OPERA IN DUBLIN
1798–1820

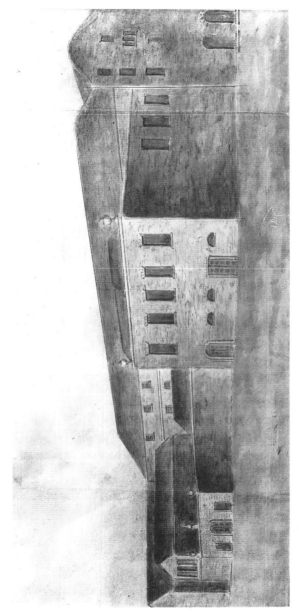

Crow Street Theatre, c. 1818, i.e. about two years before its demise. (Artist unknown)

OPERA IN DUBLIN
1798–1820

Frederick Jones and the
Crow Street Theatre

T. J. WALSH

Oxford   New York
OXFORD UNIVERSITY PRESS
1993

Oxford University Press, Walton Street, Oxford OX2 6DP
Oxford New York Toronto
Delhi Bombay Calcutta Madras Karachi
Kuala Lumpur Singapore Hong Kong Tokyo
Nairobi Dar es Salaam Cape Town
Melbourne Auckland Madrid
and associated companies in
Berlin Ibadan

Oxford is a trade mark of Oxford University Press

Published in the United States by
Oxford University Press Inc., New York

© 1993 Victoria Hamer

All rights reserved. No part of this publication may be reproduced, stored in a retrieval system, or transmitted, in any form or by any means, without the prior permission in writing of Oxford University Press. Within the UK, exceptions are allowed in respect of any fair dealing for the purpose of research or private study, or criticism or review, as permitted under the Copyright, Designs and Patents Act, 1988, or in the case of reprographic reproduction in accordance with the terms of the licences issued by the Copyright Licensing Agency. Enquiries concerning reproduction outside these terms and in other countries should be sent to the Rights Department, Oxford University Press, at the address above.

This book is sold subject to the condition that it shall not, by way of trade or otherwise, be lent, re-sold, hired out, or otherwise circulated without the publisher's prior consent in any form of binding or cover other than that in which it is published and without a similar condition including this condition being imposed on the subsequent purchaser.

British Library Cataloguing in Publication Data
Data available

ISBN 0–19–816397–5

1 3 5 7 9 10 8 6 4 2

Designed and typeset by
Boethius Press (UK) Ltd
Printed in the
United Kingdom by
Biddles Ltd, Guildford and King's Lynn

To
Jim Dundon
Angus Lee
and
John O'Sullivan
Three dear friends

CONTENTS

List of Illustrations		ix
Foreword		xi
Preface		3
1:	Interval Music 1798-1799	9
2:	Curtain Raiser 1800-1807	20
3:	Curtain up—on Catalani 1807	37
4:	Catalani Encore. 1808	63
5:	The Great God Braham 1808-1811	82
6:	Mozart Adapted 1811	105
7:	Catalani Ritorna 1812-1814	124
8:	Two English Nightingales 1815-1816	154
9:	An Irish Lark and that extraordinary Child, Master Balfe. 1817-1818	175
10:	The Original Don Giovanni 1819-1820	201
Appendix A: Cast lists of first performances of representative operas and operatic pieces performed at the Theatre Royal, Crow Street, between 1798 and 1819		233
Appendix B: List of performances of Italian Operas given in Dublin between 1808 and 1819		249
Appendix C: Programmes of Catalani's Dublin concerts in 1814		250
Appendix D: Rules and Regulations		257
To be observed by the several Performers and Persons engaged and employed in the Theatre-Royal, Crow-street, and in the Theatres of Cork and Limerick, and in all other Theatres in which Frederick Edward Jones Esq. may be interested or concerned.		
References		261
Bibliography		275
Index		278

ILLUSTRATIONS

Crow Street Theatre, c. 1818 *Reproduced by courtesy of the National Gallery of Ireland*	Frontispiece
"The Red House" (1859) — Frederick Jones' house *Reproduced by courtesy of the National Library of Ireland*	10
Section of W. Faden's *Plan of the City of Dublin* 1797	15
Charles Incledon as Macheath in *The Beggar's Opera* *Reproduced by courtesy of the Board of Trinity College, Dublin*	22
Charles Incledon as Macheath in *The Beggar's Opera* *Reproduced by courtesy of the British Library*	25
Thomas Philipps *Reproduced by courtesy of the British Library*	30
Angelica Catalani *Reproduced by courtesy of Opera Rara*	39
Angelica Catalani *Reproduced by courtesy of the National Library of Ireland*	44
Angelica Catalani *Reproduced by courtesy of the Board of Trustees of the Victoria and Albert Museum, London*	50
The contract for Catalani's Dublin performances in 1807, drawn up between F. E. Jones and Paul "de" Valabregue and co-signed by Michael Kelly *Reproduced by courtesy of the Board of Trustees of the Victoria and Albert Museum, London*	55
Rotunda and New Rooms *Reproduced by courtesy of the National Library of Ireland*	60
Angelica Catalani in M. A. Portugal's *Il Ritorno di Serse* *Reproduced by courtesy of the National Library of Ireland*	65
Angelica Catalani in *Semiramide* *Reproduced by courtesy of the Board of Trustees of the Victoria and Albert Museum, London*	67
Arias sung by Angelica Catalani in *Semiramide* *Reproduced by courtesy of the British Library Music Department*	70, 74
Nicolò Miarteni *Reproduced by courtesy of the Board of Trustees of the Victoria and Albert Museum, London*	81
Playbill for *Laugh When You Can* and *The Mock Doctor* at the Theatre Royal, June 4th, 1811 *Reproduced by courtesy of the National Library of Ireland*	87
John Braham *Reproduced by courtesy of the National Portrait Gallery, London*	90
John Braham *Reproduced by courtesy of the Board of Trustees of the Victoria and Albert Museum, London*	94
Maria Dickons *Reproduced by courtesy of the British Library*	101

Michael Kelly Reproduced by courtesy of the Board of Trustees of the Victoria and Albert Museum, London	104
"Nel cor più non mi sento" from Paisiello's *La Molinara* Reproduced by courtesy of the British Library Music Department	108
Teresa Bertinotti Radicati Reproduced by courtesy of the Board of Trustees of the Victoria and Albert Museum, London	115
Giuseppe Naldi Reproduced by courtesy of the British Library	122
Thomas Simpson Cooke Reproduced by courtesy of the Board of Trinity College, Dublin	127
Eliza O'Neill Reproduced by courtesy of the National Portrait Gallery, London	133
Camilla Ferlendis Reproduced by courtesy of the Board of Trustees of the Victoria and Albert Museum, London	136
Playbill for *Burning of Moscow* and *St Patrick's Day*, May 28, 1813 Reproduced by courtesy of the National Library of Ireland	145
John Sinclair Reproduced by courtesy of the British Library	151
Miss Hughes Reproduced by courtesy of the National Portrait Gallery, London	157
The Misses Dennett Reproduced by courtesy of the Board of Trustees of the Victoria and Albert Museum, London	164
Catherine Stephens Reproduced by courtesy of the National Portrait Gallery, London	171
Playbill for *The Castle of Andalusia* and *Lock and Key*, Theatre Royal Reproduced by courtesy of the National Library of Ireland	177
Frances Maria (Fanny) Kelly Reproduced by courtesy of the British Library	184
The Lover's Mistake (Bayley/Balfe) Reproduced by courtesy of the British Library Music Department	193
Mary Byrne Reproduced by courtesy of the British Library	202
Sir John Andrew Stevenson Reproduced by courtesy of the National Library of Ireland	213
Giuseppe Ambrogetti and Giuseppe Naldi in *Don Giovanni* Reproduced by courtesy of the Board of Trustees of the Victoria and Albert Museum, London	222
Giuseppe Ambrogetti	227
"Poney Races at the Theatre Royal, Crow Street" Reproduced by courtesy of the National Library of Ireland	231

DOCTOR TOM'S FINAL CURTAIN

We laid Tom Walsh in the earth on Friday, under a glorious Indian-summer sun, in the Barntown cemetery outside the town; that way he can sleep amid the soft green hills of his native County Wexford which he loved so much. After the requiem mass in his home church, the cortège formed up; we filled the street from side to side and end to end. Solemn robed figures walked immediately behind the hearse; easily mistaken for members of the Guild of Mastersingers, they turned out to be the entire borough council, in full fig.

The town band wasn't there; perhaps it had been wrongly thought insufficiently reverent for such an occasion. The Taoiseach, though, had sent a telegram. The flowers, piled up, made an Everest of beauty and farewell; the church was heady with their scents. We sang 'Abide With Me', and meant it.

Well, your man had done a lot for the place, starting by being born there, in 1911 (he missed his 77th birthday by a fortnight). He qualified as a doctor at Dublin University in 1944; he practised in the town from 1944 to 1955; from 1955 to 1977 he was the anaesthetist for the Wexford County Hospital. In 1951 he founded the Wexford Opera Festival, and was its director until 1966.

His worth and achievements were recognized; the University of Dublin made him first an hon MA, then a doctor of philosophy, then a doctor of literature. He was an hon. fellow of the Faculty of Anaesthetists of Ireland, a fellow of the Royal Historical Society, a Knight of Malta, a freeman of Wexford (well, I should think so). He wrote a series of scholarly books on the history of opera—this is the last of them; he was twice married and widowed; he is survived by his daughter and sister.

Facts, facts; useful things for charting the stops of life, and seeing who gets off or on; not much good at conjuring the actual man on the actual bus. That shall be my task this morning.

Tom died smiling. At least, I assume he did; he was certainly smiling when I saw him in Wexford Hospital a few days before the end. As a doctor, he could not deceive himself about his condition, and his colleagues did not try to bluff him. But there were no solemn farewells; solemn farewells were not much in his line, except to be sure, operatic ones.

Wexford knew him as 'Doctor Tom', and would call him nothing else. He had retired from active practice a decade before, but until recently he would keep his hand in by slipping over to England to do an annual locum.

When his health began to fail, some way into 1988, we devised Operation Tomplot—'we' being the group of friends who go, every autumn, to his festival. We lured him to Sussex, he all unsuspecting while we were hiding out in the hedges and ditches around him, togged up and ready to carry him off to Glyndebourne; the girls had dressed more beautifully than ever, for him. The Plot held: 'Bernard, you swindler!' he cried, as the whole gang crashed through the door. I had wondered mildly, and put the point to his daughter Victoria, what she would say if he asked why the tea-table was set for 15. 'We'll keep him out of the room,' she said, 'and anyway, Daddy wouldn't notice.'

It was perfect Glyndebourne weather that day; a cloudless sky, a breeze to cool it, the gardens beginning to recover from the devastation of the hurricane. In the interval, up on the roof-terrace, the Christies poured libations, in which we drank his health. Brian Dickie was of the company; he was general manager of Glyndebourne, but in 1967 he had had the alarming task of stepping into Tom's shoes as director of the Wexford Festival.

The Glyndebourne meeting was a moving moment; George Christie, a man who inherited a festival and thereafter dedicated his life to it, stood beside Tom Walsh, a man who created one out of nothing, and lived to see its fame spread wide. Then we went back into George's Festival Theatre for the rest of *Die Entführung*; of course it had to be Mozart for Tom, whose love for that composer was passionate and unwavering.

Not many men devote their lives to the selfless service of their fellows. Tom Walsh did it twice over; as doctor and as man of music. 'Doctor' says all that is necessary for the first part, and if you think it doesn't, ask his patients in Wexford. But 'man of music' is a feeble phrase for what it encompassed in his case. He simply decided that the quiet little town of Wexford should have an annual operatic festival to which in due course, the world would come. And the money? Tut; the ravens fed Elishah.

I often wish I had been living in Wexford at the time; I would have loved to watch the scene as he went about the town telling people of his plan, while the news went much faster about the town that Doctor Tom had gone mad. For consider: Wexford in 1951 was not only a quiet place, unheard of outside Ireland and hardly heard of even inside; it was also savagely poor. The theatre hadn't been used as such for a century (some say two); moreover it would hold only 400 people, and anyway it was now a furniture repository.

The very Muses wrung their hands and wept at so forlorn a hope, but they didn't know Doctor Tom; the iron-clad principles of rectitude and honour that guarded his life were translated into an irresistible determination to see his dream realized. The Wexford Opera Festival, with the weeping Muses engaged for the chorus as a token of forgiveness, opened its doors on time; that was 37* years ago, and they haven't shut yet. *Si monumentum requiris, circumspice.*

On Sunday morning during the festival, Tom always kept open house for his friends. Now he was adamant that he would be there to preside as usual, even if his hospital bed had to be put on wheels and pushed all the way to Lower George Street; as the week went by, though, even he had to admit defeat. But when he did, he was even more adamant that the ritual would be kept to, even if our host was from home.

Tom's Catholicism was deep, tenacious and complete; he suffered great distress when his beloved daughter married out of the faith. But there was no estrangement, and he died full of joy in the knowledge that a grandchild was soon due.**

He sought no fame, no fortune. He had got hold of the notion that he was on earth to tend the sick and spread the love of music, and he pursued both vocations with great diligence and no fuss. It pleased him,

---

* 42 (Founded 1951)
** Clara Ninette Hamer, born 1st March 1989.

as it pleased all of us, that over the years Wexford had become noticeably better off; his festival brought a good deal of money into the town.

We returned, *en masse,* to the hospital, to see him for the last time; the group was almost identical to that of the Great Tomplot. The doctors wouldn't let us in all together, but said we could go in two by two, each pair strictly enjoined to stay only a few minutes. He had been wandering a little, but he was perfectly clear with us.

He fought on for another week; death would not have dared approach his bedside until the 1988 festival was over. Last Tuesday afternoon, he fell asleep, and in sleep he left us. We who knew him will keep his memory bright, forever in his debt for the joy and friendship he and his festival have given us. We are even more blessed by having known and loved a man of such goodness, wisdom, generosity and laughter. Doubt not that he feasts in Heaven this night, with Mozart on one side of him, and Hippocrates on the other, and a glass of good red wine in his good right hand.

Bernard Levin
November, 1988

Reprinted from *The Times* by kind permission.

Frederick Jones and Opera in Dublin
1798–1820

Preface

Operatically, Dublin's Theatre Royal in Crow Street ended the 18th century with performances of such familiar works as *Netley Abbey* (played as an afterpiece to *King Lear*), *Rosina*, *The Waterman*, and *The Haunted Tower*. This type of comic or pastoral opera, combined with melodrame continued in vogue in Dublin as the popular form of musical entertainment for another thirty years or more, until it was supplanted by the operas of Bishop, Balfe, Barnett and others.

Apart from what public taste might dictate (and public taste changed much more slowly then than it does today) theatre composers and dramatists had no wish to deviate from their contemporary style. Neither did theatre managers. There were two reasons for this conservatism. One lay with the audience, already quite content with the *mélange* of plays and singing (and occasionally dancing), but above all, managers were restrained by the composition of their players. Invariably they consisted, in part, of a number of singing actors who could become acting singers as the occasion demanded, thus permitting anything from Shakespeare to a 'Musical, Rhetorical, Satirical, Mock Heroical, Vocal, Whimsical, Melo-Dramatical, Emblematical, Laughable, and Farcical Burlesque Opera'* to be performed on alternate evenings, or even on one and the same evening. The roots of this arrangement went back to the beginning of the century.** It was the manner in which the players had learned their profession, it was what they knew, and in fact, in order to earn their living, how they must remain. Moreover, actors and actresses (at least in those days) like old dogs were loath to learn new tricks, nor did their managers require it of them. Matters eventually changed with time, but meanwhile the year 1801 merely presented an example of the artistic aspect of one century overlapping the next, while continuing on its placid course.

Up to 1830 consequently, in many instances it is difficult to decide whether a work performed is in fact an opera, or merely a play with music. One reason for this, as has been noted throughout the book, is that rarely in Dublin was an exact replica of a work performed, as it had been in London. When in doubt I have relied heavily on the casting, but on occasion, as for example in *The Maid and the Magpie*, I have been influenced by information contained in the

---

* See *Saunders' News-Letter* 28 April 1814
** See *Opera in Dublin 1705-1797*, p 36

advertisements; here, the announcement of 'admired French Airs'.*

Of paramount theatrical importance to 19th century Dublin, however, is its conversion from English to Italian opera. Indeed, as the years proceeded, it was to become the century of Italian opera, commencing in 1808 with works by Mayr, Portugal, and Paisiello, and continuing with (among others) operas by Mozart in 1811 and 1819. In time the 'Golden Age' of Italian opera in Dublin would be encompassed by the years between 1840 and 1878.**

This book is a continuation of *Opera in Dublin 1705-1797*, which gives me the opportunity to make a number of corrections and additions to the first volume. Firstly, the photograph facing page 96, said to be of Nicolina Giordani, is in fact a reversed engraving of Madame Favart as Bastienne which I should have recognised since it appears in Martin Cooper's *Opéra Comique*. Moreover, I may have erred in recording that Dawson and Mahon had taken over Stretch's puppet theatre in Capel Street in 1770, instead of the theatre founded by the Smock Alley players in 1744/45, and later occupied by the Giordani-Leoni opera company in 1783/84. The latter theatre stood on the west side of Capel Street, close to the north corner of Mary's Abbey. Nevertheless, during the mid-eighteenth century, a building on the east side of Capel Street facing almost directly opposite the above, which in the reign of James I had been established as a mint, became a puppet theatre.***

The location of the Aungier Street Theatre is also puzzling. Rocque's map of Dublin for 1757 places it almost in the centre of the north side of Great Longford Street. Here it is marked O.T (old theatre) to distinguish it from N.T (the new theatre) of Crow Street, and T.R. (Theatre Royal) of Smock Alley. It is just possible that there was an entrance to it from Aungier Street, which being the main thoroughfare gave the theatre a better known address.

Mrs Muriel McCarthy, librarian of Marsh's Library, Dublin, has very kindly drawn my attention to the following extract from *A Great Archbishop of Dublin, William King, D.D., 1650-1729*, ed. Sir Charles Simeon King, Bt, (London 1906) which lends credibility to the story of Nicolini and the staff of Joseph of Arimathea. (p 17),

---

* *Freeman's Journal*, 12 December 1815
** The first Theatre Royal in Hawkins' Street burned down on 9 February 1880
*** See: William Monck Mason's 'Collections for a history of Dublin' (Gilbert Library); 'Crane Lane to Ballybough', by Timothy Dawson in the *Dublin Historical Record,* vol 27, 1974, p 134; *The Irish Builder* 1892

'Dr Berkley (late Bishop of Cloyne) after his Travels, waited upon ABp King & told him this memorable Story. When the famous Italian Singer, Seignior Nicolini was in this City [Dublin], he dined at a Gentleman's House, who had an ordinary walking Thorn Stick in the Corner of the Parlour, with a carved Head. Nicolini viewed it very attentively. The Gentleman, observing his Notice, told him that Stick had been many Years, nay he believed centuries in the family. That it was accounted a great Curiosity & some affirmed that it was Joseph of Arithmathea's [sic] Staff, who was commissioned to preach the Gospel to the Infidel Britains, & was a branch of Glassenbury [Glastonbury] miraculous Hawthorn yt buddeth every Xmas Day. The Italian earnestly requested that he wd make him a Pres.t of that Stick, for which the family recd several musical Consorts. When the Singer returned to Rome, he humbly offered this sacred Relick to the Vatican, where it was gratefully received as a venerable piece of antiquity rescued from the hands of some Irish Hereticks. The Doctor said it was exhibited to him as an inestimable Treasure.'

On reflection, the unidentified opera *Aminta* (p 54) may have been *Teraminta*, text by Henry Carey, music by John Christopher Smith or John Stanley, first performed at Lincoln's Inn Fields on 20 November 1732. Eric Walter White, in *A Register of First Performances of English Operas and Semi-Operas,* lists first performances in Dublin of two ballad operas previously unrecorded: *The Queen of Spain, or, Farinelli in Madrid,* text by James Worsdale, or James Ayres, music by J. F. Lampe, at Aungier Street on 16 April 1741, and *The Ruling Passion,* text by Leonard MacNally, music by Philip Cogan, but at Crow Street,* *not* Capel Street on 24 February 1778. Caution is advised in accepting Casanova's oft-quoted statement of meeting Tenducci and his wife, Dorothea, at Covent Garden in 1764, when he reports that Tenducci boasted of fathering two children due to having a third testicle which had not been removed during his orchidectomy. Apart from this singular anatomical abnormality, and, as has been noted, the granting of Dorothea's subsequent divorce on the grounds of 'nullity of marriage', 'void from the beginning' (p 138), Casanova had arrived in London in 1763, leaving again in March 1764, at least a year before the couple had originally met. Dan Maddison O'Brien advised me that 'the Lord C—' (fn, p 135) was more probably Lord Clanwilliam than Clanricarde, County Tipperary being in the barony of Clanwilliam.

Of a performance of *La Frascatana* at the King's Theatre, the *New Morning Post* of 27 November 1776 reports: 'Savoi has neither shake nor swell and treads the stage with the grace of an elephant' (p 190). At a meeting of the committee of the Royal College of Surgeons in Ireland held circa 31 August 1784, 'a Mr

* *Freeman's Journal,* 19-21 February 1778

T[homas] Carmichael offered to let or sell the Opera House, Capel Street, for a Hall. The College felt it would not answer that purpose.' Carmichael had performed at the Aungier Street Theatre in 1737 and had been prompter at the Smock Alley Theatre for the seasons 1750-52 and 1772.

Since the book was published more definite, although not conclusive, evidence concerning Michael Kelly's birth-date (p 263) has come to light: *A Biographical Dictionary of Actors, Actresses, Musicians, Dancers, Managers & Other Stage Personnel in London 1660-1800* gives this taken from a tombstone which once stood in St Paul's Churchyard, Covent Garden, as 'Aug.t 12th 1762', which may help some future historian to decide exactly in what year he set sail from Dublin to the Continent. Furthermore, an extract from a letter reported in *Joseph II als Theaterdirektor. Ungedruckte Briefe und Aktenstücke aus den Kinderjahren des Burgtheaters* edited by Dr R. Payer von Thurn (Wien 1920, p 70) suggests that when Kelly left Vienna early in 1787 his intention was to have returned there. It reads: 'To Count Rosenberg…Prague 29 September [1786]. Here is my opinion of the questions you ask me …I have granted O'Kelly [*sic*] 3 months leave—to include the 6 weeks of Lent, it would truly be a great loss to Italian opera if he were to go, since he is very good in many roles and spoils none.' My observation that Kelly's relationship with Mrs Crouch 'was ambiguous and has never been defined' (p 269), which was questioned, arises from a comment by Joseph Farington in his diary under the date of 3 January 1800, where he states: 'Mrs Crouch has lived in the same House with Kelly the Singer many years under suspicious circumstances, though they do not regularly cohabit.'

Joseph Kelly *was* Michael's brother (p 279) but seems to have been the rake of the family. Sir Jonah Barrington records, 'Mr Joseph Kelly of unfortunate fate, brother of Mr Michael Kelly (who, by the bye, does not say a word about him in his Reminiscences)', and makes up for the omission by devoting eleven pages to him in his *Personal Sketches of his Own Times*. Padge Reck in *Wexford— A Municipal History* discloses that the playwright-parson, Sir Henry Bate Dudley (p 168), was admitted a freeman of the borough in 1809. The 'Accompanier on the Pianoforte, the Baron de Grifft' (p 296) was an obvious misspelling for William B. de Krifft, who, born about 1765 in England, was a pianist and composer whose music was published from 1789. In 1791, while travelling through Germany, he gave a concert of some of his compositions at the Court of Koblenz. Dr Ita Margaret Hogan reports in *Anglo-Irish Music 1780-1830* that in 1804 he had moved to Waterford.

In a generous review of the book, Winton Dean offered the following enlightenment: 'Nicolini never sang in Galliard's *Calypso and Telemachus*: Mrs Barbier's last stage appearance was at Covent Garden in 1740, not in 1733; 'Hor la tromba', one of her favourite show pieces, comes from *Rinaldo*, not *Tamerlano*… Ann Catley was by no means the first woman to play Macheath; Mrs Weichsell was a theatre - as well as a concert - singer. Michael Arne was indeed the son of Thomas Augustine, but not by his wife.'

To all the above who contributed this added information I tender my sincere thanks.

I realise that in this present volume as in the previous one I have included much which does not pertain directly to opera—or to Frederick Jones. My reason for this, if reason is required, is that I believe to understand the conglomerate background of opera in Dublin up to 1820—and indeed beyond—one must have some knowledge of Dublin's Theatre Royal, as well as a passing acquaintance of events such as the Catalani concerts in 1814.

I now wish to thank both library staffs and individuals, all of whom were so helpful with the writing of this book. First comes the library of Cambridge University, where my research was carried out in surroundings not only of great scholarship, but, for me at least, great happiness. In London there were the British Library and the Music Department of the University of London Library, while in Dublin I am indebted to the National Library, Trinity College Library and the Library of the Royal Irish Academy. Among individuals I should like to thank Professor L. M. Cullen, F.T.C.D., for resolving the genealogy of Mary Byrne's family, and Patrick Fagan (whose book on Dublin's history, *The Second City*, is a goldmine for the researcher), for helping to elucidate the locations of Dublin's 18th century theatres. As ever, Catlin O'Rourke, Wexford County Librarian, and her assistant, Kathleen Lucking, were indefatigable in searching out books for me, not otherwise easily obtained. Leslie, Jennett and John Hewitt have my sincere thanks for transforming this story from manuscript into handsome print, as has Gerard Moriarty for helping to seek, and in finding, illustrations. I leave until last Helen Gregory, who, as with all I have written, has again given her generous and unstinting help from the beginning of the first chapter to the completion of the index.

This book, with the preceding volume, covers the history of opera in Dublin for its first 115 years. Nevertheless, to date, a further 170 years remain. May I recommend to some young scholar interested in opera and in search of a thesis that here is one to hand, especially for the period 1829-1878, a period when the excitement of Italian opera coursed down the years, a spectacle of melodrama or coarse comedy redeemed by music, enthralling an unsophisticated yet discerning audience.

T.J.W.
October 1988.

My father delivered the manuscript of this book to Boethius Press only a few weeks before his death on 8 November 1988. My father therefore did not have an opportunity to make any final alterations or corrections at proof stage. He made some alterations to the text shortly before he died, and I am grateful to Helen Gregory for adding the revised source references, which are given in brackets, alongside those of the author, and for providing captions for the illustrations. Boethius Press typeset the book and brought it to camera ready stage and I am very appreciative to Leslie and Jennett Hewitt of Boethius Press for their work. I am indebted to Oxford University Press for completing the printing and publication of the book.

Victoria Hamer
May 1993.

Chapter I
Interval Music 1798-1799

It might not have surprised the theatre patentee, Frederick Edward ('Buck') Jones (in his day every fashionable Dublin gentleman was a 'buck' if not a 'rake'), to know that eighty years after his death his name would still live. Yet he surely would have been astounded to learn that it had survived, not in connection with Crow Street Theatre but with, of all things, a sporting body called the Gaelic Athletic Association, for in 1914 one did not speak of attending football and hurling matches at 'Croke Park', one called it 'Jones's Road'. Jones's Road it was, for in his time it was a wooded avenue leading directly to his elegant Georgian house, now part of Holy Cross College, where it still stands as the Red House, at present the convent of the Daughters of Charity of St Vincent de Paul.

Jones was born at Vesington in County Meath about 1759, educated at Trinity College (although his name does not appear in *Alumni Dublinenses*), and later, as a man of means and position, had spent some years on the Continent. He is described as 'one of the handsomest men of his time: in stature he was above six feet, and somewhat resembled George IV when Prince of Wales, in his person, aristocratic deportment, and polished manners.'[1] He lived in princely style at his home which he rented from Lord Mountjoy. Known first as Fortick's Grove, he later restored it to its original name of Clonliffe. His career as a theatre manager commenced on 6 March 1793 when, in partnership with the Earl of Westmeath and encouraged by Dublin society (which at the time was greatly dissatisfied with Richard Daly's* management of Crow Street), he opened a private amateur theatre at the Music Hall in Fishamble Street.** Here he gave performances, even acting in some of them, until 1796.

He was friendly with the then Lord Lieutenant, the Earl of Westmorland, and so in March 1794 the Government ratified his position by granting him a licence. This was for seven years, during which time he could produce 'interludes, tragedies, comedies, preludes, operas, burlettas, plays, farces, pantomines, of what nature soever, decent and becoming, and not profane or

---

* See *Opera in Dublin 1705-1797*

** J. D. Herbert [Dowling] the artist-cum-actor in his *Irish Varieties for the last fifty years* describes the hall when taken over by Jones as 'a shell so appropriate for his plan that he could decorate it as he wished'. (p 257-8)

"The Red House" (1859)

A sad aspect of the once elegant house of Frederick E. Jones.

obnoxious',[2] provided that any new theatre did not seat more than the theatre in Fishamble Street. As a private theatre admission remained limited to subscribers only. Actors, since they were drawn from the members of the gentlemen subscribers, were not to be paid, but actresses, who in those days were rarely if ever thought of as ladies, were professionals. (Jones, nevertheless, is said to have married an actress from his company—a Miss Campion.[3]).

He was also 'empowered to eject all scandalous, disorderly, or other persons as he thought proper'.[4] His next move, in 1796, was to petition the Lord Lieutenant (now the Earl Camden) for a patent to open a second public theatre, on the plea that Richard Daly's mismanagement of Crow Street had made such an application inevitable. In this he was supported by 'many of the principal nobility and gentry of Ireland'.[5] Daly, undoubtedly realising that competition would be ruinous to himself if not to them both, decided not to oppose the application, and instead offered to retire if Jones would pay him reasonable compensation. This solution was acceptable to the Government, and so the Principal Secretary of State, Thomas Pelham, on behalf of the Lord Lieutenant, wrote to Jones on 2 November 1796, informing him that his patent would be granted. At the same time he counselled, 'his Excellency expects that if Mr Daly shall offer you reasonable terms, you will be ready to enter into accommodation with him'.[6]

A correspondence ensued. We read that on 10 March 1797, Daly has agreed to an offer from Jones of £800 a year for life, with £400 a year of that amount for the lives of his children, for which sums he requires security to be approved by the Attorney-General. On 29 May, the Attorney-General, Arthur Wolfe, (later Lord Kilwarden, ill-fatedly assassinated in the Emmet rising) wrote as follows to the Lord Lieutenant: 'My Lord, I had the honour yesterday evening of receiving your Excellency's letter with Mr Jones's new proposition on the subject of his security to Daly. I have sent to Daly, and have no hesitation to inform your Excellency that I shall certainly report the security as sufficient; and in the meanwhile your Excellency may, if you think fit, so signify to my Lord Westmeath, or Mr Jones.'[7]

The security consisted, in part, of six thousand pounds in five per cent debentures lodged in La Touche's bank.[8] In addition to the annuity reported above, Jones also became liable for an annuity of £232 payable to the representatives of the Reverend Thomas Wilson, D.D., Senior Fellow of Trinity College and late mortgagee of Smock-Alley Theatre, with which Crow Street had been encumbered at the time of Daly's take-over from Thomas Ryder. Then, on the prompting of the Under-Secretary of State, Jones agreed to pay Daly a further £300 a year from rents of the Cork and Limerick theatres, which Daly had also held under lease. Finally, Daly was to be granted a private box in Crow Street during his lifetime, with six transferable tickets nightly, and for good measure, if without any well-defined good reason, was permitted to burden the theatre with a further 30 transferable tickets. Daly assigned the Crow Street Theatre property on 12 August 1797, and it passed to Jones under the Privy Seal at St James's on 25 June 1798.[9]

It permitted him to build or hold a theatre in the city or county of Dublin for 21 years, to engage as many players as his company should require, and to pay them reasonable salaries. In return he would receive an annual government subvention of £350 (which had 'been paid for the last sixty years'[10]), and could charge the usual prices for tickets. There seems to have been one proviso only. No performances 'reproachful to the Christian religion in general, or to the Church of England in particular, nor any abuse or misrepresentation of sacred characters' were to be permitted.[11]

Jones consequently had paid a high price to indulge his theatrical bent, especially so when we learn that he had taken over the theatre 'without scenery, without wardrobe, without music, without ornament, without a manager—not even a decent staircase or convenient passage in the whole house; neither performers, nor tradesmen, nor creditors of any description were paid; it was sunk to the lowest state of degradation.' But 'immediately on Mr Jones obtaining the patent, the house underwent a total change,'[12] and before it re-opened we learn: 'The stage is lengthened to a depth of eighty-four feet. Four additional seats in the Pit, which has new entrances with a passage through the middle and sides, over which are flap seats on hinges, to put down when the press of company may require it; they are all covered with green morine [moreen]. The Boxes, Lattices, &c, have also each two additional seats in depth, covered with crimson morine. The seats of the first Gallery are covered with carpeting; and those of the upper Gallery, matted; the iron railing before it, richly gilded. The corridors round the Boxes, Lattices, &c, are six feet wide, all flagged with Portland stone—and all the stairs are also of Portland stone—There are also very convenient rooms for walking or taking refreshments—neat water closets—and all the entrances are very capacious.

The Orchestra is entirely new modeled [in mahogany]—and in it is introduced a grand piano-forte, organized. The workmanship of the Stage Doors is in the highest style of execution—the fancy painting of the panels drawn by Waldré* and painted by Philipo [Zafforini] exhibit much taste. The King's Arms over the Stage correspond with the style of the other ornaments, and deserve great praise—they cost above 400*l*. Mr Baily, an Irishman, we understand, is the artist. Nothing can exceed the elegance of the Pilastres, which exhibit in the shafts very elegant specimens of fancy painting; the capitals are richly carved and have the most striking effects. The ceiling is elegantly painted—the subject, Apollo and the Muses. The panels are gilt and painted in arabesques. The pillars supporting the gallery &c. &c. are of iron, highly gilt and varnished. The scenes, we have learned, as well as all the other arrangements are designed by Waldré and painted by Messrs Philipo [Zafforini] and Coyle.'[13]** Following the opening the same journalist reported, 'Upon the whole, we do not hesitate to pronounce our Phoenix the most elegant theatre in the British dominions.'[14]

Although this is obviously a 'puff' for the new proprietor, yet Jones is said to have spent nearly £12,000 on refurbishing.[15] The auditorium was reputed to be capable of holding an audience of 2000, and when filled at regular prices box-office takings could amount to: pit, £83; second gallery, £63; upper gallery £23, besides being 'so well constructed, it is said, that those in the remotest part could distinctly hear and see the performance.'[16] There is little to support the latter claim, at least one contemporary print suggests the contrary.

The exterior of the theatre has been described as 'singularly rude and unsightly, an irregular mass of brick defying all symmetry, and divested of any architectural ornament that should distinguish it as a public building.'[17] The theatre in fact fronted Cecilia Street, occupying the ground between Fownes Street and the north end of Crow Street,*** and was 'approached

---

* Waldré, Vincent (Vincenzo). (b. Vicenza, Italy, c. 1742; d. Dublin, August 1814). Painter and architect. Came to Ireland in 1787 in the train of George, Earl Temple, later Marquess of Buckingham on his second appointment as Viceroy. Was commissioned to decorate Dublin Castle, his principal work there being the ceiling of St Patrick's Hall. In 1804 he was living at No 12 Charlemont Street.

** Waldré had five assistant scene painters engaged by Jones. With Zafforini, Coyle and Baily there were also Farrel and Seymour. (*Freeman's Journal*, 27 Jan. 1798)

*** Crow Street, site of the suppressed monastery of St Augustine, was named after William Crow, 'chirographer and Chief Prothonotary to the Court of Common Pleas' to whom it had been assigned in 1597. Originally the street was T-shaped, later the passage running east-west was named Cecilia Street

from Dame-street and Temple-bar, by four narrow inconvenient avenues.'[18] The entrance to the Pit was in Temple Lane, the Gallery door in Fownes Street opposite which stood the King's Arms Tavern, popular with both artists and audiences. Much patronised by theatrical performers too were the Horseshoe and Magpie in Temple Bar, and the Shakespeare Tavern, which by 1795 had moved from Temple Lane (beside the Pit door) to Fownes Street at the corner of Cope Street.[19]

On 23 January 1798 the following notice appeared in *The Freeman's Journal*: 'The Public are respectfully informed, that the Theatre Royal in Crow-street will be opened on Monday the 29th instant with a play and farce.'* The same paper announced further: 'At a meeting of several of Mr Jones's Friends who have formed a Committee for promoting the interest of the Drama, the following resolutions were agreed to for the Convenience of the Public, and Regulation of the Theatre. The Committee assure the Public, that every necessary Precaution having been taken, the House is perfectly well aired for their Reception. Places to be taken at Mr Jesson's, Boxkeeper, No 12 Dame Street, and Tickets to be delivered and paid for at the time of taking Places, or otherwise they cannot be reserved. No servants to remain in the Boxes, for the purpose of Keeping Places, later than the End of the second Act. On account of the very great Expence incurred in the fitting up, enlarging, and ornamenting the Theatre—nothing under full prices can be taken. All carriages to set down and take up with the Horses' Heads directed towards Temple-lane, and all chairs to set down at the front Door;** but for the Accommodation of the Public, a door will be opened, at the end of the Play, in Fownes'-street for Chairs only. By Order of the Committee. J. Monk Mason,*** Chairman.

---

* But Jones's management had begun earlier with a season which commenced at the Fishamble Street theatre on 20 November 1797, 'on account of the unavoidable delay attending the extensive Alterations and Decorations of the Crow-street Theatre.' (*Freeman's Journal*, 14 Nov. 1797)

** *The Dublin Evening Post* of 24 August 1797 had announced, 'A new Box door will be made opposite to Crow-street, i.e. in Cecilia Street.

*** John Monck Mason (1726-1809) Member of the Irish Bar and Shakespearean commentator of doubtful scholarship, who sat in the Irish House of Commons at various periods between 1761 and 1798. Originally from County Galway, he was the founding father of a legal-literary family that settled in Dublin. He had a nephew, William (1775-1859), Irish historian, bibliophile and art collector, whose son, Thomas Monck Mason, unwisely relinquished his studies at the Bar, undertaking instead the management of the King's Theatre, London, in 1832, on which venture his losses were said to have amounted to 'upwards of £14,000.' *The Court Journal*, 11 Aug. 1832

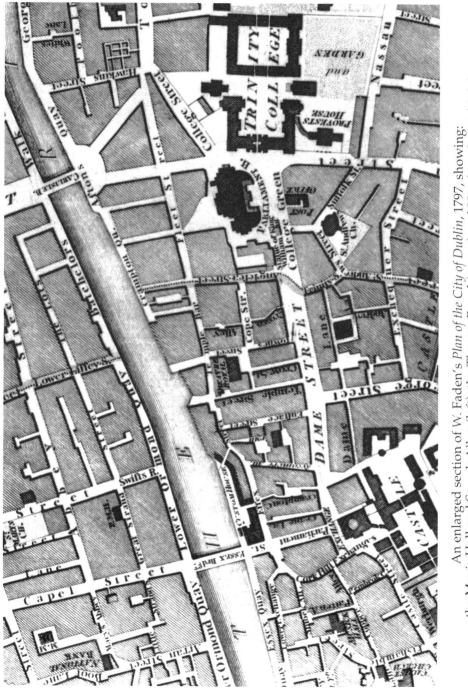

An enlarged section of W. Faden's *Plan of the City of Dublin*, 1797, showing: the Music Hall and Smock Alley (left), the Theatre Royal (centre) and Hawkins Street (right).

'It is requested that no Gentleman will appear in Boots, nor wear his Hat on in Lower Boxes,' was added on 25 January, and on 24 February, we read: 'The Boxes will be served with Tea, Coffee, Fruit, and all other Refreshments, from the Coffee-room; there are also places fitted up for Refreshments in the Pit and both Galleries.' The play for the opening was *The Merchant of Venice*, followed by *The Lying Valet*, and the evening began with the overture to Handel's *Occasional Oratorio*, played by an augmented orchestra under the direction of Mr Bianchi. It was almost like old times under Daly's management. The orchestra leader, Mr Bianchi* was not the opera composer, Francesco Bianchi (1752-1810) as is still occasionally reported. The performer in Dublin was a violinist and composer of ephemeral musical pieces. In a discordant correspondence with the violinist, Felix Yaniewicz, published in *Saunders' News-Letter* on 20 and 21 March 1799, he signs himself 'J. Bianchi'. *A Dictionary of Musicians from the Earliest Times*, ed. John S. Sainsbury, 1825, reprint, New York, 1971, gives his initials as J. M. C., and states that he 'died at Neuilly, near Paris, of a deep decline in 1802 aged only 26 years.' His 'deep decline' is confirmed by a pathetic announcement in *Saunders' News-Letter* on 16 February 1801, concerning a benefit for him, when he 'hopes his numerous friends will excuse his not being able to wait on them in person; assures them that nothing but a long and dangerous illness could have made him deficient in shewing, on this occasion that gratitude which their former generous patronage has so indelibly stamped in his breast.' *Saunders' News-Letter* of 15 September 1802 further confirms that he had died on the previous '30th of August in Paris...in the 27th year of his age.'

The performances of comic or ballad operas which took place during 1798 were, with one exception, revivals from earlier years. The exception was *The Adopted Child*, first performed on 17 December. The libretto was by Samuel Birch (dramatist and pastry cook!), the music, mostly new, by Mozart's old friend and pupil, Thomas Attwood. It was successful and *Saunders' News-Letter* reported, 'Few pieces within our recollection pleased so highly or universally.'[20]

The paucity of new productions this year was due almost entirely to extra-theatrical events, for on 25 May the Wexford Rebellion broke out, curfew was imposed, and the theatre closed for several weeks. When it

* See *Opera in Dublin 1705-1797*

re-opened on 5 July it was announced: 'The curtain will rise precisely at half past six o'clock, that the performance may be over at half past ten; as orders will be issued to the several guards that the audience may return from the theatre unmolested and unchallenged by the Sentinels until ELEVEN O'CLOCK.'[21] It was further advertised that the receipts of the first performance would be divided amongst the artists and theatre staff, 'who being debarred of their only means of support by the closing of the Theatre, have felt great embarrassment and distress.'[22] By 25 July the situation had calmed sufficiently to allow the theatre to revert to its normal opening hour of seven. Nevertheless, not surprisingly, houses were poor, and we read, 'there is scarcely sufficient money received to pay for *lighting up*…till we are entirely restored to tranquillity, the players, we fear, must continue to perform to empty benches.'[23]

In the circumstances, on 25 February 1799, Jones presented a petition to the Irish Parliament setting out all he had expended on the theatre since he had become patentee, besides declaring that because of his known loyalty, and because receipts from performances had been 'appropriated to increase the fund for carrying on the war against the enemies of Great Britain,' a conspiracy had been organised against him by 'the leaders of the disaffected to prevent the usual audience from frequenting the Theatre to his great loss.' He therefore appealed to the House for a grant to alleviate his losses. The Committee of Supply proposed a sum not exceeding £5,000, but on the second reading the House voted against the Committee's recommendation and his plea was rejected.[24] Obviously, Jones's Road, whether the connotations were theatrical, political, or social, was going to be a stony one.

1799 seems to have begun more auspiciously. At a meeting of the Dublin Musical Fund held on 4 February it was 'Proposed by Mr Giordani, 2nd by Mr Bartoli, that Madame Banti be applied to for to perform at the Ensuing Commemoration.'[25] (An annual Handel commemoration concert given under the Lord Lieutenant's patronage during Passion Week). Brigida Banti was singing at the King's Theatre, London, about this period, but does not appear to have accepted the invitation.

Equally unsuccessful was an attempt to establish 'Oratorios and Concerts of Sacred Music on Wednesdays and Fridays during Lent' at the theatre, which were advertised to 'exceed any thing of the kind ever performed in this Kingdom.

Agents,' we learn, 'have been employed on the Continent, and Dr Stephenson* has been engaged to go to London for the purpose of engaging every performer of eminence, both vocal and instrumental.' Apparently the programme was being organised by Stevenson and Thomas Ludford Bellamy, a singing actor with a 'high bass voice'[26] and Jones's deputy manager.[27] A performance of *Messiah* took place on 13 February, having been postponed from 8 February 'on account of the inclemency of the weather,' but the singers taking part were all members of the Crow Street stock company, none were performers of eminence.[28] Little wonder, therefore, that on 18 March Bellamy would 'most respectfully [beg] leave to inform the Public that he has resigned the Conducting [management] of the Oratorios, as the Speculator in those Concerts has refused to recompense him for past Exertions.'[29] One wonders was the 'speculator' Stevenson?

If so, then far more successful was a new comic opera composed by him called *Love in a Blaze*, first performed on 29 May 'with new Scenery, Dresses and Decorations. The Overture and Music entirely new.'[30] The libretto almost certainly was by Joseph Atkinson,** who acknowledged that it had been adapted 'from a little French piece without songs, called *Le Naufrage*'.[31] 'It was received with the most unbounded applause by a respectable and crowded audience', and was said to have 'every requisite for the stage…lively and sentimental dialogue appropriate to the several characters, adorned with elegant songs and choruses'. The music did honour to the taste and abilities of Dr Stevenson, while 'the scenery, dresses, and decorations, and the incidental processions formed a grand and pleasing spectacle, reflecting credit on the proprietor of our Irish Stage, who brought out this opera in a stile never before seen in this country'. It was also 'admirably well cast…the performers individually and collectively did infinite justice to the parts allotted them'.[32]

The work had eight performances so it can be considered to have been successful, and a general chauvinistic summing-up concluded, 'our Theatre is capable of bringing out entertainments which, for elegance of *dress, scenery,* and *decorations* can vie with, if not outrival, the London Stage; and the author deserves our thanks who favoured us in the first instance with the fruits of his genius, as a compliment to this country'.[33]

---

* The arranger of Moore's *Melodies*. See *Opera in Dublin 1705-1797*. The name is more correctly spelt Stevenson. See *infra*.
** See *Opera in Dublin 1705-1797*

INTERVAL MUSIC 1798-1799

Theatrically, the year ended on 16 December with a performance of *Netley Abbey** by William Pearce,[34] and two evenings later, with the first performance of *The Mouth of the Nile* by Thomas John Dibdin,[35] younger illegitimate son of Charles Dibdin.** The music of the first piece was by William Shield, with additional airs by Baumgarten, Paisiello, W. T. Parke and Samuel Howard. The composer of the second was Thomas Attwood.

It was a jejune close to the 1700s, but then for Dublin it had been a century of operatic promise rather than achievement—as exemplified by the few distinguished works produced, Gluck's *Orfeo ed Euridice*, the dramatic oratorios of Handel and some opera-buffa. Renown would come with the 1800s, at first tenuously, like a trickling brook, and then in full spate. Nor would the stimulus be artistic alone, but social and economic as well. And when, at the end of the next hundred years, the tide would recede, once again the cause would seem to have been partly social.

---

* Although announced as a first performance, the opera had been performed at Crow Street on 2 March 1796 'For the benefit of Mr and Mrs Mahon.' The cast then was: *Catherine*, Miss Brett; *Lucy Oakland*, Miss Cornelys; *Ellen Woodbine*, Mrs Mahon; *MacScrape*, Mahon; *Oakland*, Callan; *Gunnel*, Cherry

** See *Opera in Dublin 1705-1797*.

Chapter 2
Curtain Raiser 1800-1807

It is entirely coincidental that in June 1800 Crow Street Theatre should be found advertising for 'a few hundred Gallons of the best Spermaceti Oil'.[1] There was nothing symbolic of the 1800s ushering in the new industrial age about it. No doubt the advertisement had appeared in earlier years and would continue to appear for some years to come until gas lighting was installed, for assuredly the oil was for the footlights and theatre lamps. It merely indicated that then, as now, energy was important inside as well as outside the theatre, although at the time stage lighting as a rule was elementary. Spectacular effects could, on occasion, be achieved, but in general the results required much imagination to make them plausible, for in those days footlights flickered like a row of enlarged night-lights.

A musical after-piece produced on 4 July of that year was *The Turnpike Gate*, libretto by Thomas Knight, music by Mazzinghi and Reeve. This was followed on 19 July by another, *Of Age Tomorrow*. The music was by Michael Kelly* ('with the exception of the opening piece',[2] which was by Paisiello), and was considered both 'scientific and pleasing'.[3] The libretto, by T. J. Dibdin, adapted from a work by August von Kotzebue (*Der Wildfang*), was said to have 'but little novelty' and a routine plot—'a lover in disguise—a maiden aunt—and a pert chambermaid'.[4]

For the opening of the winter season the orchestra was given what was magniloquently described as 'a most formidable accession of strength' with the engagement of a 'cellist, Mr Attwood, 'brother to the composer' and 'Mr Smith,** the first trumpet in the three Kingdoms'.[5] Then on 1 May 1801, an elaborate production of a comic opera, called *The Bedouins; or, The Arabs of the Desert*, was attempted. It had been 'written by a Gentleman of this City',[6] with music by Stevenson, later to be knighted by the Lord Lieutenant, the Earl of Hardwicke, following, it is said, a convivial occasion. To have described the librettist as a 'Gentleman of this City' was scarcely correct, for he was Eyles Irwin, an orientalist and writer, who had been born in Calcutta and would die in England. It may be explained perhaps by his marriage to Miss Honor Brooke

---
* See *Opera in Dublin 1705-1797*
** Johann Georg Schmidt, born in Saxony and said to have been brought to England by the Prince Regent as solo trumpet in his private band.

of Dromavana (Co. Cork) and Firmount (Co. Longford), who, in turn, was a cousin of Henry Brooke, dramatist, and librettist of the operatic satire, *Jack the Giant-Queller*.* The Irwins may have been living temporarily in Dublin in 1801.

There was new scenery by Mr Chalmers and Filippo Zafforini, with machinery by Peter Martinelli, who would die not long after in 1803 'at his lodgings in Fownes-st'.⁷ The first performance was 'received with universal applause' by a house 'extremely crowded and brilliant', and it was given out for the following evening 'with much approbation'.⁸ If such approbation was forthcoming, it is remarkable that the work seems to have had only three performances, on May 1, 2 and 6. The last was for Stevenson's benefit, and even for that, 'the author of the Opera' (Irwin) felt 'himself impelled to promote, as far as he is able, the Interest of the celebrated Composer', entreating 'the Public, and his Friends in particular, to come forward…in compliment to the Composer, and he [would] venture to add, the native Productions of Ireland'.⁹ Nevertheless, the vocal score dedicated to the Countess of Hardwicke was published in 1806 by Goulding, Phipps, D'Almaine & Co., when it was reported both of this composition and of *Love in a Blaze*, 'Many of Sir John's warm admirers tell us he has combined the styles of Haydn, Sacchini, Storace and other modern masters with peculiar felicity'.¹⁰

This was followed on 4 July by a new musical drama called *Paul and Virginia*, taken from a story by J. H. Bernardin de Saint-Pierre, and originally adapted as an opera for the Comédie-Italienne by de Favières and Rodolphe Kreutzer. The English version (much 'mutilated and deformed',¹¹ according to Thomas Dutton) had a libretto by James Cobb and music by Mazzinghi and Reeve. For the Dublin production new scenery had been painted by Chalmers, the machinery was by Seymour and Martinelli,¹² and in addition the famous English tenor, Charles Incledon** had returned to Crow Street to take his

* See *Opera in Dublin 1705-1797*

** See *Opera in Dublin 1705-1797*. Hazlitt would report of Incledon in 1816, admittedly when he was 53: 'Mr Incledon's voice is certainly a fine one, but its very excellence makes us regret that its modulation is not equal to its depth and compass. His best notes come from him involuntarily, or are often misplaced. The effect of his singing is something like standing near a music-seller's shop, where some idle person is trying the different instruments; the flute, the trumpet, the bass-viol, give forth their sounds of varied strength and sweetness, but without order or connection.' (*A View of the English Stage*, p 244). Of his method of changing from the chest voice to falsetto, Leigh Hunt wrote about the same time: 'in his leap from one to the other [he] slammed the larynx in his throat like a Harlequin jumping through a window shutter'.(*Leigh Hunt and Opera Criticism. The "Examiner" Years. 1808-1821*, by Theodore Fenner, p 95)

Charles Incledon (1763-1826) as Macheath in *The Beggar's Opera*

original role of Paul. The work, we read, was performed with 'extraordinary eclat',[13] while of Incledon it was noted, 'his vocal powers…were never displayed to more advantage'.[14]

At the beginning of the winter season, Joseph George Holman was appointed actor-manager,[15] and the theatre was said to have undergone 'a thorough repair' with a new ceiling designed by Zafforini.[16] But for the following year of 1802 there is little to relate except an announcement that 'the Grand Piano Forte has been removed from the Orchestra to accommodate a greater number of Performers who have been engaged, so as to render the Band the first and fullest ever collected in this Kingdom, under the direction of Mr T. Cooke'.[17] The name, T. Cooke, imparts the most significant information given here, for he was Thomas Simpson Cooke, popularly known as 'Tom', the son of Bartlett Cooke, a renowned Dublin oboe player at Smock Alley Theatre. He was actually born in Crow Street in July 1783,*[18] and became a pupil of Tommaso Giordani.** He is reported to have given his first public performance at a benefit for his father at the age of eight 'when he played a concerto of Jernovick [Giornovichi] on the violoncello in a quartetto of Pleyel's'.[19] His appointment as leader or conductor of the theatre orchestra in 1802 if not earlier is consequently indicative of unusual ability and precociousness. Much more will be heard of him.

A 'new musical Farce' must also be reported. It was called *The Sixty-Third Letter*, and was first performed on 30 November. It had been written by Walley Chamberlain Oulton,** 'late of this city', and the music had been composed by Dr Samuel Arnold. Earlier in the year, it had been an unsuccessful production at the Haymarket Theatre, London. Its success in Dublin does not appear to have been any more decisive.

The year 1803 brought yet another comic opera written by T. J. Dibdin, *Il Bondocani; or, The Caliph Robber*, the music by Thomas Attwood and John Moorehead. Moorehead was an Irish-born musician who had spent almost his entire life in England. He had the misfortune to become mentally deranged at a time when this form of illness was liable to have one committed not to hospital but to gaol—as it had him. Though subsequently released, he ultimately ended the inhumanity of his life by hanging himself. *Il Bondocani* was given

---

* *Oxberry's Dramatic Biography and Histrionic Anecdotes* vol III, p 110, gives the year 1781
** See *Opera in Dublin 1705-1797*

its first Dublin performance on 4 March, and the evening's programme was completed with 'the new Dance of *Colin and Flora; or, Rural Sports*, by the four Misses Adams'.[20] A second performance on 7 March was commanded by the Lord Lieutenant and the Countess of Hardwicke, and the work seems to have been popular, for performances continued until 25 June.

It was followed on 11 July by *Ramah Droog; or, Wine does Wonders*, a comic opera written by James Cobb and set to music by Joseph Mazzinghi and William Reeve. When it was revived at Drury Lane in 1816, Leigh Hunt had found the work 'below criticism' and the music not 'much better'.[21] Nevertheless, the Dublin press (as almost ever) hailed it as having been brought forward 'with a splendor and brilliancy that reflect additional credit on the Irish stage' and goes on to report that 'the several performers exerted themselves with great success in their respective characters; Mr Munden and Mr Johnstone, the originals in London were particularly happy in the display of their rich comic abilities'.[22] John Henry Johnstone* and Joseph Shepherd Munden were favourites in Dublin from the late 1700s, and on this visit once again 'performed their principal characters highly to the satisfaction of the Dublin critics'.[23] Unfortunately for them their engagement was disrupted by Robert Emmet's abortive rising on 23 July when 'Martial law being declared...before they had their benefits, they were under the necessity of performing in the day-time, in order that the entertainments might be concluded before eight o'clock in the evening; —they commenced at one o'clock'.[24]

In 1804, theatrically the greatest activity in Dublin occurred not in the theatre, but following the anonymous publication in verse of *Familiar Epistles to Frederick E. J—s, Esq., on the Present State of the Irish Stage*. The author in fact was John Wilson Croker, born in Galway in 1780. In 1796 he entered Trinity College, Dublin, where he became conspicuous as a speaker in the Historical Society, and where Tom Moore was a year or two his senior. In 1800 he entered Lincoln's Inn. For a while he worked as a journalist in London, at the same time forming a remarkable collection of pamphlets on the French Revolution, now in the British Library. In 1802 he returned to Dublin, and two years later wrote the octosyllabic verses mentioned above. Croker later attached himself to the Munster Circuit, was elected to Parliament, and eventually appointed

---

* See *Opera in Dublin 1705-1797*. Not to be confused with Henry Erskine Johnston, born in Edinburgh, another Dublin favourite also playing in Dublin about this time.

Mr Incledon, as Macheath.

But now again, my spirits sink,
I'll raise them high with wine. [Drinks]
Beggars Opera. Act 3 Scene 3

Engraved for the Theatrical Inquisitor.

Secretary to the Admiralty. He subsequently founded the Athenaeum Club, and was instrumental in acquiring the Elgin Marbles for the British Museum.

Five editions of his 'Epistles' were printed within a year, and the affair became something of a *cause célèbre*. Teacups rattled in Dublin's drawing-rooms from Mountjoy Square to Merrion Square, but it was nothing more than a storm in one of them. Jones believed that he was being attacked personally, which Croker (anonymously) assured him was not so, declaring 'that so far from having any hostile intentions towards him, I like and esteem him as a pleasant companion and an honourable gentleman'.[25]

He then proceeds to introduce him as a gourmet:

> 'Jones, who directs with equal skill
> The bill of fare, and the play-house bill,
> Whose taste all other palates sways
> Either in dishes, or in plays'.[26]

Croker, of course, did lampoon Jones professionally through his artists, and the satire's interest here lies in the information it imparts about the singers and musicians in the company.

Jones is reproved that he:

> '…with more shameless puff will tell ye
> That Cooke is equal to Corelli;
> And liken, with unhallowed scandal
> His *noises* to the strains of Handel'.[27]

In a series of footnotes Croker expounds further on the artists he mentions. Of Cooke he writes, 'The modest and diffident Mr Tom Cooke, who played on *eight* different instruments* for his *own* benefit. I am sure it was neither benefit [n]or pleasure to any one else. This person writes *new* overtures to all the operas which are imported to our stage, beginning generally with *chords*, and ending with an Irish *jig*, and this he calls *composition*.'[28]

Next, the leading male singer in the company he portrays as:

> 'Comes Phillips [sic] writhing in grimaces,
> And tott'ring in his girlish paces.
> With feeble voice, yet sweet and true,
> (Where taste has done what taste can do;)
> ………

---

* These were reported to be flute, violin, tenor (viola), violoncello, piano, clarionet, pedal harp and trumpet. (*Saunders' News-Letter*, 3 Mar 1803)

> With labouring chest and straining throat,
> He'll seem to heave the nauseous note,
> And in *cadenze* and divisions,
> Throw up his vocal evomitions'.[29]

Of him, he explains: 'We have no professed male singer but Mr Phillips, and thus the *whole* of our musical department depends on his slender pipes'.[30] There were, however, two singing actors, and these he mocks:

> 'Or is it Lindsay's Irish howl?
> Or solemn Coyne's pedantic growl?
> 'Tis both—in dismal chaunt they join,
> And Lindsay's echoed back by Coyne'.[31]

He goes on to describe the ladies of the company as follows:

> 'But see where little Howels [sic] stands,
> And waves her supplicating hands,
> ………
> Lost in these humble ranks sonorous,
> That swell a Covent-Garden chorus,
> Thy thrilling voice, thy wond'rous taste,
> Thy beauteous person—all were waste;
> Till knowing Jones's generous care,
> Taught you breathe Hibernian air;
> And bad you lead the vocal throng,
> Unrivall'd queen of Irish song'.[32]

To Howells, Croker adds Miss Davidson and Mrs Stewart who, he avers, 'are on every account incapable of playing even secondary parts, and indeed seem…to be only fit to lengthen the procession in *Alphonso*, or swell the choruses of the *Castle Spectre*'.[33] His opinion, at least of Mrs Stewart, is partly confirmed, for, from another source we learn, 'Of the ladies who appear in the opera, Mrs Stewart is the best actress and the worst singer—yet her voice is an excellent one, and if she could divest herself of her airs and giddiness, and avoid foolish attempts at quivering and shaking, there is none would be heard with more pleasure'.[34] Croker also notes that Jones's top salary for artists was £5 a week. In London, the better performers were paid £10, £15 and even £20.

It was with this basic vocal cast, therefore, that Jones's first opera production for 1804 was presented on 21 January. It was a 'musical drama', *The Wife of*

Two Husbands, libretto by Cobb, music by Mazzinghi—'the Overture and Accompaniments composed and arranged by Mr T. Cooke'.[35] It was an adaptation from a French play (as so many English productions were then), called *La Femme à deux Maris*, by Guilbert de Pixérécourt, which had had its first performance in 1802, and subsequently a very successful run at the Théâtre de l'Ambigu-Comique, Paris. In Dublin, it would be performed until late February at least.

It was followed on 27 February by George Colman's *Love Laughs at Locksmiths*, with music composed by Michael Kelly. This too was a very free translation from the French, this time of Méhul's opera, *Une Folie,* the libretto of which was by J. N. Bouilly, the original author of *Fidelio*. Kelly rather pretentiously explains, 'The original music of both pieces [he is speaking also of Dalayrac's *La Maison à vendre*] was very good, but not calculated for an English audience; I therefore recomposed the whole of the music for them. Both pieces met with prodigious success on our stage, but particularly *Love Laughs at Locksmiths*, which is to the present time (1824) a great favourite.'[36]

The year 1805* saw a second musical farce, *Matrimony*, this time written by James Kenney and composed by Matthew King, presented on 24 January, in which 'the songs and dialogue produced much laughter'.[37] Kenney, then aged just 25, was another Irish-born dramatist. His father was manager and part-owner of Boodle's Club, in St James's, London, and had apprenticed him to a banking-house. But his interest lay with the theatre, and with the success of his first effort, a farce called *Raising the Wind*, performed at Covent Garden in 1803, he turned entirely to writing. He married Thomas Holcroft's** widow. *Matrimony* was his second theatrical venture, and for it he had adapted the libretto of B. J. Marsollier's opéra-comique, *Adolphe et Clara; ou Les Deux Prisonniers*, originally set to music by Dalayrac.

---

* It should be noted that other minor theatres existed in Dublin at this time. Besides the well-known Royal Amphitheatre in Peter Street where *The Naval Pillar*, a musical afterpiece written by T. J. Dibdin, with music by John Moorehead, was performed in November 1799, there was also the Opera-House Theatre in Capel Street, where Mrs Mountain appeared ten times between 27 July and 17 August 1805. She presented a mixed entertainment which included 'an imitation of Signora Grassini', while on 12 and 14 August of the same year, Incledon sang in *his* mixed entertainment at Percy's Private Theatre, Stafford Street (*Saunders' News-Letter*, 9, 12, 13 August 1805). Another private theatre would open in Fishamble Street on 7 November 1808 (*Freeman's Journal*, 7 November 1808)

** See *Opera in Dublin 1705-1797*

1806 went one better by staging, on 22 February, an entirely new opera '(never acted)',[38] called *The Five Lovers*, with music by Tom Cooke. It was 'written by a Gentleman of this City',[39] a certain Counsellor Swift,[40] later confirmed as the eccentric Theophilus Swift,[41] a second cousin of the renowned Jonathan. Theophilus had fought a duel with the Duke of Richmond (when Colonel Charles Lennox) and had attacked the Fellows of Trinity College for not awarding distinction to one of his sons in his examination, charging them, among other transgressions, of violating the rule which forbade them to marry. Following the first performance, *The Five Lovers* was 'given out' for 25 February 'amidst a tumult of applause from every part of the house, and some hisses from the pit'.[42] Perhaps the hisses occurred because 'there was not a male singer in the House!' [43]

Earlier in the month it had been reported, 'There never was such picking of pockets known in so short a time in this city, as took place at the Theatre on Wednesday night. There must have been a numerous gang of the nimble-fingered tribe, who had formed a plan for the purpose, as the work was carried on from the boxes and pit to the upper gallery'.[44] One way or another theatre-going cannot have been very entertaining just then, for a concurrent letter to *The Monthly Mirror* roundly condemns the standard of the company, and incidentally suggests that Philipps may have left temporarily. 'We never had a *worse* [company]…' writes a correspondent, 'We have a Mr Stephens from Edinburgh (in the room of Philipps); but surely the manager does not intend to pass this gentlemen off to us as a *first* singer… Mrs Nunn [late Mrs John Addison, née Willems] is our first singer; her voice is tolerable, but her person and acting are indifferent. Mrs Cooke [late Miss Howells] is a pretty little singer'.[45] Even the replaced Thomas Philipps left much to be desired, if William Hazlitt is to be believed. 'Of Mr Philipps we would not wish to speak' he reports from London, 'but as he puts himself forward and is put forward by others, we must say something. He is said to be an imitator of Mr Braham; if so, the imitation is a vile one. This gentleman has one qualification, which has been said to be the great secret of pleasing others, that he is evidently pleased with himself. But he does not produce a corresponding effect upon us; we have not one particle of sympathy with his wonderful self-complacency. We should wish never to hear him sing again; or, if he must sing, at least, we should hope never to see him act'.[46]

Thomas Philipps, tenor (1774-1841).

Philipps met a tragic end at Hartford railway station in Cheshire, while travelling to Dublin for an engagement. He stepped on to the train while it was in motion, but his foot slipped and he was crushed beneath the wheels.[47]

On 17 June, *The Hunter of the Alps*, a musical piece by Michael Kelly with a libretto by William Dimond, had a successful run of six performances. It was noteworthy in having the principal role of Felix played by the renowned actor-manager, Robert William Elliston, a role he had earlier taken in the first performance of the work at the Haymarket Theatre, and which in Dublin he performed 'beyond the reach of criticism'.[48] It was noteworthy also in having an 'overture *à la chasse* and a delightful hunting song, both by T. Cooke [which] received peals of approbation'.[49]

Then, on 14 August, for the first time in Dublin, the popular comic opera, *The Cabinet*, was produced, the libretto by the prolific T. J. Dibdin, the music by William Reeve, John Moorehead, John Davy and John Braham. Incledon, engaged once again, sang Lorenzo, another role he had created at Covent Garden. The opera's first performance did not have an easy passage, however, and we read that 'The curtain did not draw up until half past 8 o'clock, though the clamour of the audience at one time was so violent as to threaten the demolition of the scenes. Indeed, we never witnessed in a Dublin Theatre so complete a *row*. Instead, however, of making an apology, instead of attempting to pacify the audience, the play commenced without any explanation. No wonder that the House became almost ungovernable. Mr Foote at length made his appearance, but for a considerable time all his efforts to be heard proved unavailing—at last he succeeded, after the most commendable perseverance, in informing the audience that the delay arose from some irregularity in the delivery of the dresses from the wardrobe. The play was at length suffered to proceed'.[50]

This was not the only report of theatrical uproar at the time. Indeed, an indictment published by the *Hibernian Magazine* two years before suggests that uproar may well have been a theatrical way of life in Dublin just then.* 'How often are the public disturbed by the squalling of children…?' we read, 'to what inconveniences are the spectators exposed, from "the shot of the galleries", the occupants of which *exalted* station employ the interval between the acts

---

* As indeed it also seems to have been in London. See report in *The Times* on Monday, 17 June 1805 of a riot at the King's Theatre on the previous Saturday evening.

in assailing the orchestra with rotten apples, half-devoured oranges, fragments of gingerbread, &c, and horrid to relate, sometimes glass bottles! but which, seldom reaching their destination, fall upon the heads of persons situated in the pit, to the utter spoliation of wigs, bonnets, hats and cloaks; the annoyance of the company; the stoppage of the performance; and sometimes the loss of eye-sight. The hideous noise of the catcall, the banging of box and gallery doors, the incessant babbling and unblushing effrontery of *ladies of pleasure*, and the licentious conversation of drunken apprentices and journeymen, are all nuisances *within* the playhouse, which call loudly for a *theatrical police*; and upon *coming to* or *quitting* the theatre, the horrid blasphemy, obscenity, and rude language of the orange-women, the rush of company that is admitted in the lobbies, the herd of hackney-coachmen, livery-servants, pick-pockets, bill-hawkers, &c. &c. who surround the doors, prove the absolute necessity of a speedy reform in the avenues'.[51] Obviously nothing had changed in almost sixty years since Thomas Sheridan's remonstrance on the same problem in 1747.*

The stormy season ended with a final performance of *The Cabinet* on 20 August. It was a benefit for Philipps in which, besides the role of Orlando, he included the air, 'Let fame sound the trumpet', accompanied on that instrument by Henry Willmann,** while Incledon sang 'The Thorn' (both songs by Shield), and the two artists joined together to sing the duet 'All's Well' from *The English Fleet in 1342*, by Braham.[52] [According to a playbill the evening then ended with both artists appearing in *Inkle and Yarico*.] At the end of the engagement 'Mr Jones, in consequence of Incledon's successful efforts in behalf of the Theatre…made him a present of a most elegant snuff-box.'[53]

A voice from the previous century was resurrected when, on 3 September 1806, we read: "Mr Urbani seizes the earliest opportunity of announcing his arrival from his tour to Scotland, to the Nobility, Gentry, his Friends and the Public, and begs to inform them that all commands in his musical capacity addressed to him at 29 Nassau-street, will be punctually attended to'.[54] Pietro Urbani* had in fact left Edinburgh in 1804, following the failure of his music business and oratorio concerts there, and from then onwards appears to have

---

* See *Opera in Dublin 1705-1797*
** Whom Michael Kelly describes as 'the finest trumpet player I ever heard in any country…his execution on the instrument almost baffled belief.' *Reminiscences*, p 306

commuted between the two countries for much of the remainder of his life. On 13 September 1804 he organised a concert in Dublin at the Rotunda with the French violinist, Paul Alday, when he sang Italian arias, a Scottish and an English song, and the old Irish air, *An Cailín*. He and his wife seemed to have moved from place to place about the city, staying at one lodging after another. At this period he gives his address as 'No. 17, Eustace-st'.[55] Then in the following December, returning to Dublin from another visit abroad, he announces his intention 'to continue for the remainder of the Winter Season' there, and invites 'such Ladies and Gentlemen as wish to be instructed in the true Italian stile of Singing…to send their Commands to his Appartments, No. 5, Trinity Place'.[56] He was still at this address in March 1805, during which month, as well as in April, he sang at concerts in the Rotunda[57] and the New Concert Rooms, Rutland Square.*[58] His pieces now included an 'Italian Duet' with the tenor, John Spray, and a 'Comic Trio', with Guerini and the bass, David Weyman, both by Mozart, besides 'Total Eclipse' from Handel's *Samson*.[59] Since the comic trio was 'La mia Dorabella' from *Così fan tutte*, and since Guerini seems to have been Vincenzo Guerini, a principal violinist from Crow Street theatre orchestra, this, coupled with Urbani's falsetto, must have made it a very comic trio indeed, though scarcely what Mozart would have wished.

On 15 January 1806, he was among the singers (including Philipps and Mrs Cooke) who had offered their services for a Grand Concert 'in aid of the Fund for the intended Monument of Lord Nelson',[60] which took place at the Rotunda. It is probable that in 1806 he remained in Dublin, at least until the summer, for his name appears at a second concert held in January,[61] as well as at concerts occurring during March,[62] April,[63] and June.[64] In these his choice of songs vacillated between 'Verdi prati' from Handel's *Alcina*, and 'Molly Astore'. One of his partners in the June concert was the celebrated Mrs Salmon, then aged only 19 or 20, and at the beginning of a very successful, though short-lived, career in concert and oratorio. The concert was at the New Rooms, Rutland Square, one of several under the direction of her uncle, the clarinettist, Mr Mahon** (probably John) and of her first appearance it was reported: 'We do not remember hearing "Sweet Bird" performed with such great effect as Mrs Salmon gave it… Her tone of voice is sweet and melodious, her shake perfect,

---

* Now Parnell Square
** John, and his brother William, were violinists and viola players as well as clarinettists

with an enchanting swell and diminuendo surpassing any description'.⁶⁵

In 1807 Urbani's name again appears in concert announcements: at the Rotunda on 25 March, when he sang 'He was despised' from *Messiah*,⁶⁶ and at the Concert Rooms, Rutland Square, on 17 June, when he sang his own 'Italian Rondo, Ah frenate il pianto', 'The last time I came o'er the moor', and, joined by John Spray and Mrs Mahon, a glee arranged on Charles Thomas Carter's* popular air, 'O Nannie, wilt thou gang wi' me'.⁶⁷

An interesting opera, if only socially, called *The First Attempt; or, The Whim of the Moment* was given its very first performance on 4 March of this same year. The librettist was Robert Owenson's* daughter, Sydney Owenson, later Lady Morgan,** the popularity of whose novel, *The Wild Irish Girl*, published in 1806, 'was the means' we are told, 'of gaining her admission to the best society'.⁶⁸ Of her stage work, we read, 'Last night Miss Owenson's Opera was performed for the first time to a most brilliant and crowded audience, and received the greatest applause. Mr Owenson made his first appearance these nine years [as O'Driscoll] and met with a most flattering reception. Mr Talbot spoke an occasional address, at a short notice, in his usual elegant style, like a scholar and a gentleman. The piece was given out for Friday amidst the most unbounded applause'.⁶⁹ The music also seems to have been selected by Miss Owenson who, having been a pupil of Tommaso Giordani, was a composer of sorts as well, but it was 'harmonized'⁷⁰ by Tom Cooke, who also composed the overture. Many national airs were included, especially with her father in the cast, and indeed we read that 'his *lilts* were loudly encored'.⁷¹ The 'best society' in the persons of the Lord Lieutenant and the Duchess of Bedford (shortly to return to London) greatly honoured her by attending the third performance which was for her benefit. The role of Nicholette was taken by Mrs Tom Cooke, née Howells, who had recently married the orchestral leader. It was said that whenever subsequently she had to play a love scene, or sing the ballad, 'When I'm a widow', her husband was the butt of raillery from the galleries.⁷²

A second musical piece to be produced on 28 April was a one-act interlude, called *The Invisible Girl*. The librettist was Theodore E. Hook, the composer,

---

* See *Opera in Dublin 1705-1797*

** Her husband, Sir Thomas Charles Morgan (physician to the Marquess of Abercorn), received singing lessons from Rossini when the latter visited London in 1824. See: *Rossini and some forgotten nightingales*, Derwent, p 227

his father, James. It was of little consequence, based on an idea borrowed from a French farce, *Le Parleur Éternel* by Charles Maurice [Descombes], in which one character alone, Captain All-clack, speaks and sings for all the others, including the final chorus, with the titular heroine singing off-stage throughout. Of far greater interest was the drama taking place in the auditorium about the same time. Anyone acquainted with Dublin's theatrical history will scarcely be surprised to read such an observation as, 'Those who visit the Dublin theatre are frequently entertained during the intervals of [a] stage performance, with local sallies of wit from the upper gallery'.[73] Things got out of hand, however, (in every sense of the word) with the oyster shell incident. On Thursday, 21 May, the highly dangerous missile was thrown from the upper gallery into the pit, 'by which act of wanton barbarity a Gentleman was severely wounded'. Jones at once offered a reward of £50 (undoubtedly secure in the knowledge that no informer would be forthcoming to accept it) 'to any Person or Persons who shall prosecute the offender to conviction'.[74] An unlikely outcome.

But if a shooting war had erupted, a cold war had also broken out. On 19 April, the Duke of Richmond landed in Dublin and was sworn in as Lord Lieutenant. As a general in the army he had earlier served in Ireland, and at this period of his life, was a noted anti-Catholic bigot. It is not entirely surprising, therefore, that when he and his Duchess paid their first visit to the theatre on 30 May, 'the house was fashionably, but by no means so fully attended, as…witnessed on similar occasions. When their Graces entered, a profound silence prevailed, until her Grace* saluted the audience, when the native gallantry of the country broke through every political consideration, and some plaudits followed.' That, however, did not resolve the situation, and the report continues: 'One circumstance could not pass unnoticed: our popular air, "Patrick's Day" [a Catholic party tune; another was "Garryowen"] was frequently called for, but it was not played during the night, and this was the first time it was omitted for more than twelve months. What are we to think of this? Are we returning back upon the gloomy days of party distinctions? Is such violation of national feeling the way to harmonize this country? But we sincerely hope the omission was accidental, and that we shall not again

---

* Who by coincidence, was a sister of the Duchess of Bedford whom she had just succeeded, both being daughters of the fourth Duke of Gordon

have occasion to remark upon it. When the Duke and Duchess of Bedford first visited the Theatre Royal, Crow-street, *Patrick's Day* was ordered by those illustrious personages.[75]

Cause was given once again for comment, however, (and this time more forcibly), when on 18 July their Graces honoured the theatre for a 'second time...and for a *second* time our national air, "Patrick's Day" was omitted, though frequently called for. If the omission proceeds from Mr [Tom] Cooke, he will no doubt meet the *reward* he *merits* for such an insult to a Public that has heaped upon him many favors. If from a *higher* authority, it should not excite our wonder, though it may public indignation and contempt'.[76] Honour in fact would not be done to the 'national air' until—and even then only partially—late March 1808, when 'Patrick's Day was played for the *first time* before their Graces...at the Theatre...we must, however, observe that Lord and Lady Hardwicke, and their Graces the Duke and Duchess of Bedford, always rose and remained standing when this favourite and national air was played!'[77] Richmond, who would remain in Ireland until 1813, did not seem set to have a happy sojourn there.

The season ended on 3 August with as patriotically named a piece as any English viceroy could wish for, Braham and Dibdin's 'historic comic opera' of *The English Fleet in 1342*. Its *raison d'être* was probably the arrival in Dublin, for a short season, of Charles Incledon, who assumed his original role of Fitzwalter. But while the work would be performed at Covent Garden and Drury Lane 'upwards of 200 nights, with most unbounded applause',[78] here 'all the exertions of Incledon, Philipps, &c could not give the piece any interest with the public'.[79] It was given a final performance on 8 August.

On the whole it had been a tolerably successful season, although hardly distinguished by the production of new operas. But then, what year had been? 1807, however, would have its place in Dublin's operatic history, for within five days of the season ending, a star of the first magnitude would arrive, a star not only in her own right, but as the precursor of an entirely new style of opera performance, a style which, during the nineteenth century, would establish itself firmly in Dublin; too firmly perhaps, for the same style scarcely altered remained almost until today.

Chapter 3
Curtain up—on Catalani 1807

The prima donna has been part of opera both substantively and symbolically since the beginning of the 18th century. Such is her mystique and radiance that barring a nuclear holocaust she is bound to remain, spreading her aura and animation to the beginning of the 22nd century. Each decade creates its own star, sometimes there are two, even three, at the same time, who in today's jet age streak across the sky like refulgent meteors, descending to earth for brief spells to coruscate about the great opera houses.

In her time such was Angelica Catalani, although her mode of travel, perforce, was more leisurely. She was born in Sinigaglia (now usually spelled Senigallia) on the Adriatic coast, south of Rimini on 10 May 1780, and so varied are reports of her early life that it may be appropriate to record here some details up to the time of her arrival in Dublin. She was the daughter of Agostino Catalani, a goldsmith from the nearby village of Mondolfo, and his wife, Antonia Summi, a laundress from Ancona.[1] Agostino had a good bass voice and on the foundation of a new chapel choir in the Sinigaglia cathedral in 1778 he was invited to join, in company with the renowned sopranist, Girolamo Crescentini, then a pupil of the choirmaster, Pietro Morandi. At the same time, during the 1777-78 carnival season in Ancona he sang a leading role in Anfossi's *La Vera Costanza*,[2] while in 1782 and '83 he was secondo buffo at the Teatro San Samuele in Venice.[3]

Even as a child Angelica gave evidence of having a remarkable voice, which prompted her father to bring her to Maestro Morandi for lessons.[4] Next she was placed at school in the Convent of Santa Lucia* in Gubbio, probably for the opportunity it gave her of receiving a free education as a member of the chapel choir. The opinion is supported by the circumstance that both Sinigaglia and Gubbio were then located in the Papal States. As a member of the cappella in the former town her father would have had some influence with Cardinal Bernardino Onorati its founder, who in turn would have every influence with the Sisters of Santa Lucia.[5]

---

* The convent still exists in Gubbio but no record remains of Catalani having been a pupil there (personal communication)

About 1797 Angelica,* then just 17 years old, is said to have returned home where, it is recorded, the Venetian impresario, Alberto Cavos, recruited her for the 1797-98 season at the Fenice Theatre.[6] According to one source (regrettably not confirmed) this came about through the sudden death of a prima donna already engaged. Cavos was rescued from his predicament by his music copyist, Zamboni, who mentioned Catalani to him.[7] Her father was engaged for the Fenice at the same time. In the autumn of 1797 she made her debut there in Mayr's *La Lodoiska*, singing the title role. The sopranist, Luigi Marchesi, sang Lovinski, while her father took the part of Giskano.[8] She sang in at least four other operas there during the season,[9] and then, some months later appeared at the Teatro della Pergola, Florence.[10] She returned to the Fenice in 1800, but meanwhile, during the carnival season of 1799 she had sung the title role in Mosca's *Ifigenia in Aulide* at the Teatro Argentina in Rome.[11] In 1800, on 1 October, she had also appeared at the Teatro San Pietro, Trieste, in Cimarosa's *Gli Orazi e i Curiazi*,[12] an opera in which she had taken part earlier in Venice during January-February of the same year.[13] Next she was engaged for the Mecca of all Italian artists, then as now, the Teatro alla Scala, Milan.

Here she was invited by Zingarelli to sing the protagonist's role in the creation of his opera, *Clitennestra* at the opening of the season on 26 December 1800.[14] Having sung there a month later in Nicolini's *I Baccanali di Roma*,[15] she was next engaged for the Italian Opera in Lisbon where from 1801 until 1806 she remained the idol of both Court and public.

She made her debut there at the Teatro São Carlos on 27 September 1801,[16] once again as the protagonist, this time in Portugal's** *La Morte di Semiramide*,[17] and less than four years later, her father, (who had joined the company there as Lusignano in Portugal's *Zaira* on 7 May 1802[18]) could proudly write to Pietro Morandi in Sinigaglia: 'My dear daughter Angelica is now not alone "the

---

* It may be noted that Angelica Catalani had a sister-in-law, Adelina or Maria Antonia (married to her elder brother Benizio, a conductor); a French woman, mediocre both as singer and actress, but who, because of her illustrious name succeeded in finding impresarios to engage her and indulgent audiences to applaud her. About 1828 she appeared in Paris, where she presented herself as Angelica's cousin, and where, through Angelica's influence she was engaged at the Théâtre Italien. But the Parisian public refused to accept her. She is probably the artist with that surname who sang Elena in the première of Schira's *Elena e Malvina* at the Scala, Milan, on 17 November 1832.

** Portugal's real name was Marcos Antonio da Ascençâo (or, da Assumpçâo) but he became known by the more euphonious Marcos Portugal or Portogallo.}

Portrait of Angelica Catalani by the Cork-born artist, Adam Buck (1759-1833).

celebrated", but "la divina", and has been re-engaged for a further three years at a fee of 32,000 crociati—that is 8,000 zecchini* a year'. He then goes on to list the presents of jewellery which she had also received from the Queen.[19] But this re-engagement she was never to fulfil for in 1805 she married an official attached to the French Embassy[20] in Lisbon, named impartially Paul** and Ferdinand André Valabrègue, who now became her manager. Her last creation in Lisbon (during the carnival of 1806) seems to have been Vonima in Portugal's *La Morte di Mitridate*.[21]

Travelling through Spain she halted in Madrid, where on 22 and 24 March she gave two concerts at the Teatro de los Caños del Peral, reputedly under the patronage of Queen María Luisa. Each evening between the acts of a sacred drama (it was Lent), she sang four arias from her repertoire, and prices of admission were raised to 20, 40 and 80 reals.[22]

She then stopped at Bordeaux for another concert,[23] where seats cost six francs each,[24] and thence to Paris. She gave two concerts at St Cloud on 4 and 11 May 1806 in the presence of Napoleon, and three at the Opéra on 21 July, 11 August and 3 September. Seat prices were trebled for the latter public performances, and although this did not lessen the demand for seats,*** the Paris critics were not entirely disarmed.[25] Napoleon liked her, however, and commanded her to an audience at the Tuileries where he peremptorily offered her a place at the Opéra at a salary of 100,000 francs a year, with two months vacation. Intimidated by the brusquerie of the encounter she retired without having the courage to tell her imperial impresario that she was already committed to an engagement in London through the British Ambassador in Portugal.[26] Consequently she was obliged to set sail by stealth, crossing from Morlaix to Southampton[27] on board a vessel which had been sent for an exchange of prisoners. She relates how her baby son, then only seven months old, made the crossing with her.[28]

Having arrived in London she made her first appearance there at the King's Theatre on 13 December 1806,[29] in the same opera in which she had made her

---

* A zecchino was worth approximately half a sovereign.
** A manuscript letter in the British Library (Department of Manuscripts Egerton 2159 f80) addressed to a 'Monsieur Benelli' is signed P. De Valabrègue.
*** When she returned to Paris in 1825 to give concerts at the Salle Cléry she was a failure. Time had passed and all the brilliance had vanished with it. (*Vie et Aventures des Cantatrices Célèbres*, M. and L. Escudier, p 242).

debut at the Teatro São Carlos in 1801, Portugal's *La Morte di Semiramide*. In this opera, and those that followed, Catalani created the sensation expected of her, and Frederick Jones, hearing of her fame, set out at once for London to offer her an engagement. Opportunely the then acting manager of the King's Theatre was the Irish tenor, Michael Kelly* (he had held the post since 1796) and in his *Reminiscences* he relates how he had selected singers from among the King's Theatre company to travel to Dublin. He explains that for his help with the engagement he was to have a free benefit, while 'Madame Catalani was to have a clear half of the receipts for each night's performance; and Mr Jones the other half for paying all the performers, orchestra, &c', and concludes by recounting that 'the agreement was signed and sealed in my saloon in Pall Mall. M. Valabrique, [sic] Madame Catalani's husband, was kind enough to offer me a seat in the travelling-carriage to Dublin, which I accepted. I was their guest throughout the whole journey'.[30]

It could hardly be called a company which Kelly brought to Dublin, for, apart from Catalani and himself (he too would sing besides presiding at the pianoforte) there would be two others only in the party, Giovanni Morelli, a buffo-baritone, and Carlo Rovedino, a bass. This in turn may explain how Jones was able to offer Catalani such very liberal terms as half the gross receipts.

Of these two male singers, Giovanni Morelli was the more important. Michael Kelly tells how he began life in Florence as Lord Cowper's running footman. He was in fact born in Livorno (Leghorn) and made his first stage appearance at the opera house there. Cowper it appears had overheard him singing one night and, recognising the quality of his voice, had taken him under his protection and sent him for tuition to the celebrated *basso comico*, Francesco Marchesi. From 1770 onwards he appeared at the Teatro del Cocomero, Florence, where acquiring great fame he then proceeded to other leading Italian theatres as well as to Paris and London. Kelly, who first met him in Venice when Morelli was appearing at the Teatro San Samuele, unequivocally describes his bass voice as 'the finest I have ever heard'.[31] In April 1787 he had made his debut at the King's Theatre, London, in Paisiello's *Gli schiavi per amore*, when James Boaden had found him 'an actor such as the Italian stage has seldom witnessed...he was distinguished for neat articulation and an unremitting attention to the business of the whole stage'.[32]

* See *Opera in Dublin 1705-1797*.

Burney described him as having 'a base [sic] voice of nearly the same force and compass as Tasca's, but infinitely more flexible and pleasing. He is likewise a good actor and superior in all respects to every *buffo caricato* we have had since Morigi's first appearance in the *Buona Figliuola*'.[33] His name appears among the cast lists as late as February 1815, but apart from a season with Catalani at the rebuilt Pantheon in 1812, he seems to have retired from an active career by 1809.* He is known to have continued singing until he had scarcely any voice left, although he was always kindly received by the audience as an old favourite. He should have been singing reasonably well on his visit to Dublin, for his 'powerful assistance' is acknowledged in Paisiello's *Il Barbiere di Siviglia*, 'exceedingly well performed'[34] at the King's Theatre on 9 June of the same year.

Carlo Rovedino had made his debut at the King's Theatre as early as 13 March 1777, as Licida in Rauzzini's comic opera, *L'Ali d'Amore*. A year later in the pasticcio *Demofoonte*, Mount Edgcumbe remembered him as 'a young man with a good bass voice',[35] which Burney would confirm. The most graceful description of his voice comes from Giacomo G. Ferrari, who recalls it as 'both deep and high, warm and sonorous, vibrant and sweet: he combined the chest voice with the head voice like two fibres of a thread of silk, and sang cantabile like an angel'.[36] In 1781 he was a member of the company at the Teatro San Pietro in Trieste,[37] and during the same winter Kelly had heard him sing in Anfossi's *I Viaggiatori Felici* at the Teatro della Pergola, Florence.[38] Between 1788 and 1792 he is discovered in Paris at the Théâtre de Monsieur in the Tuileries (after July 1791 known as the Théâtre Feydeau) which had been founded by Léonard Autié (Marie-Antoinette's sometime hairdresser) and the renowned violinist, G. B. Viotti. Here in a performance of Giacomo Tritto's comic opera, *Le Vicende amorose*, which inaugurated the theatre on 26 January 1789, the *Journal de Paris* portrayed him as a '*superbe basse-taille, dont la manière de chanter a excité les plus vifs transports*'.[39]

In 1793 he returned to the King's Theatre where, by 1813, his voice seems to have deteriorated to the extent that a reviewer** in *The Examiner* wrote, it 'resembles the roaring of a buffalo who has caught a cold; it sets all

---

* Francesco Piovano reports that as early as '1806 he was still living in [London]—reduced to destitution'. (*Rivista musicale italiana*, vol xvi, Turin 1909, p 258).

** This was probably Thomas Barnes, often unduly critical of the performers and performances of his day.

accompaniments at defiance: the orchestra in fact seems ashamed to pay attention to his grumbling vociferations'.[40] A more kindly and perhaps less biased opinion appeared many years later, when it was written of him, 'as basses went then—when, in the serious opera, a few bars of *recitativo parlante* and a part in a trio, were all that was confided to them—[he was] a very fair basso cantante; had a good deep voice but occasionally sang woefully out of tune'.[41] He had lived most of his life in England and died there at his son's house in Sloane Street, London, on 6 October 1822,[42] aged 70. For the season of 1795/96 the King's Theatre fee for Morelli was £800 to Rovedino's £600,[43] which helps to establish their relative artistic ratings. With the passage of time and the deterioration of Morelli's voice, both would receive the same fee of £500 in 1808.[44]

And what of Michael Kelly, whose voice at this time must have been on the point of disintegrating? Six months later, Leigh Hunt would refer to his singing a duet with Braham as a 'knife-grinder's wheel accompanying a flute',[45] while a reviewer in *The Quarterly Musical Magazine* recalled, 'we remember his voice, which Dr Arnold used to say was like the tearing of brown paper'.[46] Even as early as 1800, Thomas Dutton habitually referred to Kelly's 'monotonous croak'.[47] But if Kelly's reputation as a singer is overrated because of his fortuitous association with *Le Nozze di Figaro*, equally under-rated is his ability as a composer. His friend, R. B. Sheridan's mocking apophthegm of 'composer of wines and importer of music' like many a witticism has little to justify it, yet the ridicule lives on. Even a cursory examination of his music shows that Leigh Hunt was much closer to the mark when he wrote, 'we think better of some of Mr Kelly's compositions than many do. If he stole them, as it is alleged, we should like to see the originals; and then we shall maintain that he has been a very tasteful thief; but till then, we must maintain that he has sometimes shewn himself a very tasteful composer'.[48] But Kelly was not only a singer and composer, he was a competent pianist as well, an accomplishment scarcely recognised. The entertainments about to commence in Dublin began on most evenings with an overture 'the Piano-Forte by Mr Kelly—and', notes the reviewer, 'we cannot but consider him peculiarly happy on this instrument'.[49]

Of course in 1807 it mattered little whether the other singers had voices at all, since Catalani's was the only voice of interest to the Dublin public. It was indeed remarkable and almost universally acclaimed, although an attempt to

assess its range and quality today is hampered by contradictory information. Her range is described by Spohr as extensive, from G below the staff to B above,[50] but he recalls some of her singing 'sounding something like the howling of the wind in a chimney'.[51] Various journals* writing of her first appearance in London credit her with a range of three octaves, a distance 'she covers with the greatest of ease',[52] quotes one, thus anticipating another renowned equilibristic act. Biographical details, published in Berlin in 1816 recount how she astonished the English public with a 'run of half tones up to high E flat, sustained for an entire three bars!',[53] but a report from Milan in the same year states that her voice 'has a compass of low G (violin key) to high B flat', and explains that 'the higher notes can appear only in runs'.[54]

We know that the range of her voice became lower with time, but although she is commonly described as a soprano, from the above references and from musical examples published of her singing, to modern ears she may have been a mezzo soprano with an extensive upper register. Confirmation of this comes from yet another London journal, where we read, 'the upper notes are entirely the produce of art and incessant practice. Of such notes she can never be quite certain, and being liable to falter, they will occasionally be out of tune'.[55] Her out of tune singing seems to have been her besetting fault and to have existed from the beginning of her career. Dr William Crotch has reported that 'she frequently sings out of tune',[56] and a reviewer in the *Quarterly Musical Magazine* concurs, observing, 'she varied from the pitch frequently'.[57] Even one of the newspapers which had awarded her her three octaves felt obliged to impart that 'she sings *frequently out of tune* and does not even appear to be sensible of the circumstance',[58] adding 'she seems in every slow movement to have a certain apprehension and difficulty in producing the exact note'.[59]

Coupled with this other blemishes are recorded. One seems to have been the not uncommon problem experienced in passing from one register to another. Spohr records that at about E or F this was quite audible and that three or four notes thereabouts were noticeably weaker than those in the very low or very high parts of her voice. Further, in order to disguise the defect, she used to sing phrases occurring on these notes in half-voice.[60] It was probably

---

* They included *The Courier*, 15 December 1806, *L'Ambigu*, 20 December 1806, and *The Daily Advertiser*, 25 January 1807, although such unanimity may create the suspicion of one reviewer merely duplicating another.

to this imperfection that the physician and musical amateur, Benedetto Frizzi, was referring when, having heard her when she was 20, he would later write that her voice had 'a certain point of weakness, due to some particular negative physical cause and though she turns it to advantage, her voice will never lose that defect entirely, never acquire a uniformity, an evenness'.[61] In contrast, however, yet another writer would report that her head voice began about E or F, 'by virtue of an almost imperceptible transition'.[62]

Then there were times when she seemed to force her voice beyond its natural power or compass until one could 'scarcely distinguish the sound she produces from a scream'.[63] This fault also seems to have been present from the very beginning of her career, for the critic G. B. Nunez, having lavishly praised a performance by her in Rome in February 1799, a month later chides her for pressing her voice until her notes are 'liable to turn into shrieks'.[64] Not that the voice lacked power; in fact one reviewer confirms that 'no band was sufficiently powerful to cover it, no nerves strong enough to resist its influence'.[65] Finally we are told, 'her vocalisation lacked legato',[66] and a London critic who had heard her during her first season there in 1807, having marvelled at 'her manner of ascending the chromatic scale by half-tones…and her skipping the double octave' has to admit that 'in the *sostenuto*, one of the great *stamina* of the vocal art, she seldom soars above mediocrity'.[67]

But, the above being said, almost all else was awe and admiration. Her voice 'was full, rich, and magnificent beyond any other voice we ever heard',[68] records one reviewer, while another writes, at moments 'her spun-out and gently swelling notes produce the effect of musical glasses'.[69] Radiciotti reports that 'Her vocalisation was truly something quite prodigious. Amid the endless embellishments that she weaved with a rare elegance, one heard clearly her chromatic scales, executed with a trill on each note which scintillated like a diamond of the first water. At times she produced a vigorous shake, imitating the highest trill of the lark; at other times she veiled her voice to produce a gentler effect'.[70] 'Her *staccati* are executed with a precision that only a well-handled bow could rival',[71] reports a French writer, who compares her coloratura singing to the high staccato notes later achieved occasionally by Adelina Patti. One of her favourite caprices of ornament is said to have been 'a sort of imitation of the swell and fall of a bell…sweeping through the air with the most delightful undulation'.[72]

Lord Mount Edgcumbe confirms this, but wishes 'she [were] less lavish in the display of these wonderful powers, and sought to please more than to surprise', adding, 'her taste is vicious, her excessive love of ornament spoiling every simple air, and her greatest delight...being in songs of a bold and spirited character...in which she can indulge in *ad libitum* passages with a luxuriance and redundancy no other singer ever possessed'.[73] As she grew older both her singing and her stage deportment became greatly exaggerated until it was declared, 'her whole manner is grown into a caricature of its former self. When she begins one of those interminable roulades up the scale, she gradually raises her body, which she had before stooped to almost a level with the ground, until having won her way, with quivering lip and chattering chin, to the very topmost note, she tosses back her head, and all its nodding feathers with an air of triumph; then suddenly falls to a note two octaves and a half lower with incredible *aplomb*; and smiles like a victorious Amazon over a conquered enemy'.[74] Perhaps the critic who described her style and career most succinctly of all was the one who wrote: 'The truth is that she could play with her voice at pleasure...she could do anything. Yet she failed to touch the heart'.[75]

What she did touch—and touch constantly and effortlessly—was the tinder of excitement which exists amongst every theatre audience, ever ready to burst into a blaze of applause when a great artist is present to provide the needful spark, and Leigh Hunt, in recounting her singing of Handel's 'Angels ever bright and fair' at Covent Garden, expresses this well. He writes: 'She occasionally throws out a note which reaches us at a distance like vocal lightning, and makes us wonder what her voice would be if she exerted it through a whole song; but this vehemence and swell she can contrast with the utmost delicacy and tenuity of warbling; her shake on the upper tones is pure, crystal quivering, like water in sunshine, and seems as if it would be perpetual; and when she suddenly springs aloft from a low note to one of inconceivable height and fineness, dropping down from thence a few still, small utterances, you might shut your eyes and fancy a fairy being, who has shot up to the music of the spheres, and with one finger after another touches them to our distant ears'.[76]

Catalani and the small group of other artists left London for Dublin on 6 August 1807[77] and, travelling via Birmingham (where she attended a play at the theatre), took ship from Portpatrick[78] for Donaghadee—either to take advantage of the short sea crossing, or in order to visit Belfast to arrange her later engagement there—arriving in Dublin six days later.[79] There seems to have been some question of the Polish violinist, Felix Yaniewicz,[80] and the tenor, Righi,[81] accompanying them, but neither artist eventually appeared. Catalani was reported to be 'too much fatigued by her voyage'[82] to appear at her first concert on Saturday, the 13th. (One newspaper ironically surmised that 'either *her notes* or the *Irish notes* were not in readiness'.[83]) Kelly relates that she 'was received and caressed in every society',[84] and there is evidence of this.

The exigencies of touring endured even by the most celebrated artists at this period are notably described in letters written to the Duke of Clarence (later William IV) by his mistress, the famous actress Miss Dorothy Jordan, when she visited Dublin some two years later. She writes as a very embittered and very lonely woman, but if she is not exaggerating, her misfortunes evoke a dismal picture of theatrical travelling at the time, even for the most celebrated artists.

First, there was the sea crossing from Holyhead, which took her ten hours,* during which she recounts, 'We were all dreadfully sick—and quite wet in the little beds, the sea having several times passed over the middle part of the Packet'.[85] Then there were her lodgings in James's Street ('the dirt of the house is beyond everything') which, two days after her arrival she declares she must leave 'or be *literally starved*', explaining, 'It is so bad that I have offered *ten guineas* to be allowed to quit them. Today, I am to draw a dinner from a *tavern*, which will be the first we have had since we came here'.[86] She then settled in Danes Street (*sic*, presumably Dame Street**) whence she writes, 'I dress at my lodgings and go to the House in a chair, and the first *time* of course drew the curtains. The wretches that followed the chair told Thomas [a servant whom earlier she had had to warn against abusing 'the Catholicks'], if I did that *next* time they would throw *stones* into the chair. I now manage by putting a large loose gown over my dress and a muslin veil and leave the curtains undrawn'.[87] She did admit rather grudgingly that 'The audiences are perhaps more *enthusiastic* than even *Bath* or *Bristol*', but that was 'the nature of the *people*'.[88] Finally, receipts were most disappointing. She had expected to receive over £2,000 for her season, but instead had to report, 'If I make £1,000 [I] fear it will be *the utmost*', adding that,

---

* She was lucky! The average crossing was said to take about 14 hours. (*The Public Ledger*, 26 September 1806.)

** Catalani was less parsimonious about her board and lodging, and so presumably more comfortable, staying at Moran's Hotel, Sackville Street (*British Press*, 21 September 1807).

'after all the *boast* that was made, I find that £200 is what is thought *here* a *wonderful stock night*. The House does not hold near five hundred* [pounds]'.[89] During the season she was most hospitably entertained by Mrs Joseph Le Fanu (R. B. Sheridan's sister), Sir Jonah and Lady Barrington and John Philpot Curran. Sydney Owenson called upon her but she refused to receive her.

Catalani seems to have been most cordially received, for when she returned the following year we read of her giving 'a grand dinner' (presumably returning hospitality) 'to several persons of distinction: among the company assembled were the Countess of Meath, Earl Harrington, Earl of Meath, Lady E. Henry, Mr Grattan, and several ladies of fashion accompanied by their beautiful daughters'.[90] There was even a report sent from London that 'Lord Castlereagh unbends his mind from the cares of the War Department in the fascinating company of Mme Catalani'.**[91] Moreover, regardless of Mrs Jordan's disillusion over receipts, Catalani suffered no such disappointment, her husband, M. Valabrègue saw to that.

Valabrègue has been described as 'a man totally untutored in music and not at all of an artistic nature'.[92] Nevertheless the age-old story (which seems to have originated with Castil-Blaze) of how he was asked by Catalani to have the piano lowered and then proceeded to have the legs sawn shorter, may be dismissed as apocryphal. The equally oft-told tale of his retort to the theatre manager's protest that the enormity of Catalani's fee would preclude the engagement of other talented artists—'Ma femme et quatre ou cinq poupées, voilà tout ce qu'il faut',[93] is perhaps marginally more authentic. He was undoubtedly a shrewd man when arranging his wife's engagements, as is evinced by a description of her singing at a reception in Devonshire House about this time. 'She had all her diamonds on', we read, 'and entirely eclipsed Lady Harrowby, who was standing by her at the harpsichord'.[94] But he was also a highly unpopular man. Rumour relegated him either to being 'a life guard man, or something of that kind, not above it',[95] or 'a stupid ignorant soldier'.[96] That he had been an officer in the French army before the revolution,[97] which was also asserted, seems much nearer to the truth, although it was possibly because of this story that Michael Kelly had to publicly protest in the London papers that he, and Catalani, were not spies of Napoleon![98] Socially he seems

---

* At prices charged for her performances, but see Catalani *infra*.
** See also later in this chapter, and in Chapter 7

Portrait of Angelica Catalani
by the Kilkenny-born artist, John Comerford (c. 1770-1832)

to have lacked *politesse* and here opinions of him varied from 'a little forward Frenchman',[99] to 'presumption and impudence double-distilled'.[100] However, since we are told that Catalani was 'very much attached to her Husband',[101] presumably the world's opinion did not really trouble him.

There seems to have been uncertainty about the terms and conditions of Catalani's engagement, caused perhaps by the fact that Frederick Jones (as will be noted later) would soon leave Crow Street for a time to manage Drury Lane Theatre, London. One early announcement reported her to be coming 'for only three performances, at the enormous price of 1,400 guineas!'[102] but even as late as June the same source declares, 'we cannot yet inform [our readers] in what manner the public are to be gratified with her fascinating powers—whether from the boards of Crow Street, or from the proxy Orchestra of the Rotunda'.[103] In the event, the public was to be gratified from both. Neither would it be for a mere three performances; in all, there would be ten. As for her fee, besides two benefits, this was subsequently announced to be 'four hundred British pounds'* for each performance,[104] while in March 1808 a report from London credited her with having earned £2,500 during her visit.[105] Was she worth these then enormous sums? A reviewer in *The Freeman's Journal* reported that it was 'not considered too much by those who have witnessed her exquisite and uncommon merits'.[106] Conversely, in a case taken some time later by Edmund Waters against William Taylor (both of the King's Theatre) over Catalani's engagement there for 1808, the Lord Chancellor (no lover of opera it seems) in giving judgment 'confessed himself wholly incompetent to the task of deciding with respect to Mme Catalani's salary. For his own part he would not give five shillings to hear her sing half a year together.'[107]

From the first, Catalani's reputation created excitement, yet the Dublin public seemed dilatory in taking seats until they could have first-hand information from their friends of her abilities. Three subscription concerts at the Rotunda were announced, the subscription for the three being two guineas; single tickets bought at the door cost one guinea. Then, on 20 August, poor ticket sales were indicated when it was advertised: 'The Subscription having closed, Tickets may be had at the door at 16s 3d** each'.[108]

---

* Irish and English currencies did not merge until 1826. In 1807 the English pound was worth £1 1s 8d Irish. Today, nearly two centuries later, *plus ça change…!*
** Irish coinage, equalling 15 shillings English.

The series of concerts had begun two evenings earlier on 18 August, when we read: 'At an early hour the great room began to fill, and at eight there was an assembly of nearly 500 persons* of the very first distinction. Mr M. Kelly, the director of the concerts, who officiated as Master of the Musical Ceremonies, led her to her place, amid general and repeated applause. It is not possible to describe the sensations which she awakened—it was astonishment mingled with rapture; and these were not a little heightened by her extreme beauty; the rapidity of her execution in such *distinct* and rapid progression of *semitones* ascending and descending, through the entire chromatic scale, was such an instance of flexibility of organ and facility of practice as was perhaps never heard from any other singer that preceded her… Signor Rovedino and Signor Morelli gave two duets in the pleasantest style. Mr T. Cooke filled the place of leader to a numerous and respectable band with a spirit and steadiness, and yet with a dependance on the singer, as evinced the possession of no common talents, particularly in so young an artist'.[109]

She conquered her audience quickly and 'At the end of the first movement of that beautiful song, "Son Regina" by Portogallo, the rapturous applauses which burst from every quarter, and the reiterated exclamations of "*Bravo! Bravo!*" through the room were sufficient evidences how her astonishing powers had electrified the audience. Her second song, *O quanto l'anima*** by the celebrated Maier [*sic*], was not less astonishing than the former; she was repeatedly applauded at every close, and a general *encore* followed at its conclusion, which she very kindly conceded to. But where we conceive she outshone her former efforts was in the last Grand Aria, *Frenar vorrei le lagrime*,*** all that passion could warm, or soul and feeling could bestow, were in this *chef d'oeuvre* called into action'.[110]

The second concert was postponed until 22 August 'in consequence of the very great exertions of Madame Catalani last night',[111] and the programme reveals that the company was joined by the tenor Thomas Philipps and Mrs Cooke from the Crow Street company. Happily, we learn that 'The room

---

* A poor attendance.
** An aria by Mayr which she used to introduce in Nasolini's *Cleopatra*, and Portugal's *Il Ritorno di Serse*.
*** No composer's name is given, but an aria beginning with these words appears both in Cimarosa's *Gli Orazi ed i Curiazi* and in Portugal's *Semiramide*. The latter composer is perhaps the more likely.

which can contain 1500 persons was nearly full… Every person of taste and fashion in and within some miles of the city, were [sic] present'.[112] So word-of-mouth communication had been effective. It is not inconceivable that the postponement was really intended to allow time to sell more tickets following the disappointing attendance at the first concert. A prior announcement of the second concert that 'the entertainments upon the whole, we understand, will be more desirable'[113] at least seems to imply that excerpts from Italian opera alone (even with Catalani) were not enough and that Philipps and Mrs Cooke had been brought forward to make the concerts more popular.

It opened with Michael Kelly singing, 'I've roam'd through many a weary round' from his opera *The Gipsy Prince* (libretto by Tom Moore), 'in which he evinced his usual judgment and finished execution. Mrs Cooke followed Mr Kelly and was very much applauded.[114] Named compositions which were performed during the evening included 'Donne, donne, chi vi crede' by Mengozzi, sung by Morelli,[115] a repeat of 'Frenar vorrei le lagrime' sung by Catalani, who also sang Sarti's charming cavatina, 'Lungi dal caro bene', 'which was so enrapturing that the company could not avoid *encoring*'.[116] Other items were 'A pleasing song from Rovedino',[117] who also sang in two duets, one by Bianchi* with Philipps,[118] the other with Morelli, and 'an elegant Duet between Mme Catalani and Mr Kelly, which was a very high treat indeed to musical amateurs'.[119]

Two further concerts were now given at the Rotunda on 25 and 28 August (one more than originally announced), the latter for Catalani's benefit. As on 22 August these commenced at 8.30. On the first evening pieces by Italian composers whose names had not appeared in the two previous programmes included an aria by Sacchini sung by Rovedino, who, with Morelli also sang a duet by Martini.** Morelli sang a recitative and aria by 'Buranello' (presumably Galuppi), while a quartette by Nasolini was performed by Catalani, Mrs Cooke, Kelly and Rovedino.[120] On 28 August Catalani and Kelly sang 'Il tuo destino, ingrata' from Nasolini's *La Morte di Mitridate*.[121] Rovedino and Morelli each sang an unidentified aria, the first by Cimarosa, the second by Mozart, 'which Morelli executed with his usual humour'.[122] For her solos, with 'O quanto l'anima', Catalani contributed the scena and aria 'Alla pompa', from Sarti's

---

* Probably Francesco.
** Almost certainly Martín y Soler.

Mitridate, besides the aria 'Per queste amare lagrime' also by Sarti, and then ended the concert with 'God save the King', singing the first verse,* Kelly taking up the second, while the other artists joined in the chorus.[123] Throughout her career this would remain a contrivance of Madame's to stimulate enthusiasm among her British audiences. Strangely enough, she may have been obliged to stimulate enthusiasm in Dublin just then, for, although it was her benefit, news received in London proclaimed, 'The attraction of Catalani, like that of all mortal excellence is now on the wane; she sang divinely on Saturday** eve last to a very respectable, though not a very crowded Theatre'.[124] If the report is true it may have been this decrease in attendance which prompted the engagement of a new artist to assist her on the occasion. He was 'the celebrated Young Spaniard (of Irish parentage[!], 12 years old)' who performed on the violin, 'his first appearance in this Kingdom'.[125] The young man of such anomalous nationality was Michael Rophino Lacy, born in Bilbao of an Irish father*** and Spanish mother, who would return to the annals of opera in Dublin some 20 years or so later with his adaptation of operas by Rossini and Weber.

As has been noted earlier, Tom Cooke was leader of the orchestra, and although a reference 'Song. Mr Cooke',[126] which occurs, is in this instance a misprint for his wife, it could easily have been correct, for subsequently he would appear on stage as a tenor. Cooke was a Proteus of the theatre if ever there was one, and years later when he had gone to London it would be

---

*Ellen Creathorne Clayton (in *Queens of Song*, Vol I, 307) the source of whose authority it has been found impossible to trace, reports that in order to help her memorise the words Catalani had them written out phonetically. It is imputed that her version of the King's English read as follows:

'O Lord avar GOD,
Arais, schaetar
Is enemis, and
Meke them fol.
Confond tear
Politekse, frosstre
tear nevise trix,
On George avar hopes
We fix, GOD save te
Kin.

** Incorrect, the concert was held on 28 August, Saturday fell on the 29th. The error raises some doubt about the report's accuracy, and especially its objectivity.

*** The *Dictionary of National Biography* describes his father as English, as does the first edition of *Grove's Dictionary*. However by the second edition of Grove his nationality had been altered to Irish.

The contract for Catalani's Dublin performances in 1807,
drawn up between F. E. Jones and Paul "de" Valabregue (the singer's husband),
and co-signed by Michael Kelly.

written of him: "At Drury Lane M. Tom Cooke is at the same time director of the music, leader of the orchestra, and actor for the role of second tenor, when there is one in the opera. If he does not have to appear on the stage until the second act, he leads the orchestra during the first act, gives up his place in the orchestra to some miserable violinist during the second act, returns later, enveloped in a greatcoat, to beat the big drum in an important passage because there is no one else to do it, or to assist the double basses'.[127] Even of the Dublin concerts Michael Kelly recounts a story of his remarkable versatility. 'One morning at a rehearsal at the Rotunda, Madame Catalani was so ill with a sick headache, that she could not rehearse her song; and as it was extremely difficult for the orchestra, she begged of me to have it rehearsed by the band. Cooke asked me for the part from which Madame Catalani sang; I gave it him. He placed it on one side of his music desk, and on the other, his first violin part from which he was to play; and to my great astonishment, Madame Catalani's, and that of all present, he sang every note of the song, at the same time playing his own part on the violin'.[128]

The company now transferred to Crow Street Theatre, and although the performances there are sometimes described as operas (Kelly himself records, 'There were two operas to be performed, *Semiramide* and *Il Fanatico per la Musica*'),[129] what in fact took place were concerts similar to those held at the Rotunda followed by *scenes* from operas, of which there were three; Portugal's *La Morte di Semiramide*, Nasolini's *La Morte di Cleopatra*, and Mayr's *Il Fanatico per la Musica*. Certainly it was to everyone's advantage to try to present it as a full scale opera season, and accordingly this was done, but reports from the London *Morning Post* describe what was performed as 'the most striking and popular scenes',[130] while the *Morning Herald* places it beyond doubt when it states, 'There was also an Abridgment of, or *Select Scenes*, as the piece was called, from the opera of *Semiramide*'.[131]

During the previous April, particulars of London fashions for the season had been published in a Dublin journal when 'Opera Dress' was described as follows: 'A round robe of pliant white satin, made to fit close to the form; trimmed round the bottom, bosom, and sleeves, with gold brocade ribband. The Curaçao turban of white satin embroidered in spots of raised gold, confined on the forehead with Indian bandeau of the same composition. Necklace, one row of fine brilliants, from whence is suspended a most curious Egyptian amulet.

Earrings and bracelets to correspond. Hair closely confined under the turban behind, and worn in irregular curls in front, divided over the left eyebrow so as to discover the temple. Rose-wood Opera fan, with mount composed of military trophies in transparencies. White kid gloves and shoes'.[132] And that presumably was what fashionable ladies were wearing in Dublin too. Indeed, the year before when attending the theatre (where 'the boxes were filled by the small remnant of our Nobility',[133] due to the Act of Union), 'The Duchess of Bedford wore a turban, with a plume of white ostrich feathers, and a wreath of diamonds, besides a most brilliant necklace, and splendid ear-rings. Her dress was white muslin, over sarsenet of the same colour, trimmed with roses'.[134]

Life and fashion were certainly influenced by Catalani during her visit. For example, Kendrick Yeates, Optician, 22 Henry-st (whose successors are still in business at the Grafton Arcade 180 years later) were importing 'a great variety of Opera Glasses, such as are used in the Opera House, London' and respectfully requested 'Ladies and Gentlemen wanting such…to apply early, as the demand is great, in consequence of the Opera commencing at the Theatre Royal in which Madame Catalani performs'.[135] Down in County Waterford at Tramore where a race meeting is still held each year about the same time, we learn that *'Madame Catalani*, Mr Scully's chestnut filly rode by Mr Weir won easy'.[136] Catalani's greatest accolade of all however was to have her 'striking Likeness…in the Character of Cleopatra' displayed by Madame Tussaud in an exhibition of her wax works at 'No 22, New Sackville-street'.[137]

Six performances in all were given at Crow Street on September 2, 5, 8, 10 (Catalani's benefit), 12 (Kelly's benefit), and 14. The first of these ended with 'select scenes' from *Semiramide*, with Catalani as Semiramide, Rovedino as Seleuco, and Morelli apparently singing both Mitrane and the High Priest (Oroe). It had been intended to perform *Il Fanatico per la Musica*, but this had to be postponed 'on account of poor Kelly being suddenly laid up with an excruciating fit of the gout'.[138] The evening's concert began with an unnamed overture by Winter, followed by arias by Paisiello[A] and Sarti[B] sung by Rovedino and Catalani respectively, a duet by Cimarosa[C] sung by Morelli and Rovedino, and an aria by the same composer[D] sung by Morelli, a trio by Sacchini[E] sung by Philipps, Mrs Cooke and Rovedino, and a duet by Winter[F] sung by Philipps and Mrs Cooke. Then, an important item since it was printed in capitals, 'Nel cor più non mi sento'[G] sung by Catalani '(with her own variations)', the concert ending with a 'Finale-Rondo[H]'.*[139] The orchestra, which, it was promised, would be 'numerous and select' continued to be led by Tom Cooke,[140] but later advertisements announced that 'Dr Cogan [a very early teacher of Kelly] will preside at the pianoforte',[141] as he had for Mara 14 years previously.** Prices of admission were Boxes and Pit, 11s 4d, First Gallery 5s 5d, Second Gallery, 3s 3d (Irish) and the anticipation of full houses is confirmed by the announcement that 'All complimentary admissions are for the present suspended'.[142]

Complimentary tickets unfortunately were not 'free admissions'. The latter had arisen through Jones's need to borrow money which he did by issuing £100 shares on the theatre at five per cent, plus 'a transferable ticket to visit the performances for the time the lenders should be out of their money'.[143] But the arrangement had been made before the engagement of so expensive an artist as Catalani was contemplated, and now Jones was refusing admission to some very dissatisfied shareholders on the evenings she performed. *The Freeman's Journal* (on Jones's side) appealed to the ticket-holders' combined instincts of *noblesse oblige* and patriotism, confidently exhorting that 'no Gentleman would make use of his free tickets to the injury of the Patentee, when such expence is incurred to contribute public amusement. Before the Union that might be demanded as a matter of right; but since having sunk into shopkeepers and reduced nobility, whose means will not allow them to reside in London, we conceive that we are peculiarly favoured by enjoying even once any of these luxuries in which the rank and fashion of England

---

* A 'Non già ancora sorta l'aurora'.
B 'Lungi dal caro bene' *Giulio Sabino*
C 'Se fiato in corpo avete' *Il Matrimonio Segreto*.
D 'Mezzo mondo aver girato'. *The Daily Advertiser* of 4 May 1808 gives Cimarosa as the composer but it has not been found possible to trace any aria of this title by him. His aria, 'Mezzo mondo ho visitato' occurs in *La Pastorella Nobile* by P. A. Guglielmi. Morelli had sung in this opera at the King's Theatre.
E ?'In quellina oh fiotto'. ?Sacchini.
F 'Me n'andrò di Giove al piede'. *Il Trionfo dell'amor fraterno*.
G A well known duet from Paisiello's *La Molinara* on which both Beethoven and Paganini composed variations. Translated as 'Hope told a flattering tale', possibly by Peter Pindar (Dr John Wolcot).
H Identification very uncertain. A reference to 'Nel cor più non mi sento' may be intended or it may have been 'Contento il cor nel seno' by Mayr or Ferrari.

An added complication in attempting to identify arias performed during this period is that the same opening line (especially when bitter tears are said to flow!) may occur in several compositions by different composers.

** See *Opera in Dublin 1705-1797*.

frequently indulge'.¹⁴⁴ This peroration may have relieved the problem for a time, but it did not resolve it.

The meretricious variations with which Catalani embroidered simple arias such as 'Nel cor più non mi sento' were continuously and forcefully criticised and have been described as 'at once her honour and her disgrace to have introduced into practice in England'. The review continues, 'We use this phrase of double interpretation because her chiefest display of agility was manifested in these efforts "O dolce concento"* and "Nel cor più non mi sento" as she sang them, are at one and the same time the most beautiful specimens of simple, pathetic, and lively melodies converted into the most exuberantly florid songs of execution… The selection of such airs for such a purpose was therefore doubly erroneous.—It degraded the vox humana to mere instrumentation, and it perverted and polluted the most exquisite specimen of genuine feeling to this vile purpose… Yet, strange to tell, it was in these very songs that Catalani drew more rapturous applause and perhaps more of the approbation of the entire mass of the public than from any other source'.¹⁴⁵ Which, incidentally, helps to explain the capital letters in the Dublin announcement of 'Nel cor più non mi sento'.

One cannot be certain of what she really did sing in the opera scenes which she chose. Her *Semiramide* excerpts included the cavatina, 'La pena ch'io sento' and the recitative 'Quel pallor' as well as the bravura air, 'Son Regina, son guerriera'. Not infrequently she (and her colleagues) transferred favourite arias from one opera to another. Moreover, she also introduced entirely new arias specially written for her by one composer into another's operas. It should be remembered that she had brought her own music to Dublin, most of it, possibly all, in manuscript. For example, on 10 September, into scenes from *Cleopatra*, she introduced 'Son Regina', and 'a cavatina which she has not yet sung in this City',¹⁴⁶ while on 12 September, to *Semiramide* was allotted 'the Grand Scena and Aria entitled 'Alla pompa' from Sarti's *Mitridate*.¹⁴⁷ 'Nel cor più non mi sento', and a final polacca, 'Contento il cor nel seno' by Mayr became part of *Il Fanatico per la Musica*.¹⁴⁸ It is also quite impossible to decide whether these scenes were performed in costume and with scenery. What little evidence there is suggests they were not.

---

* The chorus 'Das klinget so herrlich' sung by Monostatos and the slaves in Mozart's *Die Zauberflöte* with variations by G. G. Ferrari or Paer.

A 1795 print of the Rotunda and New Rooms.

Scenes from *Semiramide* were performed on 5 September in a command performance. We learn that 'Their Graces the Lord Lieutenant and Duchess [of Richmond] were received at their entrance with great marks of applause and appeared highly delighted with the fascinating powers of this incomparable woman'.[149] One aria not included in any previous concert, 'O cara mia amata figlia' by Paisiello was sung on that evening by Rovedino.[150] Scenes from *Cleopatra,* 'after which the most favourite scenes from *Il Fanatico per la Musica*'[151] were presented on 8 September. Catalani took the role of Cleopatra in the former, while Kelly sang Marc-Antonio, and the casting for *Il Fanatico* was Donna Aristea, Catalani; Don Febeo, Morelli; Il Conte Carolino, Kelly; and Biscroma, Rovedino. The version of the latter opera which was presented probably consisted of one act only, (if as much as that, since who was there to sing the other roles?) and which, when performed in an abridged manner on the previous 16 July had had a new scene 'arranged purposely to introduce Mme Catalani in male attire'.[152] It was reported that Catalani had 'attracted another brilliant and crowded assemblage...the receipts amounted nearly to £1,200'.[153] Virtually the same programme was performed when she had a second benefit two evenings later on 10 September on which occasion 'she appeared in an entire suit of *Irish* manufacture'.*[154]

Kelly took his benefit on 12 September but not without considerable backstage friction as reported hereunder. We read: 'The *Operatical Harmonists* from London it seems closed their late *musicals* in Dublin with a violent *discord.* Kelly complains aloud of the ingratitude of his *friends* Morelli and Rovedino whom he retained in the suite of Mme Catalani, and agreed to pay them fifteen guineas each per night, out of the receipts of the Theatre; but on his own benefit on Saturday last, they declined singing, unless they received the sum of twenty-five guineas each. The consequence was a refusal on the part of Kelly, who contrived to gratify a brilliant and overflowing audience without troubling his *Italian Coadjutors*'.[155] This he achieved by once again singing Marc-Antonio to Catalani's Cleopatra and Arsace to her Semiramide. Because of the contretemps, the remainder of the programme consisted of items from Kelly's operas, *The Gipsy Prince, Forty Thieves, Youth, Love and Folly* and *Feudal Times,* in the performance of which he was helped out by Philipps and Mrs Cooke'.[156]

* This information tangentially lends credence to the opinion that the performances were not given in stage costume, since one suit only (although possibly designed in theatrical style) is specified for two different operas.

There was one further performance on Monday, 14 September, when, 'notwithstanding Madame Catalani is under the necessity of being in Belfast on Wednesday next…she kindly consented to sing once more here…on the condition that one hundred guineas of the Receipts be given to the Lord Mayor to be appropriated for the Release of Prisoners confined for small debts'. Further, 'In order to gratify the curiosity of a greater number', prices were reduced to Pit, 6s; First Gallery, 4s; Second Gallery, 2s; (British).[157] The performance consisted of a *pot-pourri* of the most popular items of the season, ending with the ubiquitous 'God save the King', and on the following day it was reported that she had 'closed her engagement…in a manner honourable to her feelings as it was creditable to her talents'.[158]

She set out for Belfast on 15 September accompanied by Morelli and Rovedino,[159] where she would spend four nights before continuing on to Edinburgh, but previous to her departure she held a levée which 'was attended by the principal nobility and gentry [then] in Dublin'.[160] Before leaving she also sent the following letter to the press, or at least someone sent it for her, since we are informed at the time that 'she cannot speak English'.[161] 'I beg you will, through the medium of your respectable Paper, express my sentiments of gratitude for the many acts of kindness conferred on me by the inhabitants of this metropolis, the remembrance of which never can be obliterated from my memory'.[162] A vastly different appraisal of Dublin audiences than Mrs Jordan would have us believe.

There remains a postscript, in fact, two postscripts. The first concerns Michael Kelly, who, on 21 September, 'embarked on board the Montrose packet for Holyhead…having reaped a golden theatrical harvest in his native city'.[163] The second tells of what some might denigrate as a party of 'canary fanciers', for we learn that 'Sir John Stevenson, Michael Gavan Esq., and a select party of Amateurs have determined on visiting Belfast to have the pleasure of hearing again this Musical Phenomenon'.[164] Let it be realised that that journey would have taken place in a comfortless stage coach.

Chapter 4
Catalani Encore. 1808

It may be appropriate at this stage to correct the frequent errors which occur in Kelly's *Reminiscences* concerning his Dublin visits. In 1807, as already noted, *scenes* from the operas and not two complete operas were performed. He makes a further error in including the tenor Deville among the singers he had brought from London. This artist would not appear in Dublin until 1819. When describing the Italian company's second Dublin visit in 1808, he may also be incorrect in stating, 'I set off with them on the first of August',[1] since on 2 August it was reported, 'Mme Catalani and Mr Kelly leave London on Friday for Dublin',[2] and the first performance did not take place in Dublin until 20 August. Kelly, however, could have left with other members of the company before her. We know for example that three of the group, Siboni, Miarteni, and his wife, stopped at Liverpool on their way, to take part in a concert given by the violinist, Yaniewicz, during the week previous to 20 August.[3] He is certainly incorrect when he records, 'We had to perform two grand serious operas',[4] since three serious and three comic operas were produced. (Just how complete the versions were will be discussed presently.) He is equally incorrect in reporting: 'After performing six nights in Dublin we proceeded to give six performances at Cork*',[5] following which he recounts four performances in Limerick and a further six in Dublin.[6] The Dublin season consisted of 15 performances in all, not 12, while the Cork performances announced 'for three nights'[7] were probably extended to four, but Limerick had only three.[8] Chronologically, most astounding of all his misinformation is that, 'At the close of my engagements in Ireland I set off for London, and…arrived in Pall Mall on the 21st of September'.[9] Kelly in fact would have a benefit in Dublin on 22 September. The date he gives for his return might be considered a slip of the pen for, say, 21 October had he not cited as confirmation the 'destruction of Covent Garden Theatre the night previous by fire',[10] which occurred very early on the morning of 20 September.

---

* Although this may have been the number of performances originally intended there. *The Daily Advertiser* of 30 September confirms that the company was 'engaged to perform six nights at the Cork Theatre'.

The troupe engaged for this year at least made the mounting of a complete opera possible, provided one was prepared to accept what must have been quite outrageous cuts in the scores, and a complete absence of chorus. Since, once again, Catalani headed the company, Dublin audiences accepted readily. With Catalani there were return visits of Rovedino and, of course, Kelly, and debuts by Siboni, a tenor, Miarteni, a bass, and Signora Miarteni, his wife, who was the 'seconda donna'. Tom Cooke was again leader of the orchestra; the well-known player, Spagnoletti, had been engaged as principal second violin, and Michele C. Mortellari had been appointed in Philip Cogan's place as conductor at the piano.

Paolo Diana Spagnoletti was born in Cremona either in 1768 or 1773,* and had trained at the Conservatorio della Pietà de'Turchini, Naples. Thence he went to Milan and from there was brought to London by the famous tenor, Viganoni, where he arrived in 1799,[11] and where he would spend the remainder of his life. He joined the King's Theatre orchestra in 1804, playing with the second desks, but, from the first, he appears to have deputised for Charles Weichsel as leader. As he grew older he suffered from a succession of apoplectic strokes, eventually dying from one on 23 September 1834. His position in musical London seems to be set forth very fairly in the following opinion: 'The zealous and skilful efforts of the leader** although not absolutely meeting the *acmé* of perfection desirable in so arduous a station, have given universal satisfaction'.[12]

Mortellari was an old musical colleague of his who had joined the King's Theatre as harpsichordist in the same year of 1804. In later years he would compose ballet music for the same theatre. His reputed father, Michele Mortellari, had been an opera composer, a pupil of Piccinni, who had settled in London in 1785. Mortellari junior quit the King's Theatre in 1807 for other engagements, whence his parting was regretted both by *The Courier*[13] and *Morning Chronicle*.[14]

The relative positions of 'Conductor of the Orchestra' and 'Leader of the Band' need explanation since they may seem synonymous. The 'conductor' at the time was a keyboard player, the 'maestro al cembalo', who sat at his piano-forte or harpsichord and guided the singers on the stage, but in no way

---

* The former year seems to have appeared first in Baptie's *Handbook of Musical Biography*, and to have been copied ever since, but the Kensington (Brompton District) Parish register gives his age when he died as 61.

** Of the King's Theatre orchestra, which he became in 1816/17.

Madame Catalani, in the Character of Argenide, in the Opera of Il: Ritorno di Serse

Angelica Catalani in the first of the three operas of that name—composed by M. A. Portugal in 1797. The two later versions are by Zingarelli (1809) and Nasolini (1816).

did he conduct the orchestral players. They were led or conducted by the leader or principal violin, which arrangement not infrequently provoked disharmony both on stage and in the pit. Rarely there could be even a third hand involved. For example, an Italian company presenting burlettas at the Pantheon Theatre, London, in 1812 had Spagnoletti as 'Leader of the Band', and Vincent Novello 'At the Piano'.[15] Yet at an earlier rehearsal it was Vittorio Trento, the 'Composer and Director of Music',[16] who 'stood on the stage and directed the orchestra'.[17] Perhaps either Spagnoletti or Novello—or both, were missing from their posts on that day! This separate control of instrumentalists and singers persisted until the early 1830s, when the two musicians in charge began to be replaced by one, called in the beginning, the 'Director of the Music'.

Spohr, in a letter written on 17 April 1820 to his friend the composer-cum-banker, Wilhelm Speyer, sums up the situation in London as follows: 'The method of conducting here—in the theatre and at concerts—really is the most absurd imaginable. They bring on two conductors, yet neither really conducts. He who is described as "conductor" on the bills, sits at the piano and plays from the full score, but indicates neither the beat nor the tempi; this is supposed to be done by the "leader" or first violin, but since he has only a first-violin part in front of him, he can hardly be of any help to the orchestra, and thus contents himself with sawing away at his part, and letting the orchestra keep with him as best it can... The Orchestra [at the Italian Opera], conducted as described above, constantly staggers along, forever in danger of coming to grief'.[18]

The size and composition of the orchestra also merits attention. In Dublin in 1808 it numbered 36 players, made up of eight first violins, six second violins, two violas, three 'cellos, two double basses, two flutes, two oboes, two clarinets, two bassoons, two horns, two trumpets, one trombone, one 'Double Drums', and one piano-forte.* But as to its quality?! The pianist, Ignaz Moscheles, who, when he visited Dublin in 1826 to play a concerto complained that he had suffered 'martyrdom at the rehearsal, chiefly from the wind instruments'.[19] Later he qualified his criticism by diplomatically if unrelatedly commending

---

* The names of the players were: *Leader*, Tom Cooke; *Principal Second Violin*, Spagnoletti; *First Violins*, Barret, Cloken, Cubitt, Barton, Glover, Crozier, Wilson; *Second Violins*, Fallon, E. Cooke, Aubry, Gain, Mellish; *Violas*, Bowden, Thompson; *Violoncellos*, R. Cooke, Kelly, Robinson; *Double Basses*, Grey, Smith; *Flutes*, D. Bourke, B. Cooke; *Hautbois*, Catalani, Metheringham; *Clarionets*, Sipper, Burgess; *Bassoons*, Bond, Atter; *Horns*, Alley, Mulligan; *Trumpets*, Walker, Davis; *Trombone*, Meglier; *Double Drums*, Henley; *Piano-forte*, Mortellari. (*Freeman's Journal* 30 August 1808)

Madame CATALANI in SEMIRAMIDE,
her first Appearance in England, Dec.r 13.th, 1806.

'the musical taste and enthusiasm of the Irish nation'.[20] Catalani, the oboist, whose name appears among the Dublin orchestral players of 1808, was Madam Catalani's second brother, Guglielmo.[21] In the following November he married Miss Cranfield, a ballet dancer from the King's Theatre.[22] He died in Paris in 1819, at the early age of 33.[23]

Amongst this season's new singers were Nicolo Miarteni, who was replacing Morelli but was no substitute for him, and Miarteni's wife, Maria. Originally they are found performing with an Italian opera company at the German Theatre, Amsterdam, between 1804 and 1806, with Miarteni also acting as impresario.[24] They had been engaged for the King's Theatre by Waters, either in Italy[25] or Amsterdam[26] at a fee of £400 and £100 respectively for the season,[27] but as earlier recounted, Waters was then embroiled in litigation with the theatre's manager, Taylor, which would not be resolved until 1814. So, the Miartenis were to find themselves in London with a worthless contract and without engagements. A Colonel Greville unexpectedly came to their aid. This gentleman had lately established the Argyle Street Fashionable Institute, which edifice combined 'an elegant theatre with rooms for cards and various entertainments'.[28] It was patronised almost exclusively by the nobility, an audience 'generally distinguished by a cold reception of talents employed for their amusement, and by a constant and somewhat *unfeeling* warfare with the orchestra, to determine which should be the loudest, the *talkers* or the *performers*'.[29] In this private theatre the Miartenis were first heard in London on 22 February 1808, when the audience included 5 Dukes (including those of Cambridge and Sussex), 2 Duchesses, 4 Marquises, 6 Marchionesses, 12 Earls, 19 Countesses and 15 Peers.[30] Nicolo had in fact made an inauspicious debut there on 11 February before an invited party of Greville's numerous friends, when one reviewer contritely observed, 'We are sorry that we cannot announce him as an acquisition to our musical stock. He has a very unproductive voice'.[31] Some concerts later, what was probably a more considered judgement recorded: 'As a singer his voice is firm and remarkably extensive, but it is more as an actor that he must calculate on erecting the substantial fabric of his fame'.[32] Eventually on 31 May 1808 he reached the King's Theatre, where he sang with Catalani, once again to conflicting reviews. He does however possess one claim to distinction for he was the first to introduce Cimarosa's *Il Maestro di Cappella* (performed earlier by him in Amsterdam) to London, at the Argyle Rooms on 4 April 1808.³³ But for Maria Miarteni there can be no dispensation, since of her we read, 'though very powerful [she] is very defective, from her being perpetually out of tune'.³⁴

An artist of far greater stature was the tenor, Giuseppe (Vincenzo Antonio) Siboni, born in Forli on 27 January 1780. The best description of his singing comes from a retrospective review which appeared in 1830. 'Siboni's style', it records, 'was the best of any tenor we have heard; somewhat too florid occasionally; but, when the scene demanded it, grand and commanding. His person was above the middle size, [a less kind description was 'bulky'³⁵] and his action rather fitted for the serious opera in which he was comparatively seldom allowed to appear, than for the comic, to which he was condemned. Many of Paer's finest tenor parts were written for him. His compass reached two octaves, from B flat to B flat, and his falsetto blended so well with the *voce di petto*, that the transition was almost indiscernible. On the other hand, his tone was guttural and husky'.³⁶ Mount Edgcumbe confirms the latter, describing his voice as 'thick and tremulous',³⁷ although a third report speaks of 'a fine tenor voice of peculiar sweetness'.³⁸ Yet another report, from Berlin in 1812, declares, 'Signor Siboni's voice is actually baritone—but produces many high notes by virtue of the falsetto. The voice is no longer in its first bloom—indeed, it is in its decline, but with art and assiduity he finds the means still to charm and entertain his listeners'.³⁹ His principal shortcoming seems to have been a small voice, described as 'a voce di camera…delightful in a room'⁴⁰ but 'too feeble'⁴¹ for the King's Theatre. He had an extremely successful continental career, singing (besides Berlin) in Rimini, where he had made his debut in 1797, Florence, Genoa, Milan, Naples, Prague, Warsaw, Vienna, and St Petersburg, before he settled in Copenhagen as director of the Royal Opera and of the Conservatorium. He died in that city on 28 March 1839.

With the singers Kelly had brought two ballet dancers to Dublin on this occasion. Ballet divertissements had long formed occasional interludes in Dublin stage performances,* but seldom if ever were they a part of opera productions. The two artists engaged were Mlle Presle, originally from the Paris Opéra⁴² via the King's Theatre, and M. St Pierre. Mlle Presle seems to have been rather exceptional. On her first appearance at the King's Theatre it was recorded, 'Her person is small but well formed, and her features contain a delicacy of

---

* See *Opera in Dublin 1705-1797*.

Qual Pallor — Recit.
and
Son Regina — Air,

Sung by

Madame Catalani

In the Opera of

SEMIRAMIDE,

Composed by *Marcus Portogallo.*

Price 4/6

Printed & Sold by Mr. Kelly at his Opera Saloon, No. 9 Pall Mall.

expression, which rendered her performance very interesting. She has great agility of person, and her dancing is extremely graceful'.[43] Her salary there for the 1808 season was £650.[44] She came to Dublin as principal dancer, and her inability to perform on any evening due to illness was continually reported, until on 8 September one read the culminating mournful news: 'Mlle Presle sailed on Friday last for England—she was a charming dancer—her illness is a decline, and it is feared she cannot recover'.[45] Finally, on 21 December, it was reported: 'Mlle Presle died about three weeks ago, on board the Packet, as she was returning to the Continent…carried off in the 18th year of her age'.[46]

The unfortunate artist was replaced by Madame Nora, also from the King's Theatre. She had arrived there from the theatre in Lisbon in 1807,[47] when it was formally announced, 'Her person is well made and she treads the Stage with great ease and elegance'.[48] Her first appearance in Dublin on 8 September is reported to have been 'honoured with rapturous applause'.[49] St Pierre, yet another from the King's Theatre, where he performed between 1801 and 1803, was well known in Dublin both as dancer and choreographer. In January 1804 he is to be found dancing with 'Miss Adams and Miss H. Adams'.[50] There were in fact *four* Misses Adams, 'natives of Ireland',[51] all ballet dancers.[52] Reprehensibly, during a performance in 1807, 'an orange was thrown at one of [them] by some ill-natured contemptible fellow'.[53] Later in the same year, St Pierre married 'Miss Cramer of Crow-st',[54] and ten years later, evidently still residing in Dublin, took 'a large and commodious House in Stephen's-green South [No. 59]…to establish an Academy for Dancing'.[55] An interesting ballet in which he danced in 1815 was *Blaise et Babet; ou, La Fête au Village*, 'The music by the celebrated Rossi—the Overture and Finale by Jean Jacques Rousseau'.[56] He was considered to be a better dancer than choreographer, for it is recorded, 'We admire the talents of M. St Pierre, as a member of the *Corps de Ballet*, but as an active *composer* or *compiler* or *Maître de Ballet*, he is evidently culpably negligent'.[57]

Other dancers who appeared in Dublin during the early part of the century were Signor Angellini (? Pietro Angiolini), who also played 'select pieces of music on the viol d'amour',[58] and Madame La Costa, both of whom had first appeared in Dublin as early as 20 January 1801.[59] There was also Miss Lupino from the London theatres, who danced at Crow Street in 1807,[60] and who, in the following winter of 1808, danced in St Petersburg. Neither Mlle Nora nor

St Pierre took part in any of the operas. Their contribution comprised a *pas seul* or *pas de deux* danced between the acts or at the end of the opera.

As has been noted earlier, the 1808 season commenced on 20 August, and once again couturiers and perruquiers were courting ladies of rank and fashion. One, Baseggio, who was Hair-Dresser to the Right Hon. Countess of Harrington and Madame Catalani, was offering a 'Hair Border *Ala* [sic] *Catalani* without sewing'. from his salon at 18 Dawson Street, price eight shillings.[61] There was a change in ladies' fashions from the previous year. For 1808, 'scarlet mantles profusely trimmed with fur, were general [at least in London], and the *Mary Queen of Scots* caps made in black velvet, edged with pearls, imparted additional charms to the fair wearers. The *Cleòpatra* pendant pearl, which hangs on the forehead, has a very rich effect'.[62] There was also a slight reduction in admission prices. This year boxes were 11s 4½d; pit, 7s 6d; first gallery, 5s; and second gallery, 2s 6d.[63]

The first opera chosen for performance was Portugal's *Semiramide*, with Catalani, Maria Miarteni, Siboni, Rovedino and Miarteni. Some days later it was reviewed as follows: 'On Saturday night the fame of Mme Catalani and the promise of a complete Italian Opera (for the company, we understand, is full), invited to the Theatre Royal all the rank and fashion in town, as well as many amateurs from the country. The Opera was *Semiramide*. …[Catalani's] *entrée* was greeted with applauses, loud and reiterated; indeed, the beauty and striking majesty of manner with which she entered, would have called down a burst of applause, even on an unknown candidate for public favour;—but when the clear, pure, and finely modulated tones of her voice thrilled through the house, it would be difficult to describe the enthusiasm which she excited—impossible to convey an idea of the impression she made. She felt the opinion which the Irish Public entertain of her powers—and she repaid the plaudits she received by a display of voice, never, we will venture to say, equalled—and by a style of acting, never, perhaps, surpassed… Signor Siboni sustained Arsace with great ability…and was received as he deserved, with distinguished applause'.[64] For good measure, during the performance Spagnoletti and Mortellari performed a concerto on the violin and pianoforte between the acts.[65] *Semiramide* would be given a second performance on 15 September.

A performance of *Il Fanatico per la Musica* (with Act I of *Semiramide*) was postponed from 23 August to the 25th due to most equivocal circumstances.

Saunders' News-Letter gave the reason of 'Signor Siboni being suddenly seized with a severe cold and hoarseness',[66] but both *The Dublin Evening Post* and *The Freeman's Journal* published the far more dramatic news that it was 'In consequence of the sudden death of Signora Siboni (wife to the singer)'.[67] Since *The Freeman's Journal* retracted the following day to the more prosaic cause for cancellation, a 'severe cold',[68] and since Siboni appeared in the performance on the 25th, one is encouraged to believe that the news, like the premature report of another famous death, was much exaggerated. No reference at all is made to the affair in a review (although the report reappears in the London *Morning Chronicle* of 2 September); instead, we read that Catalani was 'distinguished by a vein of humour most admirably expressed, and possessed of an arch vivacity which places her pretensions as a Comic Actress in a very high point of view. She gave "Hope told a flattering tale"* with her own variations, in the most exquisite manner'.[69]

Il Fanatico per la Musica, which was first known as *Che Originali!*** later altered to *Il Trionfo della musica* and *La Musicomania* (from the French piece *La Musicomanie* on which it was based) would ultimately become best known under this title. One can reasonably assume that this was the one-act version performed at the King's Theatre in 1808, and again in June 1809, and retrospectively described as a 'pasticcio' for which Mayr, Fioravanti 'and half-a-dozen other mediocre composers were laid under contribution, and the child of many parents announced as a comic opera'.[70] It had six performances in Dublin, on August 25, 30, September 13, 20, 22, and October 15, and became the most popular opera of the season. So popular, in fact, that 'the raising of the prices of admission [even as early as 25 August] had the happy effect of not over-crowding the house, and yet of making the general receipts proportionate to the vast expense'.[71]

It is difficult to tell whether the receipts were very good or very bad, for there are conflicting reports. On the one hand we are told, 'The Duke and Duchess of Richmond, and all the higher classes of the nobility have patronized the Operas in Dublin; but notwithstanding the incomparable charms of Catalani's voice, they have failed of attraction'.[72] Then, on the other, we read,

---

* Presumably in Italian.

** *Che Originali!* had first been presented as a one-act opera, but *Il Fanatico per la Musica* was later extended to two acts.

'The name of Catalani operated like a talisman, and the half-guineas rolled into the Theatrical Treasury in a golden plenteous tide. She sung with undiminished effect, though labouring under a severe indisposition which required the application of *blisters*'.[73]

Then, too, there was the recurring problem of 'Free Admission Rights', and this year a share-holder named Doyle, having been barred from a Catalani performance, issued a writ of execution against Frederick Jones for the amount of his £100 bond. Jones in turn pleaded that the contract was usurious, since, besides the legal five per cent interest, the ticket usually sold for a further 15 guineas a season. In summing up, 'The Master of the Rolls said, that before he could decide whether the contract was usurious, he must ascertain the value of Mme Catalani's *notes*. Only two Masters in Court, Messrs Henn and King, were musical amateurs, and he would not trouble them with the enquiry. The case on the defendant's [Doyle's] part was no more usurious than that of a person who might lend money to Mr Foot the tobacconist, at legal interest, and get *a sneeze* from the effluvia in the shop. He could not shut his eyes against what he well knew to be the custom in this case. Persons admitted gratis were necessary appendages to the Manager, and necessary stuffing for a House, without whose attendance bad acting and worse plays would run a greater hazard than not being received with applause'.[74] Jones's motion was refused with costs.

On 27 August, *Didone* by Paisiello was produced. The cast was made up of the entire company—including even Kelly—who, up till then had 'been prevented by a violent attack of the gout from indulging in potations of claret with his convivial Countrymen'[75] from taking part in the operas. His illness may also have been the reason for having Siboni appointed joint 'conductor of the management of the stage'.[76] *Didone* had been brought out at the King's Theatre the previous January, where musically it was considered unsatisfactory. It 'abounds with spirited passages', we read, 'but [Paisiello's] genius does not appear suited to the grand character of music; there is too much playfulness for dignity, too much beauty for sublimity'.[77] It was a splendid vehicle for Catalani, however, and she 'shewed more propriety of action as Dido than she has ever before exhibited in the serious operas; her figure and animated exertions appeared to great advantage'.[78] Kelly confirms that 'La Didone was her *triomphe*, both as an actress and a singer'.[79]

In the Dublin production, 'Kelly, [as Æneal recovered from his indisposition, never appeared to more advantage for these many years', while, 'Never was acknowledgment of talent more justly bestowed than on Siboni'[80] as Jarba. It was also advertised that 'In the course of the Opera will be exhibited an entire new scene representing the Palace of Dido, painted by Signor Zafforini'.[81] The performance ended with a ballet. *Didone* would have two further performances, on September 6 and 22, and single acts would be staged on two other evenings.

Much has been written earlier about Catalani's voice and singing, so perhaps this is a suitable point at which to say something of her as an actress. In appearance, we are told, she is a brunette, 'rather inclining to *embonpoint*',[82] and 'though not tall, possesses a very animated and interesting figure; her face, which is oval, and reminds us of that style of female beauty Sir Joshua Reynolds invariably fixed upon his canvas, is full of feminine expression and softness'.[83] That she also looked attractive on stage is confirmed by an account of her performance of Vitellia in *La Clemenza di Tito* at the King's Theatre. 'Her dress on her first appearance,' we learn, 'was magnificently arranged, and the glittering diadem, the head wreathed like an antique bust, the scarlet tunic and the looped and tasselled drapery, gave the full impression of regal grandeur to a face and form filled, beyond all that we have ever seen, to the expression of dignity and grace—the softness of a woman mingled with the solemn and tragic majesty of a fallen queen'.[84] However, the assertion that 'Her dignified acting' (in the 'ghost scene' from *Semiramide*) was 'acknowledged not to be excelled by the great French actress, Madame Clairon'[85] must remain open to question.

One remarkable characteristic she possessed, referred to again and again in the literature, is the impression she gave her audience of smiling continuously. 'If a deaf man was to be present at the performance of Mme Catalani', we are told, 'he would, from her gestures and countenance, pronounce that she was performing a comic character, for joy is the only passion she attempts to express; even the dose of poison is drunk as unconcernedly as if it was a glass of lemonade, and the threatened death of the husband and children are absolutely treated as very comical circumstances'.[86] An extraordinary reason is advanced to explain this tendency. It seems that, while at a distance she appeared to have 'a smile upon her countenance in the deepest parts of her serious scenes, on coming closer one realised that what one had thought to be a smile was, in fact, only a violent convulsion of the muscles… This draws up the mouth at the extremities, and so seems to a distant spectator a cheerful expression'.[87] Nevertheless, this hardly explains her seeming lack of inner motivation.

She did seem to try to become involved, and on one occasion at least, her 'attention to the *bye play*' is commended—'the more so, as the other performers seem to consider such a thing as totally unneccesary; indeed, from their apparent anxiety to be near the prompter, they rather appear like children at their school lesson'.[88] One opinion which seems both objective and accurate has set down her standing as an actress as follows: 'This Lady if she could be divested of her foreign accent, and introduced into any principal character of English Tragedy, would probably be regarded by the few persons who really understand the science of the Stage, as a very indifferent Actress: and when connoisseurs are heard to express an admiration of her talents, *as an Actress*, they mean to be understood as speaking of her performances compared with those of the Ladies and Gentlemen who occupy the inferior situations. She certainly is a good Actress *for the Opera*: and so Master Betty was a good Actor, *for a boy*'.[89] And that would seem to have been it!

There was, of course, little to help her in the way of production since production of opera then was non-existent—when it was not farcical. Examples of the latter can be found in performances of *Didone* and *Semiramide* in London. It was asserted of the former, 'As this Opera is taken from ancient history, we have a right to expect an adherence to ancient customs… In the palace of Dido we did not expect to see a library table with inkstand, &c',[90] and of *Semiramide*, 'The ghost of Nino is admirably managed but his exit is made not a little ludicrous by his turning round to shut the door'.[91] Another performance tells of the ghost appearing to Semiramide and 'shaking at her with savage grin his grisly locks'.[92]

It was a time when theatre lighting consisted of tallow candles stuck into tin circles hanging above the middle of the stage, 'which were every now and then snuffed by some performer',[93] and when tawdry scenery and costumes faded into their murky surroundings. Some of the costumes at Crow Street Theatre may have been in reasonable condition, for we know that items from the wardrobe used either to be lent or hired out 'by the manager for masquerades'.[94] Catalani naturally brought her own, though not always to acclaim, one critic observing, 'in *Didone* her robe, or mantle, although regal, was infinitely too heavy…her other dress [in *Il Fanatico*] pleased us much better'.[95]

On 1 September, the staging of new operas continued with what was purported to be Valentino Fioravanti's *Il Furbo contro il Furbo*.* When first performed at the King's Theatre the previous March it was considered to have 'delightful music with action rather more orderly than is usual in an Italian *imbroglio*'.[96] This would seem a fair assessment of Fioravanti's and Tottola's work, until one reads on to find: 'The idea is taken from Le Sage's comedy of *Crispin rival de son Maître* and is adapted to the Italian stage by Buonaiuti—a man of rich talents. The music was originally by Fioravanti, but with considerable alterations by [Giacomo G.] Ferrari, who has prepared it for the voice of Catalani. Both he and the poet will acquire increased fame from the performance'.[97] Ferrari was a minor composer, born in the South Tyrol, and trained in Italy, who had been engaged as harpsichordist at the King's Theatre in 1808, where he composed operas and ballets. One wonders, especially with Tottola's libretto rewritten, how much of Fioravanti's music survived. Ironically, the most popular air in the opera, 'Papà, non dite no', was by Buonaiuti and Ferrari.

The opera, whoever composed it, had an excellent reception in Dublin, and Catalani enjoyed a great personal success in it both as singer and actress. Two items especially are singled out, her singing of the above aria, and 'the song she addresses to Rovedino in her grotesque and strange dress'.**[98] Kelly, Miarteni, Rovedino, and especially Siboni, were also applauded, and the *Freeman's Journal* gave almost equal prominence to 'an attendant' who walked on stage behind Catalani as she sang, for a £10 bet.—'It produced a burst of laughter'.[99]

Two further performances were given, on September 3 and 10, the latter commanded by the Lord Lieutenant. On this occasion, 'Mr Jones was prevented from being present on account of a domestic calamity [His mother had died, aged 98]. Mr Kelly attended their Graces on their *entre* [sic] and departure from the Theatre'.[100] Kelly would recall the incident with pride many years later, when he related, 'it was my duty, as director, to light the vice-regal party to their box, as they came in state. His Grace was particularly kind in his

---

* Which, in its time, existed under several aliases. It began as a cantata, *Il Furbo contro il Furbo*, with Fioravanti's own libretto, produced at the Teatro Fiorentini, Naples, in 1794, and was next retitled *L'Arte contro L'Arte*, but still performed as a cantata in Turin in 1795. It was then produced as an opera with new music by Fioravanti and libretto by A. L. Tottola under the original title of *Il Furbo contro il Furbo* at the Teatro San Samuele, Venice, on 29 December 1796. (See also *Annals of Opera 1597-1940*, A. Loewenberg, 2nd [3rd] ed, col. 530).

** 'Changed by an enchanted ring', and presenting 'herself as from China'.

conversation and remarks, and, at the conclusion of the opera when I again lighted them to their carriages, her Grace the Duchess would not permit me to attend them beyond the box'.[101] On the same evening Mlle Nora made her second appearance of the season, between the first and second acts in 'a new Ballet' choreographed by St Pierre, called *'The Mexican Festival'*.[102]

Mlle Nora had made her first appearance two evenings earlier, when Portugal's *La Morte di Mitridate* was performed for the first time. When it was produced at the King's Theatre in April 1807, the *Morning Post* reported, 'Before this an English audience never heard Madame Catalani to so much advantage. She was so astonishingly great in almost every part of the Opera, that it would be superfluous to name either the particular air, or passage, in the execution of which she did not electrify her audience and diffuse universal delight'.[103] Both *Mitridate* and *Il Furbo contro il Furbo* had engaged the entire company.

Mitridate had this one performance only, but separate acts were included on two further evenings. One of these was on 17 September, when Catalani took her benefit. It was a gala performance—of a kind! To Act II of *Mitridate* was added 'the popular aria of "O quanto l'anima", a new song by Signor Siboni and a song composed by Mr Kelly'. But this was merely the beginning, for the principal event of the evening was a performance of Paisiello's *La Frascatana*, although not for the first time in Dublin.* Into this Catalani introduced Mozart's air with *her* variations, and 'the scena, so universally admired in London...in which she displays the graces of the Shawl'. Between the acts Spagnoletti played 'a concerto on the violin', and at the end of the opera there was 'an entirely new Ballet' in which Mademoiselle Nora danced 'a Spanish Fandango with the Castanets'.[104] The 'graces of the Shawl',** were revealed to be 'some very graceful attitudes' assumed by her on stage 'not unlike those of the Roman figures which we see in the ancient paintings',[105] and which, when they were demonstrated in London, 'were admired beyond description'.[106] W. T. Parke tells us it was in *La Frascatana* at the King's Theatre on 9 January 1808 that Catalani received a double encore for an arietta by Ferrari, 'Non vi fidate agli uomini'. 'As none of the great singers who had preceded her...had ever received a similar compliment, this appeared extraordinary till the fact came out that Catalani, as a part of her engagement for that season, had stipulated to have the privilege of fifty orders nightly!'[107]

---

* See *Opera in Dublin 1705-1797*.
** Taught to her by the ballet dancer, Miss Gayton.

Three further performances took place. On 20 September, Siboni took his benefit, when he and Catalani introduced 'a new duet' into the first act of *Mitridate*, while in *Il Fanatico* he sang 'an Air accompanied by himself on the Spanish guitar'.[108] Michael Kelly had his benefit on 22 September, with performances of *Didone* and *Il Fanatico*, and, it being the end of the season, Catalani brought the curtain down on 'God save the King'. There remained one final performance, on 15 October, before the company returned to England. At first, two performances had been advertised for October 13 and 15,[109] but this was later reduced to one, for which Sarti's *Gli Amanti Consolati* was announced by *Faulkner's Dublin Journal*, *Hibernian Journal*, and *The Freeman's Journal*, prior to the performance.[110] Then, on 15 October, the latter two papers advertised *Il Fanatico per la Musica* instead. The change may have been due to rehearsal difficulties, or, more likely, it had become evident at the box-office that the new opera would not draw. To *Il Fanatico* was added *La Frascatana* and 'previous to her taking leave of the public…an admired Irish Air'[111] sung by Catalani.

On a social level, before leaving Dublin Catalani stood as sponsor with Henry Grattan for the christening of a son recently born to Tom and Mrs Cooke. The child was given the names of Henry Angelo Michael, an allusion to his two distinguished sponsors.[112] (Later in life, when he had become a well known oboist, he would be known as Grattan Cooke). She also presented Mrs Cooke with a valuable diamond ring, 'in token of her personal regard, as of the great support which she received from T. Cooke in her songs, who…accompanied her in a better style than she had been used to in any other place'.[113]

Nicolò Miarteni, bass.

Chapter 5
The Great God Braham 1808-1811

While Michael Kelly was in Dublin in 1807, he received a letter from Richard Brinsley Sheridan requesting him to pass on a second one to Frederick Jones. The gist of this was that Sheridan wished to be rid of the anxieties of Drury Lane Theatre, that he had a very high regard for Jones (who was a friend of Sheridan's son, Tom) and, in 'the most honourable confidence and secrecy', proposed that Jones should travel to England as soon as possible to discuss his becoming joint manager of the theatre.*[1] There is some evidence to suggest that Jones had crossed to London earlier in the year to plant the idea in Sheridan's mind,[2] and he is even reported to have been seeking the management of the King's Theatre,[3] lately become vacant through the death of Francis Gould.

Agreement in principle must have been reached by January 1808, when Sheridan consented to 'ratify such engagements' as Jones and Tom Sheridan thought 'right for the ensuing season, or any further term for Drury-lane Theatre'.[4] Agreement in practice was arrived at towards the end of May. This offered Jones joint management at £1,000 a year, plus a percentage of the net profits for ten years, at the end of which he would be entitled to purchase a clear quarter of the theatre for £10,000 if he so desired.[5]

A while later Jones returned to Dublin to seek someone to whom he might entrust the management of Crow Street. He needed money also, and in the

---

* The following letter, written by Jones to Tom Sheridan concerning the arrangement, is to be found at the British Library, Department of Manuscripts. (Add. 42720, f 143): Dublin, Oct 20, 1808. My dear Tom, Though I must acknowledge that I am sorely inflicted with a disease quite opposite to the cacoethes scribendi yet did I answer your first letter in due course, directed to you at Springs [?] with a copy of the memorandum between your father and me which if I had not mislaid I would now send you by our friend Mich. Kelly—at all events let it not make you uneasy as I hereby free your father and yourself and the Heirs executors assigns etc of both forever of whatever agreement it contained. I took care of Drury Lane when I parted with Mr Edwin who I have bound under a solemn promise to enter into no engagement without first consulting me—I send you by Mich. a Tragedy written by Dr H[all Hartson] the author of the Countess of Salisbury and found sometime since by his nephew amongst his papers. I only wish you to read it and to let me know your opinion as soon as possible.—Holman Jr. begs my interference with you about a comedy which he is carrying over and had written for Drury Lane. I only request that it may be speedily read, which as I know you to be a man of business, I assured him you would do. Give my best compliments to Mrs S.—and believe me, Yours very sincerely, Fred. Edw. Jones.

circumstances sold an eighth share in the theatre to Edward Tuite Dalton for £5,000, and a similar share for a similar sum to John Crampton. The latter moreover undertook the uncontrolled management of the theatre at a yearly salary of £500,[6] but, from the beginning, this seems to have been a disaster.[7] Concurrently a second disaster was to strike Jones in London, where, on 24 February 1809, Drury Lane Theatre burned down. Jones now returned once again to Dublin, and resumed the management of Crow Street; Dalton, it is said, seeing his investment decline had induced Crampton to relinquish it.[8] Crampton however retained his share.[9]

His incompetence as manager had been revealed as soon as he had staged his first new opera production on 29 November 1808. This was *The Travellers; or, Music's Fascination*, written by Andrew Cherry. According to W. T. Parke, the music by Domenico Corri, an Italian composer who had settled in Edinburgh, 'professed to describe the styles of the four quarters of the world'.[10]

Andrew Cherry, the librettist, was a Limerick man, the son of a printer and book seller. His name is still occasionally recalled as the author of two songs, 'The Bay of Biscay', written for *The Spanish Dollars; or, The Priest of the Parish*, with music by John Davy, and 'The Dear Little Shamrock'. At the age of 11 he journeyed to Dublin where he found employment with another printer and bookseller, James Potts. He then turned to acting, his first recorded engagement taking place at Newry in December 1778 before he was 17.[11] Subsequently he became one of the leading comedians of his time, joining the Drury Lane company in 1802. For this theatre he also wrote one or two plays, and *The Travellers*.

The first performance of *The Travellers* in Dublin 'attracted one of the most fashionable houses' of the season. 'Indeed there was a complete overflow', but Corri's music 'although in many passages brilliant, and even striking',[12] generally disappointed. It was Cherry's drama however, and its production which suffered most criticism. 'Accustomed as we have been to all that is mean in the *saving way*', we read, 'we were not prepared for this truly ludicrous and miserable exhibition of managerial poverty. The costume in the first scene which lies in China, was indeed observed;... But amidst this, misery in the extreme meet[s] the aching eye of criticism. The Chinese banners were bordered with a few scanty remnants of yellow worsted fringe! in mock representation of silver!! The Chinese throne (reported by travellers to be the most magnificent in the world) was covered with stuff!!!... Is it thus our new manager treats the public? Is it thus he further disgraces an already highly disgraced theatre? Our stage has become proverbial for its poorness of exhibition in spectacle; but until now such mummery as this was totally unknown: nay worse, we may defy any regular theatre in the empire to produce such an instance of parsimony as this'.[13] As for Cherry's libretto, 'the piece is vilely written, and is full of political catch-claps even to nausea'.[14] Few of the artists were commended, and one, Mrs Mason, the critic decided to ignore, adding, 'and that lady has some reason to thank us for our silence'.[15] The same critic continues, 'And is this the piece which the "jackall gazetteers" of the theatre have lavishly praised in the papers? Really, really, as far as regards the theatre we blush for the moral degeneracy of the Irish press... It was heartily hissed when given out the second time'.[16] Yet it received seven further performances,[17] and would be revived during the following November, but all in all it was not a very auspicious first production for almost everyone concerned, least of all for Crampton.

A second opera, *The Exile; or, The Deserts of Siberia*, presented by him on 21 March 1809, 'after promises in the hand-bills from day to day',[18] is interesting primarily because it was based on a well-known novel of the time. This was *Elisabeth; ou, les Exilés de Sibérie* by Mme Cottin (Sophie Risteau), which would later serve as the basis of Domenico Gilardoni's libretto for Donizetti's *Otto Mesi in due ore, ossia Gli Esiliati in Siberia*. Frederick Reynolds, the librettist of *The Exile* contemporaneously described as 'a writer of powerful talent...the founder of the modern school',[19] was prolific though mediocre. The music was principally the composition of Mazzinghi,[20] but the young Henry Bishop is reported to have supplied music for it as well.[21] *The Cyclopaedian Magazine*, so critical of *The Travellers*, on this occasion spoke highly of 'The scenery, dresses &c.', adding that 'the piece possessed merit and variety, and received the most distinguished marks of applause'.[22] Most of the cast also won high praise, although there was one exception, a Mr Duff who 'waved his arms with great pantomimic grandeur, but was no character'.[23] What presumably passed for good acting at the time was delivered by Mr Rae in the role of Daran. 'How pathetic and irresistible was he when he rushed into the presence of the Empress, threw himself at her feet, and implored the release of the exiled family; many of the audience were melted into tears'.[24]

The subject of tears in the theatre in the 18th and 19th centuries was not a matter of mawkish sentiment alone. It must be remembered that almost the entire audience in the galleries, and many in the pit, were illiterate. But those who could not read at least could comprehend, and many acquired a remarkable facility for memorising. The preface to the libretto of *Elisabeth* comments, 'Thousands who knew not the *original* [novel] have wept over the *drama*'. The importance of the theatre of the period must not be overlooked as a source of education and instruction, not merely as a place of amusement.

On regaining control of his theatre, Jones's first concern seems to have been its refurbishing, but there were also fairly extensive structural alterations made from mid-July onwards under the supervision of Mr Seymour, architect.[25] 'The Pit, which heretofore had but a gloomy aspect', was now 'lined with mahogany ornamented with brass mouldings inserted in each panel'. A second pit entrance was opened from Temple Lane, and both entrances were fitted with double doors to exclude noise and cold, 'hitherto complained of'. Both entrances would also have 'money receivers, so that the danger and crush on crowded nights [would] be considerably abated'. 'The truly grand alteration' however, was the ceiling, which had been raised by six feet' forming a groined arch springing from the back of the Boxes and coved on each side'. The proscenium arch had also been raised and, we are assured (over-optimistically, perhaps) that 'even from the back row of the Upper Gallery, everything passing on the remote part of the stage can be distinctly seen'. Then, 'all the walls of the passages, corridors and lobbies' had been 'painted in oils to prevent injury to the clothes of the audience', while in the lower tier of boxes loose cushions were provided, cased with scarlet cloth, to enable the ladies who had to attend fashionable parties after the play to come in full dress, 'as it may be fairly presumed gentlemen will not enter these boxes in boots, therefore the dress of the ladies will in every respect be as free from injury as in a drawing-room'.[26] A little later there was a renewed plea for such *esprit de corps* when gentlemen who attended the theatre were recommended 'to come dressed' and not 'to appear in boots', since it was desirable as well as in the public interest 'to preserve the cleanliness and elegance of the house as long as possible'.[27] One wonders how long it *was* possible in that time of indifferent cleanliness, even in conditions far more favourable than existed in a public theatre.

Gaetano Marinari, a scenic artist, who had earlier worked in Crow Street Theatre in 1791* had been retained to carry out the decorations.[28] Born in Bologna, he had first worked at the Teatro Zagnoni there, and between 1785 and 1804 had designed scenery for Covent Garden, Drury Lane, the Haymarket, Pantheon, and King's Theatre. He was assisted by Zafforini. The new decorations really do seem to have been lavish. The centre of the ceiling, we are told, represented Hibernia protected by Jupiter and crowned by Mars, 'supported on the right by the emblems of the linen, on the left by that of the woollen manufacture…Apollo and Fame with various embellishments' adorned the proscenium, while the panels in front of the boxes were each ornamented with a different incident from Greek mythology.[29] There was also a new drop curtain: 'Euterpe, the goddess of Music, Tragedy, &c., was depicted in the centre, supported by Hercules, conducting the infant Shakespeare, to whom she had resigned her lyre, to the Temple of Minerva, goddess of Wisdom; at her feet, on the left, Time was represented sleeping by his scythe, intimating that the works of Shakespeare will live for ever… On the right was the malevolent deity, Discordia, or Envy, in chains: at the top was Iris, or the rainbow, emblematical of the variety of Shakespeare's productions… The Parcae, or Fates—Clotho, Lachesis, and Atropos—were supported by clouds on the left; opposite on the right were Melpomene and Thalia, the tragic and comic Muses'.[30]

Marinari and Seymour together had worked to such good effect that when their Graces of Richmond visited the theatre shortly after the 1809 season had begun, the house having been 'lighted up on the occasion…presented the most beautiful and magnificent *coup d'oeil*… It surpasses the old house at Covent Garden, and we are assured by those who have seen the new one, which has been so much applauded, that for taste, arrangement and splendor, it is superior'.[31]

Jones had reasons other than aesthetic for refurbishing since Henry Erskine Johnston, a Scottish singing actor popular with Dublin audiences, had recently taken over the Royal Amphitheatre** in Peter Street (renamed The Royal Hibernian Theatre) where, at a cost of £6,000, he too had affected 'material Alterations and Improvements under the inspection of Messrs Kelly and McKenna, Architects' with Filippo Zafforini, late of Crow Street Theatre, as

* See *Opera in Dublin 1705-1797*.
** Originally Astley's Theatre. See *Opera in Dublin 1705-1797*.

THEATRE-ROYAL.

This present TUESDAY, JUNE the 4th, 1811;
By Command of His Grace *THE LORD LIEUTENANT*,
Being the Anniversary of HIS MAJESTY'S BIRTH-DAY;
WILL BE PERFORMED THE COMEDY OF,

Laugh When You Can.

Bonus Mr. FULLAM,
Mortimer Mr. YOUNGER,
Sambo Mr. YORK,
Charles Mortimer Miss MOORE,
Mrs. Mortimer Miss WALSTEIN,
Emily Miss LOCKE,

Gossamer Mr. KENT,
Delville Mr. N. JONES,
Costly Mr. KING,
Farmer Blackbrook Mr. ROWSWELL,
Miss Gloomley Mrs. M'CULLOCH,
Dorothy Mrs. WESTON,

To which will be added a FARCE, called,

The Mock Doctor.

Gregory Mr FULLAM,
Leander Mr. KING,
Welch Davy Mr SLOMAN,
Dorcas Mrs. M'CULLOCH,

Sir Jasper Mr. N. JONES,
'Squire Robert Mr. CARROLL,
Doctor Hellebore Mr. ATKINS,
Charlotte Mrs. KING.

*** *BOXES FREE FOR THE LADIES,*
Who are respectfully informed that Tickets will be delivered Gratis by application at the Box-Office, between the Hours of Eleven and Three o'Clock, and to prevent any abuse of a Custom most respectable in its Origin, no Person will be admitted without a Ticket, nor any, even with whose Appearance shall not be suitable to the part of the Theatre allotted for Free Admission.

Henry Fielding's adaptation of Molière's *Le Médecin malgré lui* was cast in the form of a one-act ballad opera and first staged at Drury Lane Theatre (1732) under the full title of *The Mock Doctor or The Dumb Lady Cur'd*. It contained ten musical numbers.

his chief scenic artist.³² Both men would launch their seasons within a week of one another, Johnston on 1 November with a 'Grand Operatical and Serious Ballet Company',³³ drawn mainly from the King's Theatre, Jones on 6 November.

Celebrations had taken place in Dublin, as throughout the United Kingdom, on 25 October to commemorate the entry of George III into the 50th year of his reign. Crow Street Theatre had 'A transparency of his Majesty, most elegantly executed, and the windows festooned with variegated lamps had a fine effect, as the whole could be seen from Dame-street'.³⁴

Jones had particular cause for celebration just then, for he had a star to head his playbill, the great John Braham, an artist with the drawing power of Catalani. With dramatic imagery, combined with self-confessed racial prejudice, Charles Lamb would write of Braham: 'There is a fine scorn in his face, which nature meant to be of—Christians. The Hebrew spirit is strong in him, in spite of his proselytism. He cannot conquer the Shibboleth. How it breaks out when he sings, "The Children of Israel passed through the Red Sea!" The auditors for the moment are as Egyptians to him, and he rides over our necks in triumph'.³⁵ Presumably Lamb is referring here to Handel's *Israel in Egypt*, though, as the music historian H. Sutherland Edwards in an excess of pedantry has pointed out, 'There are, however, no such words in the tenor part of the Oratorio'.³⁶ The passage he had in mind is probably the one recorded by Joseph Heywood of Manchester as follows: 'A little thick-set man with a light brown wig all over his eyes, a generally common appearance, and a most unmistakably Jewish aspect, got up to sing one single line of recitative. He stood with his head well on one side, held his music also on one side and far out before him, gave a funny little stamp with his foot, and then proceeded to lay in his provision of breath with such a tremendous shrug of his shoulders and a swelling of his chest that I very nearly burst out laughing. He said, "But the children of Israel went on dry land", and then paused, and every sound was hushed throughout that great space, and then, as if carved out upon the solid stillness, came those three little words, *"Through the sea"*!* And our breath failed, and our pulses ceased to beat, and we bent our heads as all the wonder of the miracle seemed to pass over us with those accents—awful, radiant, resonant, triumphant!'³⁷

* Who does not seem to have got the phrase right either! According to a programme book Braham's words should have been '…but the children of Israel went on dry land, in the midst of the sea'.

THE GREAT GOD BRAHAM 1808-1811

He was born either in 1774 or 1777, the youngest child of John Abraham and his wife, Esther Abrams who, about that time was living in Goodman's Fields in the East End of London. There, as a child, he endured the poverty of ghetto immigrants. His father may have been a musician by profession, and Braham, while still very young, acquired a love of music from the ceremonies at the Great Synagogue in Duke's Place, Aldgate,* where he was befriended by the renowned singer, Michael Leoni, who is said to have been his uncle.** On 21 April 1787 he made his debut at Covent Garden as Leoni's pupil, singing 'The Soldier tir'd of War's Alarms' and James Hook's 'Ma chère amie'. Two months later, billed as 'a little boy' he was appearing at the Royalty Theatre, Wells Street, Wellclose Square. About 1788 Leoni had become bankrupt and had to flee to Jamaica, but Braham, now left to his own resources, and with his voice about to break, was befriended by a leader of the Jewish community and generous patron, Abraham Goldsmid. Then, once his voice had settled, he set out for Bath to become a pupil of Venanzio Rauzzini, who taught him for three years.

In 1796 he was engaged by Drury Lane Theatre to sing in Stephen Storace's *Mahmoud*. His success was complete, *The Morning Herald* unreservedly hailing him as 'the first public singer of the present day'.[38]

During the following season he joined the King's Theatre company, being one of the first English singers to win acclaim there. Shortly afterwards he commenced to sing in oratorio. In spite of his London success he now set out for the Continent accompanied by Nancy Storace, with whom he had recently formed a relationship. They travelled first to Paris, where they gave some concerts (at the first of which Napoleon and Josephine were said to have been present), and then on to Italy, to Florence, Milan (where both sang at the Teatro alla Scala), Genoa, Venice, and other cities, and finally to Vienna, where, receiving offers from London, they returned to England via Hamburg. Not far from Hamburg, in the village of Ottensen, they gave a concert on 1 August 1801, 'in Herrn Rainville's beautiful Concert Room' where 'there was a large audience despite the fact that admission was 1 ducat…receipts amounted to about 700 ducats'.[39] Braham now continued to sing at Covent Garden until 1804, when he had a disagreement with John Philip Kemble concerning a benefit

---

* Not far from where today there is a Braham Street, named after the singer in 1922.
** See *Opera in Dublin 1705-1797*.

John Braham, singer and composer (1774-1856)

for Nancy Storace, alleging discrimination by Kemble against them both. The dispute was resolved, however, and he returned there for the following season, appearing also at the King's Theatre in 1805 and 1806. Here, in the latter year, he sang Sesto in *La Clemenza di Tito*, the first Mozart opera to be performed in London. From 1806 until 1815 while in London he would perform no longer at Covent Garden but at Drury Lane, and it may have been the burning down of this theatre earlier in 1809 that enabled Jones to engage him to perform at Crow Street at a salary of 2,000 guineas for 15 nights. He was then 35 years of age and reaching the height of his powers.

The quality of his voice seems to have been unusual, perhaps unique. 'The oboe, clarinet, or bassoon, are the instruments of whose quality it partakes', we learn, 'though certainly it has far more richness, brilliancy, and refinement, than any of them'. His use of the falsetto, which he could take on any note from D to A and manage so skilfully that 'it was impossible to distinguish at what point he substituted the falsetto for the natural note', gave him an extended range of 19 notes, from A in the bass to E in alt. Indeed he may have inaugurated a new style of tenor singing by assimilating 'the two in their general colour and bearing with much more success than any other performer'. He used to increase the volume of his voice by 'a very peculiar alteration in the parts adjacent to the upper region of the throat and the back of the mouth, so that the voice proceed[ed] more from the head than the chest', but at the cost of quality, for the tone became disagreeably reedy, or what 'in trumpet playing is called *overbroke*'. However, he is described as being 'eminently articulate; not a syllable is lost to the hearers… His recitative is particularly masterly'.[40] A conception which Charles Lamb confirms, writing, 'He sings with understanding, as Kemble delivered dialogue. He would sing the Commandments and give an appropriate character to each prohibition'.[41] Apart from his shake, which was 'too hard, close, and rapid, and wanting [in] liquidity', it was said, 'nothing can be finer than his execution generally and most particularly when we consider it in relation to divisions. He flies through the whole compass of his voice with the smoothness and speed of light…there was no imaginable combination of notes, however various and protracted— no sort of grace in use among singers, which Braham could not…have expressed in almost any manner that was desired'.[42] His prolonging of 'vowels as the Italians are wont to do upon such syllables as mio, Dio'[43] was frowned upon, yet another source suggests that this deserved praise, recalling that the Italians '(not forgetting their Donzelli)' had paid him the tribute, 'Non c'è tenore in Italia come Braham'.[44] Undoubtedly his time spent in Italy must have influenced his style, and in England this may have elicited criticism as well as approval.

Leigh Hunt describes Braham as 'the first living singer in England, perhaps in the world' yet adds much in disparagement. He reports: 'not only is the general style of his singing meretricious, but the ornaments with all their exuberance are frequently misplaced, and so far he wants the common taste even of floridness. He wears bells on his toes as well as rings on his fingers. He will run divisions upon the most insignificant words, and trill, quaver, and roll about at you without remorse. He lights up, as it were, fifty wax candles to exhibit a nutshell; or resembles a fantastic fellow, who instead of approaching you by the ordinary path, and in a strait manner, should come up with all sorts of fluttering gestures and meanders, and accompany his concluding bow with a shake of the head and cheeks of five minutes duration'.[45]

J. E. Cox makes one important point in his favour when he writes, 'Whether he sung or shouted—and he could do both—the ear was never offended by the voice being either above or below the true pitch'.[46] Nevertheless, he shares the opinion commonly held that towards the end of a very long career Braham used to sing out of tune. Perhaps Mount Edgcumbe best of all sums up his case with discernment and impartiality when he records: 'The fact is, that he can be two distinct singers, according to the audience before whom he performs, and that to gain applause he condescends to sing, as ill, at the playhouse as he has done well at the opera'.[47]

Braham's appearance on stage was unimpressive for he was just over five feet in height, and with his short legs, was said, at least in his later years, to have 'shuffled in his gait worse...than any other man who ever trod the boards'.[48] As a Dublin reviewer expressed it, 'Of Mr Braham's acting little need be said, for in truth there is little scope for observation'.[49] But there was another factor which hampered his performing competently, the trashy libretti then in common use as dramatic clothes-horses on which the songs and concerted pieces of English operas might be hung. To again quote the above reviewer, 'The dialogue he has been compelled to utter is so very mean, the sentiments put into his mouth are so paltry and commonplace, and the situations which occur in the opera are so little interesting or dramatic, that Mr Kemble himself could make nothing of them. The most judicious way therefore, for an actor in such operas…to acquit himself, is to repeat the part without any emphasis or exertion, and cautiously to avoid any attempt at rendering it a *character*. This plan Mr Braham has prudently adopted'.[50]

As earlier mentioned, Braham's season opened in Dublin on 6 November 1809 when he played the Seraskier in *The Siege of Belgrade** in which his success was immediate and universal. *The Correspondent* reported: 'There is nothing *mechanical* in his performance; it is not mere *execution* by which we are struck but that which gives life and spirit and animation to harmony, and without which melody itself is devoid of soul, EXPRESSION, in a word, is the great characteristic of his voice. It would be impossible in words to convey an idea of this amazing power which varied as occasion required, while it appeared to be the spontaneous effusion of the moment. This was strongly exemplified in "The Austrian Trumpet", where the contest between the vocal and instrumental powers of the performers seemed to suspend the very faculties of the audience, until the conclusion of the song, when they were resumed in a universal burst of the most rapturous applause… Mrs Nunn [as Catherine] never sung better, nor with more effect, and the energy of her action in several instances was striking. Upon the whole, nothing was wanting to the fullest gratification of the audience, and the only thing we remarked likely to give pain to Irishmen, amongst whom such a thing was least likely to occur, was that a number of beautiful females were kept standing in the rere of the lattices, while *their* places in the front were occupied by men'.[51]

Curiously, there is no mention in this review of Mrs Mountain from Drury Lane, Braham's leading lady for the season. No matter what the newspapers announced about it being 'her first appearance on this Stage',[52] she had appeared there 20 years earlier,* and more recently had sung at the theatre in Capel Street. She was born Wilkinson about 1768, the child of circus performers and christened Rosomon or Rosemond**.[53] Having trained as a singer under Charles Dibdin she made a few unimportant juvenile appearances at the Haymarket Theatre in 1782. Between 1784 and 1786 she was a member of Tate Wilkinson's company (he was no relation) touring the Yorkshire circuit, but although she made some progress, especially as a singer, she seemed to have

---

* See *Opera in Dublin 1705-1797*.
** On 5 June 1787 she signed herself Rosomon in the St Martin-in-the-Fields marriage register.

been one of those artists who, either through poor personality or lack of discipline can never rise above the second-rate in their profession. About 1798, after appearing at Covent Garden for more than a decade, she had singing lessons from Venanzio Rauzzini, and this may have resulted in the remarkable improvement she displayed when she reappeared in London in 1800. But it was not until she had sung Polly in *The Beggar's Opera* at Drury Lane that year that she achieved her first real success, 'bursting upon London like a new character having made such a wonderful advancement in her profession'.[54] What seems to be a fair assessment of her talents comes from Leigh Hunt when he writes, 'Mrs Mountain is a pleasing singer in the upper rank of the mediocre, and is a better actress than most of her profession'.[55] Moreover, he had earlier paid her the compliment that she had (with Maria Dickons) 'considerably the advantage over all the other female singers in gentility of appearance'.[56] She had married the Irish violinist, John Mountain, in 1787 and had paid a second visit to the Capel Street Theatre in the summer of 1806. On that occasion her entertainments consisted of musical monologues, *The Lyric Novelist* and *The Travellers at Spa*, both arranged for her by Andrew Cherry. Into the latter she introduced 'The Dear Little Shamrock' set to music by William Shield. In both entertainments she was accompanied by her husband and a pianist.[57]

At Crow Street she was well received as Lilla in *The Siege of Belgrade*, and throughout the season 'acquitted herself very handsomely in the parts she undertook. Her voice…is clear, particularly in the higher tones; but she sometimes runs it beyond the compass which is equally distressing to the audience and to herself'.[58] In time she became a firm favourite of the London public, and eventually when she gave her farewell performance at the King's Theatre in May 1815, 'she retired, or rather was borne off the stage, amidst the fullest testimony that the occasion admitted of public respect and esteem. The pressure was so great that much of the iron railing in the passage to the pit was broken away, and many persons were in imminent danger for some time, but happily no serious accident occurred'.[59] She died on 3 July 1841, and William Robson wrote for her epitaph, 'Poor Rose! she who was one of the sweetest singers of the sweetest melodies'.[60]

Mrs Nunn, in comparison, who *was* mentioned by *The Correspondent*, and described as never having 'sung better', seems to have been surrounded mostly by malicious enemies. Of the defenceless artist we read: 'This lady's knowledge of music is unquestionable, her voice is a very fine one, and although her articulation is vile, or rather she has no articulation at all when singing, she is beyond a doubt a creditable member of the opera. But [and the But is ominous] she is unpopular; why, we shall leave others to enquire. As an actress, Mrs Nunn we think has powers, but the want of popularity to which we have just alluded, prevents her from making any progress. [And then, more ominously, since having been a member of the company from 1806, she presumably wished to remain so.] We may return to a more particular investigation of this lady's character and capacities hereafter'.[61] It would have been interesting to have seen how resolutely Mrs Nunn performed the night after she had read that personal onslaught with the threat of more to come.

Although Braham's engagement had been for 15 performances, he in fact appeared 19 times up to 8th December. The operas in which he sang were revivals, well known from earlier years. Besides *The Siege of Belgrade*, there were also *The Duenna*,* *The Castle of Andalusia*,* and *The Haunted Tower*,* as well as *The Cabinet* and *The Travellers*. Into all of these he interpolated favourite songs from other works and by other composers. They included a duet sung with Mrs Mountain from Winter and da Ponte's opera *'Il trionfo dell'amor fraterno*; 'On this cold flinty rock' from his own and William Reeve's opera, *Kais*; 'Love unperceived' from *The First Attempt*, the words by Miss Owenson, the music by Tom Cooke, as well as a new Polacca by Cooke, and a manuscript ballad, 'Keep those tears for me', written and composed by Tom Moore.[62]

His popularity seems to have been immense and we find reports such as: 'Last night, the Theatre Royal overflowed in every part—and at an early hour presented a most brilliant appearance. We observed much of rank and fashion which added to the general splendour of the Theatre, brought to memory the time when the Nobility of the land resided for the Winter Season in Dublin [prior to the Act of Union]. We understand many families of distinction have come to town for Mr Braham's nights, nor are we surprised that it should be the case, for perhaps the lapse of a century may not produce a performer endowed with such an extraordinary combination of musical talent, with so much natural ability and scientific acquirement'.[63] A week later we read: 'Mr Braham brought an overflow last night. Her Grace of Richmond and daughters

---

* See *Opera in Dublin 1705-1797*.

were in the Patentee's Box'.⁶⁴ Some nights later both the Viceroy and Duchess attended in state,⁶⁵ and by the end of the month it was announced, 'The overflow at Crow-st last night [for *The Cabinet*] exceeded every preceding appearance of Mr Braham. Even the lobbies were filled'.⁶⁶

During this period, 'the Proprietors, with much liberality',⁶⁷ gave the artist Gaetano Marinari a free benefit. Sheridan's play *Pizarro* was announced for it, and it was hoped, 'for the honour of the country…that a Scholar and a Scenic Painter of the first order will experience the fostering aid of an Irish Public on the occasion'.⁶⁸

Braham's success 'occasioned a renewal of his engagements for a few nights'⁶⁹ on 11 December, but these few were eventually extended to 13, bringing the total number of his appearances to 32. Mrs Mountain's engagement had not been extended, since presumably it was recognised that Braham, on his own, was all that was needed to bring full houses. He continued to appear in the operas noted above, but to these was added Arnold and Colman's *Inkle and Yarico*.* He also persisted in interpolating extraneous songs, among them another ballad by Tom Moore, 'Oh soon return', 'Lively Kitty' from his own opera *The English Fleet in 1342*, Felice Giardini's glee, 'Beviamo tutti tre', with Payne and Holland, and 'O thou wert born to please me', an English adaptation of the duet 'Pace, caro mio sposo' from Martin y Soler's opera, *Una cosa rara* with Mrs Nunn. This introduction of unrelated music was not universally approved, and during the season, at least three notices appeared which were to some extent critical of the practice. The writer of one having declared that he had found the plot of the opera incomprehensible, added 'Another cause…might be assigned, namely, the introduction of those supernumary songs, which have for the most part, no more connexion with the subject of the piece, than with the Koran or the Talmud'.⁷⁰

Two operas in which Braham appeared were presented for the first time in Dublin. They were, *Thirty Thousand; or, Who's the Richest?*, by Braham, Davy and Reeve, with libretto by T. J. Dibdin, and *False Alarms; or, My Cousin*, by M. P. King, for which James Kenney had written the libretto. The plot of *Thirty Thousand* was taken from one of Maria Edgeworth's *Popular Tales*, 'The Will'. The opera was given its first performance on 13 December, and although 'the house was fashionable and crowded', it was a complete failure. 'We know not

* See *Opera in Dublin 1705-1797*.

what to say', continues the reviewer, 'to attempt to describe what we do not understand, would be absurd and impertinent, and we must confess our total ignorance of the matter. If there is a plot we could not discover it, and the unanimous condemnation of the audience renders it unnecessary for us to make any enquiry on the subject'.[71]

An incident occurred during the performance as a political claptrap. To understand its significance one must know that just then there was considerable opposition by the public to the raising of prices at Covent Garden by the joint managers, John Philip Kemble and Thomas Harris, combined with a demand to have subscribers' private boxes abolished, (They were represented as 'receptacles of vice, purchased by the profligate for 400*l* a year'[72]) so that ordinary ticket buyers could have additional and better seating. A barrister named Henry Clifford had apparently provoked the box-keeper, James Brandon, who had seized him and with the help of some of his 'chuckers-out' had taken him to Bow Street police station. Clifford thereupon sued for assault and false imprisonment, and the case was heard before Sir James Mansfield, the Lord Chief Justice. A jury found for Clifford.

At Crow Street 'in the course of the opera, [the character] *Mrs Notable* introduces some friends to a gallery of paintings, she mentions the subject of each and names the artists.—When she came to a very striking portrait of John Kemble, she said "this is a celebrated actor by *Opie*". The House immediately took it as "O. P." [Old Prices] and the roar of applause was general and continued for a considerable time. Boxes, Pit, and Galleries were unanimous. The cheering and plaudits were loud and general, and thus has a Dublin audience by accident and without any previous concert [?] expressed a decided approbation of *Covent Garden extortion and Private Box immorality*. Will Sir James Mansfield say that this honest and honorable expression of the Public Will was a *riot*? Would Brandon and his ruffians have dared to drag from our Theatre last night any of those who espoused the just cause of Morality and O. P.'.[73] Apart from Braham's singing, this seems to have been the only spark of excitement during the entire evening, but the opera 'was heard with patience to the end, and upon the falling of the curtain, with all the cool solemnity of justice, consigned to the oblivion it deserved'.[74]

Nor did the second new opera, *False Alarms; or, My Cousin*, performed on 29 December, fare better, the opinion of one critic being, 'without absolutely denying it every species of merit, it appears to us entitled to nothing more than the comparative praise of not being so bad as other things that have been performed *with the most unbounded applause*, at the Theatres Royal of Drury Lane and Covent Garden'.[75]

On 22 December Braham appeared in a performance of *The Haunted Tower*, apparently without fee. This evening's entertainment had been allotted by Jones 'with his accustomed liberality' in aid of the Meath Charitable Loan Society.[76] But on the following evening Braham took part in an unscheduled presentation when he may well have given his finest performance of the season. It was a concert consisting of 'a Grand Selection of Sacred and Miscellaneous Music' in which Braham, assisted by members of the theatre company, 'a numerous chorus' and with the orchestra seated on stage and led by Tom Cooke, sang works by Handel, Haydn and others.[77] The audience was numerous and brilliant though not crowded, but 'on no occasion perhaps since his arrival in this city, did Mr Braham's powers produce such astonishing effects as upon this occasion'. His programme included 'Deeper, and deeper still' and 'Waft her, Angels' as well as 'Comfort ye, my people', and the reviewer goes on, 'It is impossible to convey to those who have not heard him, a clear and distinct idea of the ever-varying graces of his voice, and difficult to speak of them without a degree of enthusiasm in the mind and hyperbole in the language. Hitherto we had been accustomed to hear him, in general, utter the tones of mere sentiment or passion; and from the exquisite skill and animation which he evinced in this department of the art, were led to apprehend that he would be less successful in assuming the more solemn character of what is denominated Sacred Music… His voice, upon the present occasion, was grave, clear, copious and flexible like the melody of Termosiris [?], imparting a degree of passion to every subject. This was felt by the house and the feeling penetrated to the heart… Upon the whole, notwithstanding inequalities among the other performers, and some very perceivable defects in the Orchestra, the entertainment was highly satisfactory, and we should regret if this were the last specimen that Mr Braham is to give of his talents for Sacred Music'.[78]

Braham's first Dublin engagement ended on 2 January 1810, when he played the Seraskier in *The Siege of Belgrade*.[79] Thirty years later he would still be returning there for yet another season of six nights.

On 4 January 1810, just two days after Braham's departure, 'A Grand Operatic Romance called *The Forty Thieves* which has been for several months in preparation, with entirely new Scenery, Dresses and Decorations',[80] was staged.* The music was by Michael Kelly, with an overture by Tom Cooke. When first produced at Drury Lane, the libretto had been by George Colman the Younger and Charles Ward (R. B. Sheridan's brother-in-law), but other hands may have altered it for Dublin.[81] The production seems to have been lavish, but, although the opera had had a very successful run in London, surprisingly we read, 'After the departure of Mr Braham, *The Forty Thieves* and Mr Bannister** were "got up" to fill the vacuum he had caused. Both have failed—not, as it should appear, for want of attraction in either, but because the taste of the town had taken another current, and it could not speedily be altered... *The Forty Thieves* is indeed a brilliant exhibition—but it is an exhibition merely. The Arabian tale is infinitely better told in the story book than on the stage'.[82] This report is misleading, for between January and the end of March, the opera would have 22 performances—scarcely a failure.

Just before Christmas Frederick Jones had sued a Mr Corbett for a libel published in *The Hibernian Telegraph* impugning his integrity in dealing with artists. He was awarded 500*l* damages, which reparation must have been diminished by Corbett's counsel describing him in court as 'a great turbot-eating public functionary and very litigious. "There is not," said he, "a court in this hall in which he has not *performed*, although not for his *benefit*".'[83]

The year 1811 began with what at first would seem to be a work of great importance, an English version of Cherubini's *Les Deux Journées*, here called *The Escape; or, The Water Carrier*.*** The composer was announced as 'Thomas Attwood—with some of the music of Cherubini',[84] but since only two or three

---

* According to *The Correspondent* (3 January 1810) 'the properties, Machinery &c.' for this production were 'by Mr Pobje'. W. G. Strickland records that he was a wax modeller who 'lived in Whitefriars Street'. (*A Dictionary of Irish Artists*, Dublin/London 1913, 2 vols, II, pp 247/8), while Gilbert (II, p 237) reports, 'Among the marvels produced by Pobje and other mechanists of the theatre, is mentioned an imitation peacock which, worked entirely by internal machinery, strutted across the stage in the manner of the real bird'.

** John Bannister, an actor engaged to perform in Garrick's play *Isabella* on the same evening. He recounts that while on a visit to Dublin: 'I was robbed...of 109*l* in gold and twenty-five guineas in silver, stolen out of my bed-room in a small portmanteau'. *Memoirs of John Bannister* by John Adolphus. London 1839, 2v. II, 183*, 184*.

*** From its title in Germany (where it was even more successful than in France) of *Der Wasserträger*.

"Maria" Dickons, née Martha Frances Caroline Poole (c. 1770-1833), here seen as Rosina in *The Barber of Seville*.

pieces of Cherubini's music remained, while in every way the score had been drastically altered, one must consider the composition to have been essentially the work of Attwood alone. Both Thomas Holcroft and T. J. Dibdin have each been credited with the libretto, but it seems that Dibdin acknowledged the songs only, so presumably the dialogue was written by Holcroft.[85] John Fawcett, singing actor and author, who had played Michelli (The Water Carrier) in the original production at Covent Garden, may have adapted it for Dublin, where he was a member of the Crow Street company in 1810.[86] The first performance there took place on 11 January.[87]

It was followed on 20 February by *The Gay Deceivers*,[88] an adaptation from another French opera, Grétry's *Les Événements imprévus*, by George Colman the Younger, and set to music by Michael Kelly. Kelly tells us that 'it had many comic incidents, smart dialogue, and some sweet songs'.[89] (The latter of course may have owed more to Grétry than to Kelly.) He also relates, but without anything to confirm it, that the author of the original French text, Thomas d'Hèle, was really an Irish officer in the French army named Hale.[90]* The work seems to have created little stir at Crow Street.

On 2 May, Sir John Stevenson and an unknown author combined to produce 'A Grand Musical Drama, Founded on W. Scott's celebrated Poem of *The Lay of the Last Minstrel*',[91] called *Border Feuds; or, The Lady of Buccleuch*. If the music in any way resembled the 'Grand Serio Operatic Drama' called *The Patriot; or, Hermit of Saxellen*, which Stevenson had composed for the Royal Hibernian Theatre the year before, the outcome can have been of little importance, for the librettist of this work (an unknown Irish littérateur named H. B. Code) records, 'It is a fact strikingly illustrative of the facility with which Sir John composes, that the greater part of the melodies in *The Patriot* were written in one evening, while he and the author were sitting familiarly together'.[92]

Quite intriguing news appeared in both the Dublin and London newspapers in August, when it was reported that Richard Daly, who had assigned his theatre patent to Jones in 1797, was now about to enter into joint management with him. It was announced as follows: 'Mr Daly's friends will learn with pleasure, that on the opening of the Theatre next season, he will recommence his Theatrical career, and become a joint proprietor and Acting Manager.

---

* The *Dictionary of National Biography* gives his name as Hales, his ancestry to a good English family settled in Gloucestershire, and his early career to service in the English navy.

Preparations upon a splendid and extensive scale are making for the occasion. Mr Crampton and Mr Dalton will remain sleeping partners, to the extent of their shares which they have purchased. Mr Jones resigns the Cork and Limerick Theatres entirely to Mr Daly'.[93] But the take-over seems never to have taken place, nor can one consider it ever to have been a likely proposition.

It may have been Michael Kelly's imminent arrival in Dublin with another Italian Company which prompted the production of his opera, *Gustavas Vasa*, on 30 July.[94] More probably it was due to the presence just then in the Crow Street company of Mrs Dickons and Incledon, for both had earlier sung the same roles at Covent Garden which they now assumed in Dublin.

Maria Dickons had already sung with Incledon in Dublin in 1795* under her maiden name of Poole. She was now at the zenith of her career, although never an outstanding singer; her middle register was said to be 'harsh, hollow, unpleasant-sounding', and she was considered to be 'more comfortable in a high-lying part'.[95] Yet she appeared with success as the Countess opposite Catalani's Susanna in *Le Nozze di Figaro* at the King's Theatre in 1812. It was then reported that she gave 'a gratifying instance of good taste in abstaining considerably from that excess of ornament in which she too often indulges, and which, like the hoop and embroidery of a court dress, serves only to encumber and disfigure the beauty it is intended to adorn'.[96] In the circumstances it is not surprising to find that *Gustavas Vasa, the Hero of the North* was received 'with unbounded Applause'.[97]

It may be of passing fiscal interest to read that 'The Expenditure of Ireland' for the year 1810/11 was 'upwards of ten millions, and her ordinary Revenue for the same year only 3,658,985*l* 0s 8d which [was] about 700,000*l* less than that of the year 1810'.[98] The odd '8d' tells of more than unsophisticated accountancy. Whatever else, these were unsophisticated times.

---

* See *Opera in Dublin 1705-1797*. But *correctio*, she had made her debut at the King's Theatre on 1 May 1787 in a single performance of Venanzio Rauzzini's *La Vestale*, and not as there reported as Emily in *The Woodman* which took place later.

Singer, composer, theatre manager and music publisher (1762-1826).
This portrait of Michael Kelly by A. Wivell is dated c. 1825.

Chapter 6
Mozart Adapted 1811

In August 1811 Michael Kelly, in arrangement with Frederick Jones, brought yet a third group of Italian artists to Dublin. This was the first engagement of a rather extensive tour, for on leaving Dublin early in October, the company (without Kelly) proceeded first to Belfast,* where they gave a concert 'under the patronage of the amiable Marchioness of Donegall,[1] thence to Scotland, where they performed both in Glasgow and Edinburgh, and then to Manchester and Liverpool, before returning to London. Kelly and his company of singers, all entirely new to Crow Street Theatre, were reported to have set out from London on 8 August.[2] They were drawn mainly from the company at the King's Theatre, the most important of them being the soprano, Teresa Bertinotti-Radicati, and the baritone, Giuseppe Naldi.

The former was born in the Piedmont region of Italy in 1776,** moving with her family to Naples two years later. She is said to have begun taking singing lessons at the incredibly early age of four, and made her first appearance on stage at the Teatro San Carlino when she was 12, probably with the 'Ragazzi Napoletani' in an 'azione sacra per musica', *L'Apparizione di San Michele Archangelo nel Monte Gargano* by Giuseppe Coppola, during Lent 1788. As she grew older she showed promise of great beauty. Having begun her career in Naples, she is said to have later set out on a tour of Italy, visiting Turin, where, in 1801, she met and married Felice Radicati, a first-rate violinist and fifth-rate composer. She was in Livorno during September 1802, when she was described as 'an excellent singer'.[3] Between 1805 and 1808 she had recurring engagements both at Vienna and Munich. In the former city she made her first appearance in Federici's opera *Zaira*, but 'was not particularly liked'.[4] In Munich, her debut in Cimarosa's *Gli Orazi ed i Curiazi* was much more successful. 'Mme Bertinotti is a warm sensitive singer', we read. 'Her voice is not particularly strong or brilliant, but it is always very pleasing, pure and melting'.[5]

A long description is extant of her singing in Ferrara in 1805 in the opera *Virginia*. Here it is recorded: 'For roles such as Zaira and Virginia, she is

---

* *The Morning Post*, 10 October 1811, erroneously gives Cork as the venue.
** The *Enciclopedia dello Spettacolo* reports that a manuscript of her memoirs is deposited in the library of the Bologna Conservatorio.

already a little too mature [at 29?] and too heavy. She was undoubtedly once very attractive, and while she still has some *beaux restes*, she is not graceful enough; she gesticulates wildly, throws her arms into the air, presses them onto her chest—in runs and coloratura passages, her hand flies to her heart, etc. All without sense or meaning. Her voice is decidedly peculiar: recitatives she sings in contr'alto, in her arias she changes to soprano—and her runs? a veritable cock crowing. This varying of registers creates an astonishing effect—were one not to see her, one would believe oneself to be hearing two different people. The voice is very worn out, strained, and only rarely pure. Her method is that of all Italian singers of today, and consists of three or four vocal ornaments which are indiscriminately adapted to all recitatives, cavatinas, arias and finales. Her trill is very poor, and is in fact no more than a wobble and shake on one and the same note'.[6]

In 1808/09 she had gone to Amsterdam,[7] on the invitation of Louis Bonaparte, and in 1810 reached London, once again making her first appearance there in *Zaira* on 22 December. But by then the Scottish-born memoirist, Pryse Lockhart Gordon, found 'the formerly pretty face and *svelte* figure of my fair friend' much changed. 'She was indeed so altered as to be hardly recognisable. She had become fat and unwieldy; and instead of the clear and transparent skin which she once possessed, was coarse and swarthy…she relieved my embarrassment by saying, "my dear friend, you find me an old woman: a severe indisposition two years ago, and a long residence in the vile climate of Holland have destroyed me".'[8] Henry Robertson, the London *Examiner* critic, confirmed both Gordon and his Ferrarese colleague when he recorded that her role was 'by no means adapted to her figure and face, which…have too much plumpness—too much the appearance of self-enjoyment and good living, to accord with our ideas of a tragic heroine, who is generally doomed to endure all the vicissitudes of life, in a perpetual round of starvations, imprisonments, swoonings, and all the train of operatical miseries. This lady, however, makes no pretensions to acting, and being perhaps more skilful in wielding a knife and fork than a dagger, wisely avoids attempting what must appear ridiculous, and contents herself with walking on the stage or off, lifting occasionally an arm or an eye, and frowning or smiling as in duty bound'.[9]

The same London critic differs greatly however when considering her 'voice and taste'. These he found to be 'of a very superior kind; the former is powerful, melodious, and, what we rarely find, perfectly in tune. Her singing is enriched and decorated with a variety of turns and graces, as tastefully conceived as they are correctly executed. Her only defect, and that but a slight one [?], is a piercing shrillness in her upper notes which produces rather an unpleasant sensation on the ear'.[10] Had her voice and singing improved so much in the intervening five years since Ferrara, or, as so frequently happens, did different ears hear different sounds? Gordon reports that at the King's Theatre, apart from 'an opera of Mozart's (*Così fan tutte*) 'she was so ill supported, and the parts were so ill cast, that her rival [Catalani] effectually succeeded in keeping this delightful singer in the back-ground'. Bertinotti herself was more objective, however, explaining to him, 'the fact is…that I am *passé* in person and I could not expect the applause I received in my younger days'.[11]

Her husband, Felice Alessandro Radicati, a pupil of Gaetano Pugnani, accompanied her to Dublin as 'Leader of the Band'.[12] As a violinist he had toured Italy, France and England with great success, as a composer he had written arias for his wife to be included in other composers' operas. Moreover, on his visit to Dublin, as a quasi-impresario, he had brought with him 'all the best music which has lately appeared on the Continent'.[13] This was manuscript music, copied, presumably by him, and now being pirated. For example, during the 1810-11 London season, we read that the ballet master, Vestris, 'has bought from Mr Radicati the score of Mozart's celebrated opera, [*Così fan tutte*] that he may give it on his night'.[14] Composers' rights or royalties had still to be introduced. Eventually husband and wife settled in Bologna, she as a singing teacher, he as leader of the orchestra there, as well as violin professor at the Liceo Filarmonico. His death, at a relatively early age, was tragic, for he was thrown from his carriage when the horses took fright. Bertinotti survived him by over 30 years. Luigi Zamboni, Rossini's original Figaro, was her most famous pupil.

By coincidence, her compatriot, Giuseppe Naldi, came from the city of her retirement, Bologna, where he was born on 2 February 1770. He was educated for the law, but, it is recorded that with the formation of the Cisalpine republic he believed he would have no future under the new régime. Fortunately he was naturally gifted as a musician, being singer, pianist, and violoncellist. He sang in Milan in 1796 and 1797, in Florence in 1800, and in Lisbon from 1803

"Nel cor più non mi sento" from Paisiello's *La Molinara* (1790), originally staged as *L'amor contrastato* (Naples, 1788)

to 1806.* In December 1805, when announced for the King's Theatre, London, he was expected from that city, coincidentally being described as 'the most eminent *buffo-caricato* in Europe'.[15] Curiously, on his first London appearance in P. C. Guglielmi's *Due Nozze ed un sol Marito* we read in *The Monthly Mirror*, 'His voice is a tenor'![16] Perhaps it was the same reviewer who had concurrently misinformed *La Belle Assemblée* as well.[17]

W. T. Parke affirms, however, that he was no tenor but a 'baritone'** of great power, compass, and flexibility' as well as 'an excellent actor'.[18] Mount Edgcumbe concurs concerning his acting but has reservations about his vocal quality, agreeing that he was 'as far as his powers allowed, a good singer' but adds that 'his voice was weak and uncertain'.[19] We also read of 'defects of faulty intonation, nasality, and a constant lagging behind the time'.[20] Although in contrast, one admiring Scottish critic refers to 'the mellow and beautiful tones of his low notes'.[21] Most reports, in fact, suggest that Naldi was a better actor than singer, and it was mainly the actor which attracted the reviewer of *The Morning Chronicle* when he wrote: 'He is a model which every Comedian ought to study. The tasteful drollery, the restrained licence, and the prompt wit with which he enlivens the subject that he had to delineate—the moderation that keeps him always within the limit of his powers, and enables him to finish with grace whatever playfulness he attempts. The gestures which, without any seeming effort gave more force to expression than the most boisterous declamation—and the musical science with which, after his most eccentric recitative, he executed the most difficult air, make him altogether a Performer of the most valuable distinction'.[22] This opinion was shared by another London critic, who, some months later, reported: 'Mr Naldi is what the Italians call

---

* F. da Fonseca Benevides, in his history of the São Carlos Theatre, Lisbon (Lisbon 1883) mentions also a *primo buffo* Antonio Naldi appearing there in 1804 during Catalani's engagement. There were also other buffo singers named Naldi singing during the latter part of the 18th century.

** The dual nomenclature may be explained as follows: 'Initially baritone roles were largely taken by bass voices of extensive range, such as Ronconi, Tamburini, and Cartagenova, who—as late at 1840—were called bassi cantanti; but a little later a certain influx of tenors, too, were to take on baritone roles. The tenor of the early 19th century was, in fact, inclined towards a timbre which today we would term baritonal—(Crivelli and Donzelli one occasionally finds referred to as *bari-tenore*)—and for the top A he would adopt the falsetto. When Rubini managed a chest voice B, and Duprez, the C, the tenor became clearly established by virtue of his top notes, and many singers who at one time would have been classified as tenors, retreated to baritone roles'. (*Enciclopedia dello Spettacolo* vol I, col 1512). See also infra pp 116-7.

the *Buffo nobile*, as he never descends to excite the laughter of the audience by tricks that accompany our *low* comedians on the English stage'.[23]

On coming to Dublin, Naldi brought with him his wife, a ballet dancer, and his daughter, Carolina. Neither took part in any of the operas performed, but Signora Naldi danced in some incidental ballets, while Carolina sang at least one duet with her father during an interval. Having been born in 1801 she was then only ten years old.[24] When she first appeared in public in London at a Philharmonic Society Concert two-and-a-half years later, it was reported that 'she executed her part [in a quintette by Cimarosa] excessively well, and promises to become a fine singer when her voice is thoroughly formed'.[25] A reference in the same announcement that she had 'hitherto only been heard in private parties chiefly at Carlton House'[26] is very probably correct (if one disregards her Dublin appearance), for Naldi was very popular as an artist among London society, and a letter written by Princess Charlotte relates, 'Naldi, the Prince [her father, later George IV] sent for. He sang to us *delightfully*, as well as the Prince, who sang several duets with him & has settled with Naldi to come & give me lessons in singing on my return from the sea side'.[27]

Carolina's professional career was of brief duration. She sang with her father in *Così fan tutte* at the Théâtre Italien, Paris, in the autumn of 1820,* when she was described as 'still young and quite attractive; she has a small, thin little voice, can appear as the innocent but not much more; she gives proof of much that is negative in her training, without any positive basis and significance to support it. She has been engaged as prima donna'.[28] She was engaged there again in 1822, partnering Giuditta Pasta in *Tancredi*,[29] and in Zingarelli's *Giulietta e Romeo*.[30] But her face, not her voice, was her fortune, and in 1823 she married the Count de Sparre and retired from the theatre to enter society in earnest. It was a wise choice, as the following opinion expressed prior to her 1822 engagement makes clear: 'Mlle Naldi, who, it is said, is about to be engaged (*horresco referens!*) for 18,000 fr., has not enough voice even for the very small auditorium of the Théâtre Italien, her intonation is unreliable, her pronunciation very English, and her acting, non-existent. She could at best fill only secondary roles'.[31] Society was indeed her true ambiance, as a report from Paris in *The Court Journal* indicates. We read: 'The pleasantest evening I

---

* Giuseppe Naldi is recorded as having made his first Paris appearance there in *Così fan tutte* on 19 September 1820. *The British Stage*, October 1820.

remember to have spent for some time [was] on Monday last chez la Comtesse Merlin. The Countess is, as you know, the very first dilettante singer in Paris… Mme de Sparre too sang the duet from the second act of the *Elisir d'Amore* with *ce bon* Lablache'.[32] Obviously a much better milieu for Mlle Naldi than the Green Room of Crow Street Theatre, or, for that matter, the Théâtre Italien.

Unhappily, her father would not end his days amid such comfort, for tragically, like his colleague Radicati in the same 1811 Dublin season, he too had a violent—indeed bizarre—end. While investigating an early version of the pressure cooker at the home of his friend, the singer Manuel Garcia, the utensil exploded, killing him.

Other singers in Michael Kelly's company were Carlo and Angiolina Cauvini. In February-March 1808 (during the reign of King Louis Bonaparte) the former was singing at the Italian Opera in Amsterdam as a first tenor. He was there described as 'this modest, and by connoisseurs, much admired singer…[whose] voice attracts the greatest attention of every listener, and [who] with his real artistry in performance, can touch his listeners' hearts in even quite trivial music'.[33] On his arrival at the King's Theatre, not long before his Dublin visit, it was reported, 'the new tenor singer…promises to be an acquisition to the Theatre: his voice is sweet, but possessing little compass; and, indeed, in his exertion to attain the latter, the only deficiency we could see in him, that of an occasional defect of execution, was visible'.[34] Subsequently *The Times* would note that he 'was little inferior to [Tramezzani], except in power of execution'.[35]

But 'of Signora Cauvini we cannot speak so favourably', observes the reviewer who had so warmly greeted her husband. 'She deserves little praise; and we should certainly allot her a place in the chorus as the just sum of her ambition, were we not aware how impossible it is to form a fair estimate of comparative merit under the influence of Catalani'.[36] 'Neither her comic nor musical powers are of the conspicuous order',[37] records a second reviewer. Yet, to be impartial one must add that, when she sang Cherubino at the Pantheon in 1812, her 'Non so più cosa son' and 'Voi che sapete' 'were executed with much science and expression…and drew down repeated plaudits',[38] while as Despina at the King's Theatre, *The Times* found she 'had much archness and readiness in the various transformations of her character'.[39] What Angiolina Cauvini also had was a very pretty face. Both Mount Edgcumbe and Kelly comment on this, the former adding that she was a pleasing actress.[40]

The remaining singers of the group, Signori Balassi and Manni were of a much lower artistic standard. Francesco Balassi, a tenor, had sung with Miarteni's company in Amsterdam,[41] and had joined the King's Theatre during the 1810/11 season. There he sang minor roles, such as Oronte (A Priest) in *Die Zauberflöte*, and Basilio in *Le Nozze di Figaro*. Nicola Manni's name first appears in London in June 1810, as a singer in a series of subscription concerts at Willis's Rooms. In these he sang in ensembles only.[42] He is described as 'a bass'[43] and 'from Russia',[44] where presumably he had been singing previously. He was among the artists singing in a concert at the Argyle Street Rooms in 1812,[45] and in the same year amongst those who entertained the guests with music at a dinner given by the Duke of York to a distinguished company which included the Prince Regent and Princess Charlotte, the Duke of Cambridge, Lord Tyrconnell, and R. B. Sheridan.[46] These six singers made up the entire vocal company which, although advertised 'to perform Italian Operas complete in all their parts',[47] presented some that were very incomplete indeed. There was one other member of the company, the conductor of the orchestra, or maestro al cembalo. He was Haydn Corri, youngest son of the composer Domenico Corri, who, not long after, would return to Dublin to settle there and become very involved in the musical life of the city. It may be noted that Michael Kelly did not sing in any of the operas during this visit, but instead acted as 'stage manager'.[48]

Kelly began his season at Crow Street on Saturday, 17 August. It was advertised to run for 12 nights, the performances to take place on Mondays, Thursdays and Saturdays, 'in the course of which a great variety will be produced'.[49] As ever, he again errs in his *Reminiscences* in stating that the company was engaged by Jones to perform two Italian operas, *Il Furbo contro il Furbo* and *Così fan tutte*. As will be noted hereunder, four operas were given, *Così fan tutte* being one, but not *Il Furbo contro il Furbo*. He is also incorrect in writing, 'we performed very few nights'[50] since 18 performances (which included benefits) were given in all.*

For an Italian opera season a sense of occasion was required and, as in previous years, advice on Opera Dress was published in the fashionable periodicals. Recommendations included 'A blue satin robe, worn over a slip of white satin, let in at the bosom and sleeves (which are short) with silver

---

* Not including a further special benefit concert for Kelly.

Moravian net work. A tunic of Egyptian brown sarsnet or crape, confined on the shoulders with diamond studs, and trimmed round the bottom with silver net, separated in small divisions by spangled open work balls. A chaplet wreath of green foil, placed twice round the hair, which is disposed in long irregular ringlets. Earrings of silver open work studded with brilliants, resembling in form the bell of a child's coral. Shoes of brown satin, bound and sandalled with silver braiding. Long gloves of white kid'.[51]

The opening production was Pietro Carlo Guglielmi's *Due Nozze ed un sol Marito*, which had been produced at the King's Theatre in 1806 for Naldi's debut (six years earlier he had appeared in the premiere in Florence*) and presumably it was now considered the most appropriate opera in which to introduce him to Dublin. It seems to have been a suitable choice, for a review published two days later is laudatory. 'No Opera could have been more ably sustained', we read, 'Madame Bertinotti is the most exquisite, certainly the most mellifluous singer we have ever heard. She has not perhaps the bold compass, but she undoubtedly has a finer quality of voice than Catalani; there is a continued stream of melody, clear, distinct, and affecting, which throughout her singing delights the ear and reaches the heart, and all this without apparent effort. When executing the most difficult passages, her features, which are extremely pretty, are composed, yet perfectly intelligent and expressive of the subject of her song. She was rapturously applauded throughout the Opera. Signora Cauvini was not less applauded than Bertinotti; she is a most interesting comic Actress, playful, arch, and entertaining in her manner; she makes her points with much vivacity and ease, and her songs are given with considerable effect; she is likely to be a great favourite with the public. Naldi, as a Buffo performer has been celebrated in the first Theatres of Europe; he is considered one of the best Comedians of the present day, and from what we observed on Saturday night, he is also deserving high praise as a serious singer. The Signors Cauvini, Man[n]i and Balassi sustained their characters with much effect, and were repeatedly applauded. The orchestra is of a very superior order, and the whole arrangement, while it proves the liberality of the Patentee, also reflects credit on our countryman, Kelly, who is the stage manager'.[52] Whatever the attendance may have been, no management could have asked for (or subscribed

---

* As had Manni, also in the same role, Farfalla, which he was to sing in Dublin. (Piovano, Francesco, 'Notizie storico-bibliografiche sulle opere di Pietro Carlo Guglielmi', *R.M.I.*, vol 16, 1909, p 490).

to!) a better notice than that. In the event, it proved the most popular opera of the season, receiving six performances.

Due Nozze was followed on 24 August by a revival of the already favourite *Il Fanatico per la Musica*, and on this occasion, (unlike the 1808 performances), the two-act version, however incomplete, was performed. Naldi, who 'personated the musical enthusiast, displayed all the eccentricities of the character and its irritability with all the effect which it so highly admits of, without the smallest approach to extravagance or over-acting. His performance upon the violoncello astonished as much as it delighted the audience, who were unprepared to hear from such an instrument the fine and delicate harmony which he extracted from it; at his conclusion he was greeted by the plaudits of all parts of the House. Madame Bertinotti in the Air of 'Nel cor più non mi sento'…evinced great knowledge of the science of music, with an excellent taste; this Air…was uncommonly well received, and concluded the performance'.[53] When first produced at the King's Theatre in 1806, this opera was said to have 'been greatly heightened by many introduced pieces by Signor Naldi'.[54] Later it will be shown that he made much the same kind of insertions in Dublin. The introduction of 'Nel cor più non mi sento' had, of course, originated with Catalani.

What must be considered a watershed of sorts in the history of opera in Dublin was now reached with the first performance, on 31 August 1811, of a Mozart opera. Because Bertinotti had had the courage to present *Così fan tutte* for her benefit at the King's Theatre during the previous May (on which venture she lost £100)[55] *Così fan tutte* was the Mozart opera which Dublin audiences first heard, since it was already in the company's repertoire. Earlier they had had the opportunity of hearing excerpts from a few of his operas over the years. The opening trio, 'La mia Dorabella' from the same work sung by Urbani and others, as well as the 'Italian Duet' with John Spray have already been noted, while there was also 'Prenderò quel brunettino' also from *Così fan tutte* sung by Mme Dussek and Miss Tyrer,[56] besides the chorus, 'Placido è il mar' from *Idomeneo*, here arranged as a quartette and sung by Mme Dussek and Messrs Phelps, Spray and Weyman.[57] At last, however, Dublin had a complete Mozart opera—or had it?

To consider the question one must return to the first King's Theatre performance of the previous May. The cast was then: Fiordiligi, Bertinotti;

Teresa Bertinotti Radicati (soprano).

Dorabella, Collini; Despina, Cauvini; Ferrando, Cauvini; Guglielmo, Tramezzani; and Don Alfonso, Naldi. In the above casting it will come as a surprise to find the baritone role of Guglielmo sung by the company's principal tenor, Tramezzani. It will probably come as a greater surprise to learn that at that period, and even much later where Mozart was concerned, type of voice seemed to bear little relation to the casting of parts. This is manifestly exemplified in a letter written in April 1825 by William Ayrton, musical director of the King's Theatre, to John Ebers, the then manager. It should be noted that *all* the singers Ayrton mentions were tenors.

'Yesterday morning,' he relates, 'I had the first rehearsal of *Così fan tutte*, an opera that has not been performed for many years, and allotted the two characters, Ferrando and Guglielmo, to the Signors Curioni and Garcia. To my surprise the latter refused that which was assigned to him, claiming the other; and when *pro forma*, I requested Signor Curioni to attempt the rejected part, he declined it...finding it as much too low for his voice, as that of Ferrando is, I fear, too high for Signor Garcia's. The ground on which the latter refuses to take the part given to him is, that it is written in the base clef. When first this admirable opera was produced in London in 1811, Signor Tramezzani, a tenor, and one of the finest singers and actors that ever graced the lyric stage, took the character of Guglielmo. Signor Crivelli also, another tenor of the highest class, accepted the same in 1817, and M. Begrez, with a much higher voice than either of the former, has likewise performed it. *Except in one instance, the part has never been sung at the King's Theatre but by a tenor*, and with the few alterations that have always been made in it, is well suited to Signor Garcia, whose compass is extending downwards, and will not permit him to sing with due effect the part of Ferrando, which he chooses. But how can Signor Garcia justify his refusal to take the character given to him, after he has so repeatedly sung, both in Paris and London, in that of Don Giovanni, which is quite as low as the other, and is also written in the base clef?—What he denominates a base part (which is in fact a barytone) is in his compass when it suits his own purpose, but it is out of his voice when the interests of Theatre alone are concerned'.[58]

To the above extraordinary situation it may be added that in 1812 Cauvini sang the Count in a shortened version of *Le Nozze di Figaro*,[59] but when the same role was offered to Tramezzani, he seems to have refused it almost more because he considered *Figaro* to be a "comic opera" than because—as a baritone role—it lay too low for him.⁶⁰ Later he relented, however, appearing as the Count in a performance for Naldi's benefit on 3 June 1813. But not without calling attention to his celebrity, declaring that 'Signor Tramezzani with a view to obliging the public…will endeavour to adapt the music to his voice'.⁶¹

For some further surprises, still at the King's Theatre, one can examine the roles of Fiordiligi and Ferrando. In the former part Bertinotti omitted both 'Come scoglio' and 'Per pietà', and 'introduced in place of the first, the cavatina "Porgi amor" from *Figaro*, and, for the second, robbed poor Collini of Dorabella's rondeau "È amore un ladroncello".'⁶² Then Carlo Cauvini 'found the two majestic songs, "Tradito, schernito dal perfido cor" and "Ah! lo veggio, quell'anima bella" too much for his elegant but very limited voice, singing only the cavatina "Un' aura amorosa" and introducing "Voi che sapete" out of *Figaro*'.⁶³ He may have made some amends for the latter incongruity by having 'executed his songs with peculiar pathos and an elegance of ornament that did great credit to his taste',⁶⁴ though still leaving much to be desired as far as Mozart's artistic intentions were concerned.

To return now to Dublin: while Bertinotti and Naldi retained their London roles of Fiordiligi and Don Alfonso, Angiolina Cauvini took over Dorabella, and Carlo Cauvini, Guglielmo (which seems to have been considered more important than Ferrando), leaving the latter role to Signor Manni, who, as noted, is described as a bass! Finally, Despina changed both name and sex becoming Vespino and was sung by Balassi, a tenor!!⁶⁵ Then, to demonstrate how much further Mozart could be degraded, we learn that at the third performance on 21 September, 'In the course of the opera will be introduced the Favourite Duet from the Opera of *La Griselda* by Paer; likewise the celebrated Canone [? 'Das klinget so herrlich'] from the much admired Opera of *Die Zauber Flöte* (The Enchanted Flute) composed by Mozart with an accompaniment on the Angelica'.⁶⁶ In spite of this, Mozart escaped more lightly in Dublin than in other places. In Edinburgh, where the company performed *Così fan tutte* some six weeks later, it was announced: 'between the acts, by particular desire, Signor Naldi will sing a Grand Song in which he introduces his celebrated Imitation of a PARROT'!!!⁶⁷

How was the opera performed in Dublin, and how was it received by the audience? Regretfully, no review can be discovered. The admittedly not very important choruses were, of course, omitted, but two factors may be surmised from London performances. The first concerns Bertinotti's singing—'perfectly in tune and given with feeling that evinced a proper estimation of Mozart's merit. No meretricious ornaments were added, but every grace was introduced in its proper situation, and executed with a precision that no other singer but Mrs Billington* could equal'.[68] The second refers to Naldi's acting as Don Alfonso— 'more like an impudent valet or *major-domo* of an hotel. We never saw any one so much at home, who seems so little conscious of the existence of any one but himself, and who throws his voice, his arms and legs about with such a total disregard of *bienséance*'.[69] All we can be certain of is that the opera received five performances, but the lengths that were gone to (as detailed above) to ensure its success, indicate that the artists as well as Michael Kelly, and probably Frederick Jones, had little confidence in Mozart if left without adaptation.

The final opera to be produced on 12 September was the favourite Bertinotti vehicle, 'the Grand Serious Opera of *Zaira* [here chosen for her benefit] composed by Chevalier Federici'.[70] Confusion exists concerning the said chevalier, since he probably was not Vincenzo Federici as was for long assumed, but one Francesco Federici, composer also of the opera *Virginia*, in which Bertinotti appeared at Ferrara in 1805. The evidence for this identification rests mainly on Beniamino Gutierrez: *Il Teatro Carcano 1803-1914*, which book displays a reproduction of a theatre bill announcing the opening of the Nuovo Teatro Carcano (Milan) with '*La Zaira*. Musica del Maestro Francesco Federici'. This would have been more conclusive had Signor Gutierrez not recorded in his text on the facing page: 'On 3 September 1803 the new theatre…was inaugurated with the "dramma serio" *Zaira*, music by the maestro Vincenzo Federici'.[71] A clear case of the right hand not knowing what the left was doing! However, Gutierrez still helps to confirm that it was Francesco, when in a footnote quoting G. Chiappari as his source, he first corroborates that the composer was 'of Genoa',[72] and then, a few pages later, quoting from an anonymous *Lettera encomiastica*, relates that the composer of *Zaira* was 'the never sufficiently appreciated or praised M[o.] Federici of Genoa'.[73]

The history of the opera itself is almost as confusing. *La Zaira, o sia il Trionfo della Religione* seems to have been performed first at the Teatro Santa Cecilia in Palermo in 1799,[74] next in Reggio Emilia in March 1802,[75] next—as an oratorio—in Naples in 1802,[76] and then at the Teatro Carcano, Milan.[77] Bertinotti

* See *Opera in Dublin 1705-1797*.

seems to have considered this work a sort of talisman, perhaps because of the interpolated arias which Radicati had composed for her. Having succeeded in it in London she chose it again as her first opera seria at the Théâtre Italien, Paris, in 1817. But the Parisians were not impressed and the music was pronounced, 'the usual Italian "Lirum-Larum", a totally characterless inferior piece without the slightest suggestion of tragedy—and quite insipid'.[78] Dublin seems to have anticipated the melancholy verdict for, although the opera was bolstered up with an overture by Mayr, a canone for three voices and a glee for four, from Radicati's opera *Phedra*, an air from Paisiello's *Il Re Teodoro in Venezia* sung by Naldi, and Giuseppe Nicolini's 'admired Prayer of Giusto Dio' sung by Bertinotti,[79] yet it was granted two performances only.

Toward the end of the season the benefit performances continued, the detailed advertisements of which on this occasion are both informative and revealing. Naldi's took place on 16 September, when *Il Fanatico per la Musica* was performed, but by no means was it performed alone. Also included were, 'In the First Act Signor Naldi and Miss Naldi' in a favourite duet, the act to conclude 'with an Allemande, danced by Madame Naldi and Miss Naldi, with an accompaniment by Signor Naldi on the "Angelica", an instrument hitherto unknown in this country. In Act the Second will be introduced a new Duet between Madame Bertinotti and Signor Naldi, with an accompaniment by him on the Violoncello, in addition to that already in the Piece. In the course of the Opera Signor Naldi will sing an air with an accompaniment on the French Guitar. After the Opera will be represented *for that night only*, a Divertissement called *La Précaution Inutile* from Dauberval's celebrated Ballet of *La Fille mal gardée* by Madame Naldi and Monsieur St Pierre, who has very kindly offered his services on this occasion'. Tickets could be had of Signor Naldi who, together with his wife and daughter, was lodging with Haydn Corri at 2 Upper Sackville Street, or of Mr McNally ('Gentleman Usher to the boxes'[80]) at the theatre.[81]

Carlo Cauvini's benefit followed next on 19 September, with *Due Nozze ed un sol Marito*. On this evening the interpolations were fewer. During the first act Bertinotti sang the favourite air of 'La mia crudel tiranna' with variations by Radicati, accompanying herself on the French guitar. (This aria was said to have been adapted from the then popular Venetian tune known as 'The Maid of Lodi'.),[82] Cauvini sang the popular air, 'Sul margine d'un rio' by the Belgian harpist and composer, François Joseph Dizi, with variations by Radicati, and with Bertinotti, the Grand Duet from Paer's *Griselda*. Between the acts there was a grand Overture by Haydn, and in act two, an air by Cimarosa sung by Cauvini. His address and that of Bertinotti and Radicati is given as 26 Westmor[e]land Street.[83]

On 24 September there was a replica of this performance of *Due Nozze* for Angiolina Cauvini, and then on 28 September there was a final performance of *Così fan tutte*, a joint benefit for Radicati and Haydn Corri. It was advertised, 'In consequence of the unbounded approbation with which the truly celebrated Opera…was received', but to it were added two excerpts from *Die Zauberflöte*, a canone and 'The manly heart', the latter sung by Bertinotti and Naldi, besides 'the celebrated scene from the Opera of *Il Fanatico*, in which Signor Naldi will accompany Madame Bertinotti on the Violoncello' concluding with 'Nel cor più non mi sento' sung by Bertinotti. And that was not all, for 'in the course of the Evening' 'Madame Naldi [had] most kindly consented to dance the much admired Shawl Dance with a…Violin Obbligato by Signor Radicati'. Further, it was promised, 'Other entertainments will be introduced'.[84] Once again, quite obviously, the artists were leaving nothing to chance—least of all Mozart—to assure full houses.

There was still one last semi-staged performance by the company when Michael Kelly (lodging at 46 Dame Street) took a benefit on 1 October. The evening included the overture to *Artaxerxes*, after which Kelly sang 'I've roamed thro' many a wearied round' from his *The Gypsy Prince*. Next, the Cauvinis sang a duet by Farinelli,* which was followed by another duet, 'Destined by Fate' sung by Naldi and Kelly, from the latter's opera, *The Royal Oak*. A 'Cavatina' was sung by Carlo Cauvini, followed by a duet from '*The Enchanted Flute*' (almost certainly 'The Manly Heart') by Bertinotti and Naldi. Kelly then sang 'a new song "The Bard of Erin" accompanied by himself on the pianoforte'. This was followed by Radicati's aria, 'Grata di tanto dono' (introduced into *Zaira*) sung by Bertinotti. Penultimately another duet was sung by Naldi and Kelly, this time, 'The Sailor Boy' from Kelly's *The Wood Demon*, and the curtain eventually fell in patriotic fashion on 'God Save the King'.[85]

* Most likely Giuseppe Farinelli (1769-1836), and not Carlo Broschi (1705-1782), known as Farinelli.

Before assessing the season's programme, it may be noted that in his *Reminiscences* Kelly has stated: 'On the 5th of September 1811, I made my last appearance on any stage, on the stage where I had made my first appearance when a boy in 1779'.[86] Apart from the incorrect date which he gives of his last performance, he is equally wrong in stating that he had made his debut in Crow Street Theatre. This had occurred in a performance of Piccinni's *La Buona Figliuola* on 17 May 1777 at Fishamble Street Theatre.*

The introduction of popular airs from other works into the different operas had been so extensive during the 1808 season that it is unnecessary to explain its repetition in 1811. Yet in that year it was perpetrated so immoderately that obviously its sole purpose was to attract audiences. Apart from the mutilation of the operas themselves, the inclusion of so much alien material must have made it impossible for an audience to follow any sort of story line in what they were witnessing on stage. Perhaps they were not expected to, or, perhaps they were interested only in a string of unrelated vocal items which titillated the ear. Moreover, there are indications that full houses were not anticipated from the beginning, since even before the season started it was advertised 'Notwithstanding the great expense attending this undertaking the prices of the Theatre will not be raised'.[87] Yet, the slightly later announcement that 'Complimentary Free Admissions to the Theatre are suspended until further notice'[88] reflects some optimism. Admittedly the Lord Lieutenant, the Duke of Richmond, was touring 'the North of Ireland'[89] at the time which exempted the nobility and gentry from having to be bored by the performances at which etiquette and self-interest would have required their presence had he been in residence. Indeed, it was reported on 23 September that because of his return there would be one or two performances that week.[90] He certainly commanded one of *Il Fanatico* for the evening of 25 September, which presumably took place.[91] There were four in all, if one includes Michael Kelly's concert, but three of these were benefit performances which had been agreed long before at the time of the artists' engagements, nor is there evidence that the Lord Lieutenant attended any of them. Neither could inadequate casting be blamed for the failure, since four of the six artists were principal singers from the King's Theatre. In a few words, the real cause can be summed up by the absence of a star, such

---

* See *Opera in Dublin 1705-1797*.

Giuseppe Naldi, bass (1770-1820)

as Catalani.* Bertinotti may have had a sweeter voice, but she lacked Catalani's aura. She was good, but not good enough to fill the theatre, and for once Kelly may be presumed to have been correct when he wrote that the houses did not answer 'the expectations either of ourselves or Mr Jones'.[92]

In spite of poor receipts at Crow Street, Jones could take consolation from the impending financial collapse of Henry Johnston's venture at the Royal Hibernian Theatre. At the time that the Italian company was ending its less than successful season, sworn affidavits from disgruntled stage carpenters and scene painters headed by Filippo Zafforini were landing like white birds of ill omen on Johnston's desk. Zafforini's affidavit tells us something of the life and hard times of stage workers in Dublin at this period. In it he 'maketh oath that he had been for eleven years and upwards in the employment of Frederick Edward Jones Esq., Patentee of the Theatre Royal, Dublin; during the whole of which time he was regularly paid his salary; that about the close of that period, he was seduced from his employment, under the promise of Mr Henry Johnston, of two pounds nineteen shillings and threepence per week more than he received at the Theatre Royal; that he withdrew from the service of the Patentee of the Theatre Royal when occupied in painting scenes which he left unfinished; that after some time he discovered Mr Henry Johnston's inability to pay him, and therefore surrendered his situation, at which period Mr Johnston was indebted to this deponent in the sum of one hundred pounds and upwards for which he received a small part in cash and two bills accepted by Mr Johnston, one for thirty guineas at three months, the other for fifty pounds at nine months'.[93]

The unhappy affair ended in February 1812 when, ominously, it was announced, 'the Creditors of Mr H. Johnston of Peter-street Theatre, are most earnestly requested to meet at Twelve o'Clock This Day at the House of Mr James Duignan, Carlisle Tavern, Bachelor's Walk, on Business of most serious importance to their interests'.[94] One can only add that whatever it was (and still is) that attracts humanity to become involved in the theatre, it can hardly be a conviction that an assured livelihood is to be found there.

---

* In August it had been 'most positively' reported from London that four months earlier, Catalani had refused 'propositions which [had] been made to her to perform in Ireland this season' but had promised to appear there during the summer of 1812. (*The Courier*, 22 August 1811)

Chapter 7
Catalani Ritorna 1812-1814

It is not generally known that Ireland's premier melodist, Tom Moore, was also involved in writing two comic operas. On reflection not so surprising when one remembers that he was one of the best actors (and singers) at the Kilkenny private theatre early in the century. For example, a report of his performance as Tom in John O'Keeffe's *Peeping Tom of Coventry* records: 'The delight and darling of the Kilkenny audience appears to be *Anacreon* Moore*. The vivacity and *naiveté* of his manner, the ease and archness of his humour, and the natural sweetness of his voice, have quite enamoured us'.[1] The only surprise from the foregoing is that his dramatic efforts failed.

The first, *The Gypsy Prince*, mostly set to music by Michael Kelly has been mentioned earlier; the second, for which Moore also 'composed and selected the music',[2]** was called *M. P.; or, The Blue Stocking* and had its first Dublin performance on 19 May 1812. When first produced during the previous year at the Lyceum Theatre, London, Leigh Hunt had expressed disappointment. 'The house is filled,' he reports, 'expectant congratulation runs from bench to bench, the most rigid and critical faces thaw in the general smile; the overture begins—why is it not over?—the curtain rises, the actors come forward, and lo, instead of an opera worthy of its poet, a farce in three acts of the old complexion! ...An unambitious, undignified, and most unworthy compilation of pun, equivoque, and clap-trap!'[3] Nor did the music please another reviewer who writes: 'The songs in the Opera are uniformly flat, feeble, and monotonous. The fire of the Composer seemed to have passed away... The Opera was received with very dubious favour. Some of the songs were hissed, some of the jokes deserved all the hisses which they could get'.[4] Amongst the music selected by Moore was a trio 'This is love' arranged from Rode's celebrated 'Air with Variations', an Irish air, 'Come lads, life's a whirligig', and a second trio, 'Girl dost thou know me', said to have been taken from *Così fan tutte*.[5]

Perhaps the most remarkable fact about the opera was Moore's own lack of faith in it, both as drama and music, for in a letter to Edward Tuite Dalton

---

* An allusion to his principal work while an undergraduate at Trinity College, Dublin, a metrical translation of Anacreon's odes.

** C. E. Horn orchestrated it and composed the overture.

(who had married Sir John Stevenson's eldest daughter, Olivia) he wrote: [*M. P.*] 'was written quite as a hasty job, and therefore gave me nothing but sickness in my stomach from beginning to end... I was very much flattered by Stevenson's favourable anticipation of my music, but I *know* he has been disappointed. It was the first time I ever composed airs premeditatedly (for I need not tell *you* that they have always come by chance); and the idea of a task disgusted and disabled me. Again I made an effort to compose for dramatic effect, which took me out of *my own* element without naturalising me in any *other*'.[6] The only advantage he seems to have gained from the Dublin production was the publication of the vocal score by William Power of 4 Westmoreland Street.

During 1812 ballet occasionally remained a featured part of opera performances. When the romantic musical, *One o'Clock; or, The Knight and Wood Daemon*, by Michael Kelly and M. P. King (libretto by 'Monk' Lewis) was performed on 2 April, 'a ballet incidental to the Piece [was] introduced into act two and danced by the Misses Adams and Miss Rock'.[7] Jones evidently had made his peace with the Misses Adams and reinstated them in the company, for, like Zafforini, they had defected to Johnston's Royal Hibernian Theatre, both complaining, as Zafforini had, of losses sustained 'by Mr Johnston's absenting himself'.[8] Nevertheless, ballet does not seem to have been popular in Dublin just then, for later in 1812 it was announced, 'Another company of Dancers have arrived, the Misses Dennett. Reinforcement is more wanting in the Drama, than in Dance: we don't see any reason for Mr Jones, or his young Manager, Tony Rock, conceiving that the town is *dancing mad*'.[9]

On 9 September it was reported by *The Morning Chronicle* that 'Madam Bertinotti, Signor Tramezzani, Rovedino and Madam Bianchi are in treaty with Mr Jones, Proprietor of the Dublin Theatre to perform Italian operas in the course of this and the next month'. Oddly, *The Freeman's Journal* of the 15 September carries a similar story, but with the significant distinction that the singers are in treaty with the proprietor of Covent Garden. In the event Bertinotti travelled instead to Lisbon, where she appeared at the Teatro São Carlos in November.[10]

Throughout the following year of 1813, the only new opera of any significance to be performed in Dublin was *The Devil's Bridge*. This had a libretto by Samuel Arnold, with music by Braham and C. E. Horn, and was given there first on

12 November, with the ubiquitous and all-embracing Tom Cooke as Count Belino, and his wife in the role of Countess Rosalvina. It was considered to be 'altogether the most interesting and best managed entertainment, that has been got up here, for a long time. In fact', we are assured, 'it wants no attraction, other incident, dialogue or music, and its characters even appear to be well cast. The House was pretty full, and it was given out for a second representation on Wednesday evening amidst the loudest acclamations'.[11] For good measure, 'at the end of the Play' there was dancing by the Misses Dennett, making their first appearance this season.[12]

Tom Cooke had earlier demonstrated his versatility yet again, when in 1811 he had announced himself as attempting to sing the tenor part of the Seraskier* in *The Siege of Belgrade*. His appearance as a singer had then 'excited universal surprise...no one had any conception that he would sustain the part altogether in such an able manner'.[13] So able was it, that he migrated to London where he appeared in the same role at the Lyceum Theatre in July 1813, winning the following commendation: 'His voice is not of great compass, but he sings with taste and feeling. His shake (which he can use through the whole range of his voice) is unquestionably to be ranked among the most delightful of that ornament. It is grand and lively on his lower tones, and upon the upper inexpressibly delicate and pleasing'.[14] Naturally Cooke had some less approving critics. 'His knowledge of music almost compensates the inherent disabilities of a voice entirely artificial, and which he cannot trust to itself for an instant',[15] wrote one, while another opined, his 'singing would be infinitely more pleasing if he did not suffer the words and notes to run out of his mouth in a flux'.[16] There was no dissent about his acting, however. All were agreed with the reviewer who wrote, 'It is in a spirit of real good-will towards him that we say he is among the very worst we ever saw'.[17] His wife seems to have received a similar assessment, for it was said of her: She 'has a sweet, melodious, and without any pretensions to excel in the higher range of her profession, a cultivated and well-managed voice...there is a naiveté and beauty and infantine delicacy in the tones that are equally attractive. As an actress she is below mediocrity—totally inattentive to the business of the scene, and with an eye

---

* As early as 1804, while 'Leader of the Orchestra' at the private theatre in Kilkenny where he played each year until 1810, he had also appeared on stage in after-pieces. (*The Private Theatre of Kilkenny*, Kilkenny 1825)

Thomas Simpson Cooke, singer, instrumentalist, "conductor" and composer (1782-1848). Here, as Don Carlos, in *The Duenna* (Sheridan/Linley).

wandering about the boxes, she prattles off her part like a little miss running through her French verbs'.[18]

Other featured singers of rather less eminence who were singing in Dublin at this time, were 'the celebrated' Madame O'Moran who made her first appearance there on 14 April in a performance of *Messiah* at the Rotunda,[19] and Madame La Finilarde, who may have been the same as a Madame Feuillade (whose husband was a professor of dancing[20]), and who on 13 May 1814 took part in a concert 'for the Benefit of Distressed and Decayed Military Musicians',[21] or even Madame Philadare, who was among the artists taking part in the annual sacred concert for 'the Benefit of the Incorporated Musical Fund Society' at the Rotunda on 22 March 1815.[22] Of more renown, at least in Dublin, was Signora Rosa, who had been engaged for Crow Street by Jones, and was reputed to have been a native of Milan,[23] (She *may* have been Rosina la Rosa, who appeared as prima-donna at the Imperial e Real Teatro, Pavia, during Carnevale 1822.[24]) We are told 'she possesses high vocal powers, and is an excellent professor of music: her figure is small but delicate: she has never yet appeared on the Stage, but she has performed with universal applause at several Concerts in London'.[25] She took part in a number of musical pieces at Crow Street, having her benefit on 29 June 1813 when she sang Victoria in *The Castle of Andalusia* '(with additional songs)'. These included, 'Madame Catalani's favourite cavatina of "Vittima sventurata", adapted to English words from the serious opera of *The Vestal* [Pucitta]—the favourite Ballad of "Robin Adair"— the admired Duet with Mrs Cooke of "Ah tu sei che" to English words from the opera of *The Virgin of the Sun**, and "Hope told a flattering tale", accompanied on the Harp'.[26] It is obvious from this that Signora Rosa's theatrical engagement (possibly impeded by a lack of English) was very much that of singer rather than actress.

Following a posting of over six years, the Duke of Richmond ended his tenure as Lord Lieutenant in 1813. When the theatre closed for the season on 7 August, he and the Duchess 'were present and there was a full house—but there was not a syllable uttered in the shape of an Address'. Reprovingly, the report continues, 'This omission of all common courtesy is really unaccountable. It is practised at no Theatre of consideration in the United Kingdom, but at Crow Street. There is a disorder in it, a village-like unseemliness, a want of

---

* 'Ah! tu sei che stringo al seno' from *La Vergine del Sole* by Gaetano Andreozzi.

metropolitan completeness and regularity, which are ill-suited to the consequence and discernment of the description of audience which this city affords'.[27] From his first visit to the theatre, when, it will be remembered, he refused to have 'St Patrick's Day' played, until his last, little had gone well for either Viceroy or country. Perhaps it would have been better for both had he acknowledged the Catholic party tune.

He was succeeded by Viscount Whitworth, who was married to the Duke of Dorset's widow. (Through some quirk of courtesy or of protocol, the lady continued to be addressed as the Duchess of Dorset.) It was hoped that Whitworth, as anti-Catholic as his predecessor, but an experienced diplomat, would stem the movement towards Emancipation. He failed, although ironically, as one of a triumvirate with his precursors Richmond and Hardwicke, his name is still known to an older generation of Dubliners through a section of a hospital now called officially (though rarely outside official circles) St Laurence's.* With the Duchess, he paid his first visit to the theatre on 20 November, where, happily, they 'were received with all the warmth and…*hospitality* of the Irish character'.[28]

Richard Daly, long in retirement, died on 8 September 1813, and at last all the theatrical threads with the 18th century were broken. And what of the present early 19th century? According to a report published in *The Freeman's Journal* some optimism prevailed, for it stated: 'We are firmly convinced that Dublin is, notwithstanding the Union, well able to maintain two Theatres. Nothing can be more absurd than speculation and opinions of a contrary description…notwithstanding [that] Dublin has lost a number of wealthy residents and occasional visitors'.[29] Yet in the challenge a sense of doubt seems to obtrude. It has continued so for the intervening 170 years. Even today, the same challenge, the same feeling of uncertainty remains in theatrical Dublin.

The year 1814 ushered in two musical works by an English opera composer of transient renown. He was Henry, later Sir Henry, Bishop, today known principally as the composer of sentimental songs such as 'Lo, hear the gentle lark', 'My pretty Jane', and 'Home Sweet Home'. From July to October 1820 he acted as *maestro al piano* at the then temporary theatre in Dublin, receiving the freedom of the city in the same year. The first of the two Bishop pieces,

---

* St Laurence Hospital was closed down in 1987 when it was superseded by Beaumont Hospital.

performed on 4 January, was a 'new Melo Drame' called *The Miller and his Men*, the libretto by Isaac Pocock. It was repeated on the following evening 'with increased effect', and we learn that 'the scenery [by Marinari] is beautiful...the performance throughout much to be praised, and the general interest of the piece, which is indeed great, can only be equalled by that of the State Lottery to begin drawing on the 18th of this month'.[30] As noted, this work of Bishop's was not called an opera but a melodrame, which has been described as *'sui generis*, an *olla podrida* of tragedy, comedy, opera, farce, and pantomime, partaking more or less of any of the qualities of these as the whim and judgment of the writer pleases'.[31]*

There were no theatre performances in mid-January due to 'the inclemency of the weather, unparalleled for half-a-century', and 'the immense quantity of snow', which had to be 'taken away in carts'.[32] However, by February the weather had returned to winter normalcy, and the 24th having been appointed by the Lord Lieutenant the day to celebrate Queen Charlotte's birthday, the boxes at the theatre *'were free for Ladies*! but the *Gentlemen made free* also!—a more disgraceful scene could scarcely be witnessed—why is this custom preserved when 'tis known to produce but impropriety and riot? A collection of wretched and misguided females, principally servants, were admitted—not a single line of the Performance could be distinguished from the screaming of the *Ladies* in the Boxes, occasioned by the *obstrusive* attention of the *Gentlemen*... A party of the Police interfered, which gave the *Gentlemen* an opportunity of "Kicking up a Row!" Several of the Youths were however taken into safe keeping!'[33]

The Miller and his Men was followed on 1 March by a second melodrame by Bishop called *The Hungarian Cottage; or, The Brazen Bust*. The librettist of this piece was the distinguished actor, Charles Kemble, then playing in Dublin, who had already taken the role of Frederick when the work was first presented at Covent Garden. It seems to have been much less 'melo' than 'drame' expressly for Kemble, since none of the cast were noteworthy as singers, and Bishop's music seems to have consisted of an overture and 20 passages of action music only, without any vocal numbers.

---

* For a more academic description of melodrame (and *The Miller and his Men*) see *The Athlone History of Music in Britain*, vol V. *The Romantic Age 1800-1914*. Ed. Nicholas Temperley, London 1981, p 301.

Over the years little was changing in the theatre where uproar and the throwing of missiles was concerned. On 3 March a reader living in Capel Street addressed the following letter to Frederick Jones in *The Freeman's Journal*: 'Sir. At the desire of several Gentlemen who were in the Pitt of the Theatre last night, I beg to make known a wicked outrage that occurred at the time the audience were standing and uncovered. A weighty broken tumbler was flung from the Upper Gallery and struck me in the breast; had it struck me on the head it might prove of serious consequence (although an individual in humble life) to my family... I conceive that a prohibition to the admission of bottles and tumblers from the refreshing Box to the Gallery would in some measure tend to prevent a repetition of the outrage; for, if the person who threw the tumbler had it not so convenient, I think the cruel thought of flinging it amongst the unoffending spectators would never have occurred to him'.

Then on 29 May, yet again, politics entered. During a performance of Shield and Pearce's opera, *Netley Abbey*, 'a ridiculous contest' arose 'between Fullam and some jackeens in the Upper Gallery... They took mighty offence at his *Orange* coat! (by the bye it was a *brown* one!). He was desired to change it. He most heroically exclaimed, "*I won't*!"—They assailed him with loud cries—yells—and orange peels! The actor shook his stick at them, stamped his foot, and danced in a perfect paroxysm of agony!—The boxes laughed and cried out "*Bravo*!"[34]

A musical production which seems to have been an opera, at least in intention, and not another melodrame, was presented on 22 April. It was *The Corsair; or, The Pirates' Isle*, composed by a London organist and conductor named Jonathan Blewitt, who had come to Ireland in 1811 under the patronage of Lord Cahir. He became organist at St Andrew's Church, Dublin, and succeeded Tom Cooke as musical director at Crow Street when the latter left for London in 1813. The libretto of *The Corsair* 'founded' (rather loosely one would imagine) on Byron's poem, was at first said to be by that familiar Dublin dramatist—'a Gentleman of this City',[35] but who, in this instance, was not, for we later learn that he was 'a Mr Sullivan of Cork'.[36] He was, in fact, Michael John O'Sullivan, born 1794, died 1845, who wrote poetry (one volume of which was dedicated to Tom Moore) as well as plays. A barrister-at-law of the Middle Temple, in 1818 he relinquished the law to become editor of *The Freeman's Journal*, a post he held until 1825. He subsequently edited other newspapers and periodicals including *The Theatrical Observer*. Shortly before he died he was 'in conjunction with Jonathan Blewitt the author of the Musical Critiques in *The Weekly Freeman's Journal*'.[37]

As well as conductor and organist, Blewitt appears to have been a passable composer, for we read, 'tis but justice to say, that the Overture and Music bear no little incitement of attraction from its general style of judicious arrangement and originality'.[38] In act two, 'A Turkish Dance incidental to the Piece' was danced by the Misses Dennett,[39] and one other point in connection with the performance deserves mention. The role of Medora was taken by the young Eliza O'Neill. She was the daughter of John O'Neill, the manager of an Irish strolling company and had made her first appearance in Drogheda. Later she had joined the Crow Street company, making her debut with them in October 1811. On 6 October 1814 she made her first London appearance at Covent Garden as Juliet, and thence went on to fame and fortune until her triumphant career ended prematurely in marriage. This was a romantic affair in every way. The groom was William Wrixon Becher, M.P. for Mallow, Co. Cork, where he possessed considerable estates, and who would be created a baronet on William IV's coronation in 1831. He was also a leading amateur member of the Kilkenny theatre, where he took part in over a dozen Shakespearean productions between 1807 and 1819, playing everything from Shylock to Sir Toby Belch to the Ghost in *Hamlet*, and by his imposing stage presence and voice, continually earning the approbation of the press. In October 1819 Miss O'Neill rejoined the company (she had already appeared there in 1812), when she played Juliet, and he Friar Laurence, both also appearing in *Othello*, and in Otway's *Venice Preserved*.[40] Two months later, on 18 December, they were married in the little church in Kilfane near Kilkenny, (the seat of the Power family, leading patrons of the theatre) by the Dean of Ossory, Joseph Bourke,[41] who, with his wife, regularly attended the performances. Unquestionably it was the theatrical wedding of the year in Ireland.*

On 27 June 1814 the illuminations 'in celebration of the Peace were uncommonly brilliant; amongst the most distinguished [were those of the] Theatre Royal: Front of Crow-street a beautiful transparency, painted by Signor Marinari. France, represented by a female figure, in a robe covered with

---

* Thackeray is said to have based his character of Emily Costigan, better known under her theatrical name of Miss Fotheringay, in *Pendennis,* on Eliza O'Neill. He was related by marriage to Eliza; his mother was a Becher. Moreover, in the seventeenth century there was a Becher forbear, Oliver, 'of Fotheringay and Howberry', Bedfordshire. (*Burke's Irish Family Records*. 1986)

Eliza O'Neill, actress (1791-1872).

Fleur de lis, appears sinking under grief and oppression, at the foot of Britannia, who, in the act of raising her, points to Peace, represented by a colossal female figure with the olive branch in one hand, and the French lily in the other'.[42] At last Europe was at peace, but it would be short-lived. In less than a year Napoleon would escape from Elba to land once more on French soil, and the battles would have to be fought all over again.

Meanwhile, it was announced the Peace would be celebrated in much more elaborate style on 12 August, when the Lord Lieutenant planned to review the troops in the Phoenix Park, lay the foundation stone of the new post office (in O'Connell Street, then Sackville Street, after the patronymic of his wife's first husband), give a State Dinner, and afterwards at nine o'clock proceed 'to St Stephen's Green to view the most splendid and magnificent display of Fire Works ever exhibited in this kingdom'.[43] The festivities were in celebration of not only 'the honourable Peace achieved by the wise Councils and victorious arms of his Royal Highness', the Prince Regent, but also 'the happy coincidence of the Centenary of the reign of the Illustrious House of Hanover' coupled with the Prince's birthday.[44] But to many the most important events of all were the 'Grand Musical Festivals [which were to] take place at some of the Churches in the mornings, and at the Rotunda in the evenings during…the Festivities. The principal Singer, Madame Catalani'.[45]

The Musical Festival had as president, the Duke of Wellington, and as patrons, 24 distinguished personages, who, with the Lord Lieutenant and Lord Mayor, included a number of the Irish peerage headed by the Duke of Leinster. It commenced with *Messiah* performed at St Werburgh's Church on the morning of 8 August, and was followed by two further concerts on the mornings of 10 and 13 August. On the evenings of 9, 10, 11 and 15 August, concerts of vocal and instrumental music were given in the large room of the Rotunda.

Catalani's supporting artists included the singers Mme Ferlendis, Signor Chiodi, Mrs Bianchi-Lacy, Miss Hughes, Master Stansbury, and Messrs Lacy (Bianchi-Lacy's husband), Spray, Garbett, Tinney and Tett, while the principal instrumentalists* were Luigia Gerbini, violinist; Loder, leader of the orchestra;

---

* The names of the complete orchestra are given as follows: Bowden, viola; Willman, clarionet; Bond, bassoon; Weidner and B. Cooke, flutes; Mulligan and Burgess, horns; H. Willman and Walker, trumpets; Gray and Cubitt, double basses; Tomkinson, trombone; Jenkinson, double drums; Smith, organ. They were assisted by Barton, Mahon, Barrett, R. Barton, Saunders, Gale, Nelson, Daly, Glover, J. Barton, Bird, Kavanagh, Glover jun., Metherington [given as Metheringham in 1808], Smith, Broad, Robinson… (*Freeman's Journal*, 2 August 1814)

Mori, violin; Ferlendis, oboe; and Lindley, violoncello. Of the supporting singers the last five gentlemen were probably all local. Mr Tinney is referred to as

> 'Fat, jolly, and brazen, why man thou'rt a ninny.
> Your name should be *Brass*-y, it cannot be *Tin*-ney'.[46]

By far the most important singer appearing with Catalani was Ferlendis. Little is known of her prior to her marriage to Alessandro, younger son of Giuseppe Ferlendis, a renowned oboist. Both sons followed their father by becoming distinguished oboists or cor anglais players. Concerning his wife, Choron and Fayolle's *Dictionnaire Historique des Musiciens* of 1810, and all editions of Grove, identify her as Madame or Signora Barberi, an Italian contralto who had made her debut in Lisbon and had married Alessandro Ferlendis there in 1802. We know that Alessandro Ferlendis and his father were performing in Lisbon in 1801.[47] We know too that a Camilla Barberi sang at the São Carlos Theatre during the same season,[48] which confirms her identity.

In 1805 Signora Ferlendis made a reasonably successful appearance in Fioravanti's *La Capricciosa pentita* in Paris at the Theatre Louvois, by then renamed Théâtre de l'Impératrice. She was then described as having 'a fine stage presence; though not fully endowed with the desirable embonpoint, her face is quite pleasing—perhaps not at first glance, as her features have something irregular about them… The Signora's waist on the other hand, is slim and slender: many of our celebrated French young ladies would be very happy to possess such a figure'.[49] Passing next to St Petersburg about Easter 1812 we learn, 'A first appearance here was that of Mme Ferlendis, an Italian, and her husband, who is a younger brother (or relative) of our court musician the cor anglais player'. The review continues, 'Mme Ferlendis has a good, though far from perfect contralto'.[50]* Her first London appearance took place at the King's Theatre on 13 May 1813, in *La Dama Soldato* by Ferdinando Orlandi (from which opera she had earlier sung a buffo aria to 'well deserved applause'[51] in St Petersburg). It was reported, she 'will prove an acquisition to this stage in its present impoverished state. Her voice is a low soprano, limited in its compass and not very powerful, but the tone is full and she sings without any painful effort. Her power of execution is not great, nevertheless she seems

---

* In the same series of concerts, though not necessarily in the same concert, one of the artists taking part was John Field. (*Journal des Luxus und der Moden*, July 1812, p 452)

Camilla Ferlendis née Barberi

to know the extent of her own powers and does not often attempt any thing she is not conscious of being able to perform... Her acting is graceful, and abounds with the most agreeable *naiveté*, and in the serious part of the opera she proved that her talent may be occasionally employed in that line of acting in a very useful manner'.[52] Mount Edgcumbe confirms this when he writes: she 'was I think less liked than she deserved, for she had a very good contralto voice, and was far from a bad buffa'.[53]

Another singer from the King's Theatre of much lesser significance than Ferlendis, was Giuseppe Chiodi. He is first discovered singing in Amsterdam in 1808, and was still appearing as late as 1824. In both 1808 and 1812 he was a member of the Italian Opera Company then performing there, and in the latter year is said to have 'previously been a bass singer in the Royal Chapel'.[54] By 1814 he was appearing in concerts and is described as 'formerly basso at the Italian Theatre'.[55] In 1814 he sang at the King's Theatre, London (the only time he would do so) appearing in Pucitta's *Adolfo e Chiara* and as Il Conte Robinson in *Il Matrimonio Segreto*.[56] Later that year he travelled to Dublin. During the seasons of 1816 and 1817 he became a member of the Théâtre-Italien, Paris, perhaps through his meeting with Catalani in Dublin, for she was then titular director of this opera house, although at that particular time she had placed it under Ferdinando Paer's control, and had left for Munich. From 1820 to 1824, Chiodi's name is to be found only among the Winter Concerts in Amsterdam, although it is possible that he was appearing in opera in other places.

During his early years in Amsterdam he is said to have had 'a powerful but not very deep voice. His singing is too monotonous'.[57] 'He is certainly not a great singer', comments another reviewer, 'his voice and method are most suitable for comic roles, which indeed he performs quite well'.[58] In Paris in 1816 it would be said of him, 'The buffo, Chiodi, who has come to us from Amsterdam with some reputation, possesses an attractive, fairly flexible voice, is a very good actor and is full of very estimable artistic intentions. And yet— in spite of all these merits, to which could be added his occasional displays of great liveliness—his performances are inclined to leave if not his audiences in general, at least the connoisseur, somewhat cold. His performance simply lacks the true Italianate comic force, that romantic ardour, with which it should be imbued and which distinguishes the Italian buffo so favourably both from the French comic actor, and the prosaic comic German buffo... Like many Italian buffo singers, Chiodi tends to speak rather than sing the recitatives… One virtue for which Chiodi does deserve praise is an excellent diction'.[59] And what did London think of him? Of *Il Matrimonio Segreto* it is reported: 'He possesses a base voice of good compass and of a clear tone, and is a singer of some taste… It is, however, very obvious at first sight, that Signor Chiodi has not profited much by his lessons from the posture-master, for his unvaried attitude is much more ludicrous than even the most comic actor could wish, and his action is deficient in ease and grace'.[60]

Miss Hughes (later Mrs Gattie) announced as 'from the Nobility's Concerts, London',[61] was a pupil of James Bartleman, and had made her first concert appearance in London in March 1807. She then 'displayed considerable vocal powers and executed some of Handel's most difficult music with good effect, her style of singing being excellent, and reflecting much credit on her master'.[62] She did not achieve artistic maturity however until she sang in opera in Dublin in 1815, when we learn, 'it was at the pressing instance of her intimate friend, Mme Catalani, that [she] was induced to appear on the stage'.[63] Then there was Master Stansbury, aged 13 or 14, also from the Nobility's Concerts, London, who in later years as George Frederick Stansbury, returned to Dublin as musical director at the *new* Theatre Royal in Hawkins Street. There in 1834 he conducted a season of Italian Opera, and a few years later, on becoming musical director of St James's Theatre, London, adapted with rather less artistic integrity Adolphe Adam's *Le Postillon de Lonjumeau*.

Mrs Bianchi Lacy also requires comment, especially since she is frequently confused with an Angiola[64] Bianchi, a contralto who sang at the São Carlos Theatre, Lisbon, in 1807/1808,[65] and at the King's Theatre between 1809 and 1814. The artist who sang in Dublin was almost entirely a concert singer. She was born Jane Jackson in London in 1776. In 1800 she married the composer Francesco Bianchi, (the real Bianchi of *La Villanella Rapita* who did *not* appear in Dublin) becoming his widow in 1810. Two years later she married the English bass singer, John Lacy (who, as already noted, accompanied her to Dublin) thereafter combining his name with Bianchi's. This dual identification is placed beyond doubt by finding both ladies appearing in the same concert on a number of occasions. For example, at the New Rooms, Hanover Square, on 6 May 1813, 'Signora Bianchi' sang a duet by Mayr with Tramezzani, while 'Mrs Bianchi Lacy & Mr Lacy' took part in a performance of 'Cherubini's celebrated Oratorio'.[66]

Among the instrumentalists in the company, the leader, John David Loder, born in Bath in 1788, besides being a violinist later became a music publisher. He ended his days as a professor at the Royal Academy of Music, London, and as leader of the Ancient Concerts. His deputy, Nicolas Mori, was about ten years his junior. He was a pupil of François Barthélemon* and at the age of eight performed as a prodigy, playing at one of his master's concerts. Alessandro Ferlendis, as already mentioned, was a renowned oboist, born in Venice in 1783. Then there was the famous English violoncellist, Robert Lindley, born in Rotherham, Yorkshire, in 1776, who, from the time of his appointment as principal 'cellist in the King's Theatre orchestra in 1794 until his eventual retirement in 1851, remained the leading player of his instrument in the British Isles.

Perhaps the most interesting instrumentalist among them was Luigia Gerbini, who had quite a distinguished career both as violinist and as a singer. Mount Edgcumbe tends to dismiss her singing, commenting 'Gerbini, a bad singer without voice, but a good player on the fiddle, having been educated in a conservatorio at Venice'.[67] She was in fact a pupil first of Pugnani, and later of Viotti. She nevertheless cannot have been entirely 'without voice', for in November 1790, she appeared both as singer and as violinist in a pasticcio called *Il Dilettante* at the Théâtre de Monsieur, Paris.[68] (An appropriate engagement since her teacher, Viotti, had been a co-founder of the theatre). Reviews of her singing in *Il Dilettante* were on the whole tepid, one commenting: 'Signora Gerbini's voice is in general, of a very beautiful timbre, yet her entire register is not equally cultivated. Her voice tightens at the top—and is liable to produce harsh, dry, shrill notes. She ascends to the top with facility, but not always with accuracy: she needs altogether to work on her intonation'.[69] In 1798 and 1799 we find her singing in Italian opera at the Teatro de los Caños del Peral, Madrid.[70] Then, until 1801, she sang regularly at the Teatro São Carlos, Lisbon, where, on 1 February 1801, she took the role of Mandane in Cimarosa's *Artaserse*** in honour of a visit by Prince Augustus Frederick, George III's son.[71] During her stay in Lisbon she also gave successful violin recitals at the São Carlos during the intervals of the operas.[72] From December 1802 until the end of the 1803 season she appeared in six different operas, including *Il Matrimonio*

---

* See *Opera in Dublin 1705-1797*.
** With additions by Portugal.

Segreto,[73] at the King's Theatre, London. It was reported of her that 'she made a very favourable impression upon the audience by her singing, in which she displayed science and a clear full-toned voice'.[74] Other reviewers, however, remained less enthusiastic. 'Her singing pleased rather less than her violin playing',[75] is an assessment of a concert in Berlin. In fact, almost all were eulogistic concerning her violin playing. In Paris she was described as 'an excellent violinist, who combines almost masculine power and precision with feminine grace';[76] in Amsterdam, 'musical connoisseurs applauded her—particularly in Viotti's Concerto'.[77] But in Stockholm there was some retraction, for although 'she revealed much technical skill, good intonation and power: one could wish for a little more feeling—and less portamento (that notorious miaow of violinists)'.[78]

But Catalani, like Juliet, was the sun around which all other planets revolved when she sang in *Messiah** on 8 August. 'The band consisted altogether of more than a hundred vocal and instrumental performers'.[79] The chorus having been augmented by the celebrated Lancashire Singers'.[80]** Tickets costs 10s 10d. The Lord Lieutenant and his Duchess (who on 4 August had given a splendid entertainment at the Castle to which Catalani and Valabrègue 'were honoured with an invitation'),[81] having indicated their intention of honouring 'the entire of the musical performances with their presence…a splendid throne and canopy suitable to their rank and high station, [was] prepared for their accommodation at St Werburgh's Church'.[82] The second performance by the company consisted of 'a Grand Concert of Vocal and Instrumental Music' on the following evening, 9 August, at the Rotunda, when 'the Vice-regal party did not leave the room before twelve o'clock'.[83] For the second concert at St Werburgh's Church, among her arias Catalani sang Crescentini's '"Ombra adorata aspetta" with a combination of vocal power, a spirit and expression indescribably fine, and almost electrically effective… His Excellency the Lord Lieutenant and the Duchess of Dorset, not only attended, but did so punctually to the notified time for the commencement of the Oratorio'.[84] A further concert at St Werburgh's and two at the Rotunda then followed. At the Rotunda on 11 August, the audience, we learn, 'considerably overflowed the circular room. Upwards of *sixteen hundred persons* were collected'.[85]

* For programmes of concerts taken from press: See Appendix C.
** Also advertised as 'the Manchester Chorus Singers' (*Saunders' News-Letter*, 11 July 1814)

The company next set off for a short engagement at Cork, whence it was reported, 'Mme Catalani and Mme Ferlendis sung the other night a celebrated duet…to the enchantment of a crowded audience. Twice was the encore'.[86] Having travelled on to Limerick, the company then returned to Dublin for six further concerts, this time at the Theatre Royal, Crow Street, and one other Grand Sacred Concert at Denmark Street Chapel, 'in aid of the Fund for completing the New Roman Catholic Parish Chapel of St Michan, North Anne-street'.[87]

The first of the Crow Street Concerts took place on 7 September. It followed the arrangement of those given at the Rotunda in August, but included as well 'the much admired Scene from the serious Opera of *Semiramide* by Madame Catalani; in which she will sing the favourite Bravura Song of "Son Regina"'.[88] The 'much admired Scene' may suggest an entire scene in costume from the opera with other singers taking part. Newspaper announcements such as 'The Concert and Opera at the Theatre',[89] and 'Grand Concert and Italian Opera',[90] lend some emphasis to this assumption, as does the fact that the performances were taking place in the theatre and not in the Rotunda concert room. In or out of costume it is now impossible to be certain how they were performed, except that there was undoubtedly little production about them. Very likely they were replicas of the scenes as presented at the theatre when Catalani first sang there in 1807. The doors of the theatre were announced to open at seven, the performances to commence at eight precisely, and prices of admission were: Boxes 7s 6d, Pit 4s 2d, Gallery, 2s 11d, Upper Gallery, 1s 8d;[91] which must have made it hard on the ticket seller in handing out change.

The second concert on 8 September had for its principal items, 'a Scene from the Comic Opera of *Il Calzolaio** with Ferlendis and Chiodi', and 'the much admired Scene' from *La Morte di Mitridate* sung by Catalani.[92] Friday, 9th September, introduced a scene from *Due Nozze ed un sol Marito*, with Catalani and Ferlendis, 'to conclude with a Comic Scene by Mme Catalani and Signor Chiodi in which she will sing the admired air of "Papà non dite di no".'[93] The

---

* *Il Calzolaio o la Bacchetta portentosa* by Portugal. Also known as *Il Ciabattino ossia il Diavolo a quattro* and *Le Donne cambiate*. Under this latter title with libretto by Giuseppe Foppa it was performed at the Teatro San Moisè, Venice, as a farsa giocosa per musica in one act on 22 October 1797. (*Il Teatro moderno applaudito* XVII, 3, Venice 1797). Grove (6th edition) records this as the first performance and states that the libretto was based on Charles Coffey's *The Devil to Pay*. See *Opera in Dublin 1705-1797*.

following evening brought a 'Scene from a favourite Comic Opera' and concluded with a 'Comic Scene' from *Il Fanatico per la Musica*, both sung by Catalani and Chiodi.[94] A scene from an unidentified 'Comic Opera' *Il Tutore Accorto*, with Chiodi and Mme Ferlendis '(in male attire)', followed by scenes from *Mitridate* and *Il Fanatico*, into which Catalani introduced 'Nel cor più non mi sento', was the principal entertainment for the 12th,[95] and lastly, on 14 September, Catalani took her benefit, when she sang 'five of her most celebrated Scenes and Songs assisted by Mme Ferlendis and Signor Chiodi and Miss Cheese' (an expatriate from Manchester, whose curtailed Saxon name intrudes strangely amidst her Italian colleagues). Madame Catalani also 'had the honour of taking leave of the Audience in the National Air [in Ireland!] of "God Save the King"'.[96]

As mentioned earlier, on Tuesday, 13 September, there was a concert in aid of the new Roman Catholic Chapel in North Anne Street. A distinguished Catholic Committee had been formed headed by Lord Trimlestown, Earl Fingall, Viscount Killeen, Sir Edward Bellew, Bt, and others,[97] and tickets were to be had of the Reverend Gentlemen of Mary's-Lane Chapel, the Sub-Committee, and at the principal Music Shops'.[98] Preparations were made in Denmark Street 'Chapel' 'to render every possible comfort and accommodation to the crowded auditory which may be anticipated'—at one o'clock in the afternoon—and the attention to the concert was solicited 'of those over whom charity, taste, or the love of science, have an influence'.[99] Moreover, we read that the whole of the receipts of 'this day's performance are to be applied to the assistance of the funds for completing the New Chapel' which, it was believed, would 'not tend a little to elevate this accomplished Lady in the mind of a country like ours'.[100]

On the following day, 'the Committee' expressed 'their most grateful acknowledgments to Mme Catalani…to Signora Gerbini, Miss Cheese, and the other eminent Performers who were kind enough to assist on the occasion… To the Nobility, Gentry and liberal Public' they felt 'highly indebted' and 'to the Rev. Mr McGouran, for his unsolicited offer of Denmark-street Chapel, and great exertion, as well as the other Rev. Gentlemen for their polite attention'.[101] At the time, reasons religious, social, and political, dictated the absence of the Lord Lieutenant from this particular concert.

After such a strenuous engagement in Dublin, Cork and Limerick (undoubtedly following a provincial tour through Britain) one cannot be surprised to learn that at her benefit Catalani 'was rather indisposed', although 'her voice could not fail of its usual attractions'.[102] Having left Dublin she may have continued her tour (Sig. and Mme Ferlendis are reported to be travelling on to Belfast, although not necessarily with Catalani[103]), for she did not return to 'her villa, the Hermitage, Brompton, from a most successful tour in the country' until the end of October.[104] In less than three weeks she had set out again, this time for Paris, 'where she proposes giving a series of concerts'.[105]

But what had Dublin thought of the singers? Reviews on the whole are sparse and uninformative. Mme Ferlendis is reported as having 'a very extraordinary voice, canteralto [*sic*, contre alto] which we never heard before. This Lady possesses qualities in her musical talents, independent of her very superior acting, which never fail to make a great impression on the public… The comic scene between this Lady and Signor Chiodi far surpassed any thing we have ever witnessed before'.[106] 'We did not think much of Signor Chiodi's effort', wrote *The Freeman's Journal* critic in August. 'His voice is not very commanding, and there is much mischief done to any powers it possesses by his striking affectation'.[107] Yet in September the same newspaper reported, 'Signor Chiodi has a fine bass voice; his comic style of acting surpasses that of the celebrated Naldi',[108] which postulates either a rapid change of mind in the part of the critic, or a change of critic on the part of the newspaper.

Of Catalani, predictably perhaps, there is little recorded that is not a pointless panegyric. Naturally she was 'divine' and, on 11 August, 'she condescendingly attended to the encore which resounded from all parts of the room'.[109] There was also 'the bewitching archness and grace with which she performed a scene'[110] from *Due Nozze ed un sol Marito*. One has to go outside Dublin, in fact to the Continent, to discover how Catalani was singing at this time in her career. What is pre-eminent is that the voice had grown lower with the years since she had last sung in Dublin. In 1816 it was reported from Hamburg that it was 'more contralto than soprano. True,' the reviewer continues, 'she does also sing up to $\bar{\bar{B}}$ and $\bar{\bar{C}}$, and down to F., but *these* rather effortlessly produced notes did not, I confess, sound wholly pleasing to my ear. Her voice is otherwise strong and powerful: excellent in the octave from $\bar{e}$ to $\bar{\bar{c}}$; above this her voice—at *full* power—already becomes somewhat shrill'.[111] This opinion is confirmed by a critic from Leipzig. He too describes her voice as 'a powerful and sonorous mezzo-soprano', and significantly adds, as indicated by her theatre performances in Dublin, that 'her singing would seem to be more suitable for the merry, light type of music which she sings with attractive playfulness—rather than for the more serious sustained airs'.[112] The days of 'Son Regina' and 'Frenar vorrei le lagrime' were passing. Her bewitching archness and grace in *Due Nozze* and *Il Fanatico* were beginning to take precedence over the voice. But her voice was in decline even before she visited Dublin in 1814, for as early as 1811 it was reported from the King's Theatre of her Semiramide: "to our mortification and disappointment we find it is new modelled this season. All the airs are transposed; and Catalani sung her 'Son Regina' in A flat instead of B flat; and, indeed, all the other songs were one whole note lower than last season, by which they lost much of their brilliancy and effect'.[113] She would have yet a third vocal period in her career, but Dublin would not experience that for another nine years.

An old dog and a young lion had also come to Dublin in August, although in the beginning at least they had appeared not at Crow Street but at the theatre in Fishamble Street. They were Charles Incledon and a young Scot, John Sinclair, and they were both tenors. Sinclair was then 22 years old, having been born in Edinburgh on 9 December 1791. He had had instruction in music as a child, and while still young had joined the band of a Scottish regiment as a clarinet player. He managed to buy his discharge from the army, and by then, discovering that he had a fine tenor voice, set out for London and to fame and fortune. His first stage appearance is said to have taken place anonymously at the Haymarket Theatre on 7 September 1810* and with such success that Tom Welsh accepted him as a pupil. He next appeared under his own name at Covent Garden on 20 September 1811, when he played Don Carlos in *The Duenna*.

It was then reported: 'Mr Sinclair [Carlos] the new male singer, is a slight youthful figure, feebly made. His voice was full of the tremors of a first appearance, but it has some fine capabilities; it has some notes of rich and delicate sweetness—some of its turns are exquisitely tasteful; and nothing is

---

* When he may have taken over the role of Captain Cheerly in Shield's *Lock and Key* from Incledon. Incledon was announced to sing the part (*Morning Chronicle* and *The Morning Herald*, 7 September 1810) 'with the song of "The Bay of Biscay O" ...and "The Storm" (in character)'. *The Alfred, The Sun* and *Morning Chronicle* of the following day all reported that he had sung, *The Alfred* recording: 'After the play an apology was made for Incledon, who being just returned from Ireland, feared that the fatigue of so long a journey might prevent his exertions being as effective as he could wish; but an apology was superfluous, he sang delightfully and was encored in all his songs', but no reference can be found of his appearing in *Lock and Key*.

(For the Last time this Season.)

This present FRIDAY, MAY 28th, 1813;
Will be performed (11th time) a NEW GRAND HISTORICAL OPERA, called, The

BURNING of MOSCOW

(Written by a Gentleman of this City—The Music composed by Sir John Stevenson)

With New Scenery, Dresses, and Decorations.

Rostopchin (Governor of Moscow) Mr. CONWAY,
Platoff (Hettman of the Cossacks) Mr. THOMPSON,
Roffenhoff (a Tartar Chief) Mr. O'NEILL, Jun.
Archbishop of Moscow Mr. YOUNGER,
Petrowitz Mr. WILLIAMS,
Ivar Mr. NICHOLS,
Paul Mr. SHAW, Michael Mr. REID,
The Vaywode Mr. FULLAM,
Yermach Mr. CORNELLYS, Crier Mr. NORMAN.
Scroftonhoff Mr. W. FARREN,
Barney Mullahedder Mr. LEE,
Tartars Messrs. DIXON, TURNER, &c.
Cossack Chiefs Messrs. ROWSWELL and CARROLL,
St. Clair (Cheff de Brigade) Mr. FOOT,
Caulincourt Mr. FULTON,
French Soldier Mr. BURGESS,
French Officers Messrs. SAUNDERS, TOOLE, &c.
Alexina Miss O'NEILL,
Mary Mrs. COOKE,
Catherine Mrs. BURGESS,
Nuns, Peasants, &c. Mesdames JOHNSON, BARTLETT, SMITHS, EREBY, &c.

In the Course of Piece, the the Following SCENERY:—

The Banks of the River Moskwa by Moonlight;
View of Platoff's Castle ;---An Apartment in the Castle ;
Grand Hall and Banquet ; --- A Forest near Borodino ;
City of Moscow, Bridges, Boats, &c. Street near the Kremlin;
Royal Cemetry ; THE CITY ON FIRE.

The SCENERY—Painted by and under the Direction of SIG MARINARI.

To which will be added the Farce of

ST. PATRICK'S DAY!

Lieutenant O'Connor Mr. LACY, Justice Credulous Mr. FULLAM,
Doctor Rosy Mr. W. FARREN, Corporal Breakbodes Mr. NORMAN,
Serjeant Trounce Mr. LEE, Drummer Crackscull Mr. CARROLL,
First Countryman Mr. BURGESS, Robin Mr. ROWSWELL,
Second Countryman Mr. REID, Soldier Mr. GOOD.
Mrs. Bridget Mrs. BURGESS, Laurette Miss S. NORTON,

To-morrow, *THE REVENGE.*
Zanga Mr. KEMBLE, being Positively the Last Night of his Engagement.

Tickets and Places to be had of Mr. M'NALLY, at the Theatre, from Eleven 'till Three each Day.
Tickets for the Pit and Gallery will be Sold in the Office at the Entrance to the Boxes, every MONDAY, TUESDAY, THURSDAY, and SATURDAY, from Twelve to Three o'Clock.

The libretto of *The Russian Sacrifice, or, The Burning of Moscow*, to give it its full title, was by H. B. Code (= the "Gentleman of this City") and was published in Dublin in the same year (1813)

wanting to make this young person a captivating singer, but the practice of the stage'.[114] A month later it was recorded more poetically of his performance of Orlando in *The Cabinet*: Rizzio, the unhappy favourite of the Scottish Mary, it is said, was the person who first added the softer Italian graces to the sweetest of the Scots' pastoral airs, and to him the musicians of that country account themselves indebted for the elegant little ornament which embellishes without loading their favourite tunes.—To our ear Mr Sinclair's execution of the lovely song, "When away from my beautiful Maid", was a fine illustration of this combination of style; it had all the simplicity and melody of the Scots, with enough of the grace and adornment of Italian harmony. Nothing could be more sweet and enchanting'.[115]

Incledon had brought a concert party with him to Dublin which, besides himself and Sinclair, consisted of a Master Williams, a boy soprano, Mr Claremont, an actor who recited, and, as accompanist, Mr Horncastle. They presented an entertainment called '*Mirth and Minstrelsy; or, Two to One Against Care*, Consisting of Interesting Oral Matter, embracing a variety of Subjects, and introductory to a number of the best selected Songs'.[116] Prices of admission, especially for the Fishamble Street Theatre, were not low, Boxes were 5s 5d, Pit, 3s 4d, and Gallery, 2s 1d. They were advertised to play, positively for four nights only, commencing on 23 August, but then, most remarkably, two nights later, both are to be found performing in *Love in a Village* at Crow Street![117] It transpires that Jones had engaged them on what appears to have been extraordinary favourable terms—to him—for their benefit and final performance at his theatre on 2 September was 'the only source of remuneration which their engagement with the Patentee [afforded] them'.[118] Within nine days at the Crow Street Theatre they appeared in *The Cabinet*, *The Castle of Andalusia*, *The Beggar's Opera* (Macheath, Incledon), *Midas*, (Apollo, Sinclair) and a second performance of *Love in a Village*.

Of their joint first appearance in this last opera with Incledon as Hawthorn and Sinclair as Young Meadows, *Saunders' News-Letter* observed: 'Mr Incledon was in excellent voice, undiminished in either the harmony or the energy of his song; but he requires not our panegyric… Of Mr Sinclair however, the Irish public knows little except by reputation. This is his first visit among us, and it will be therefore necessary to say something of him more particularly than of his colleague in the performance. Mr Sinclair to a good stage stature unites an engaging exterior and gentlemanly deportment. He is a respectable actor, obviously understanding the character of which he is the representative, as well as the forms and business connected with stage appearance. His voice, without being very powerful, has a considerable range of compass and comprehension; it is clear, sweet, flexible, and susceptible of great modulation. His shake is distinct and perfect, and he has great rapidity in running the divisions of the musical scale, and at the same time very good articulation in executing them. To these capacities he unites excellent taste and a good manner of delivery, and on the whole is a singer of very first rate talents… The performance concluded with strong manifestations of satisfaction from the audience'.[119] *The Freeman's Journal* concurred with this approbation, appending, 'If personal advantages and easy and natural acting can be of any assistance to his vocal efforts, he enjoys them in a very considerable degree. He is very unaffected and gentlemanly under all circumstances, and is consequently already a favourite… In the course of the evening he introduced many songs in which he was highly applauded. His "Pray, Goody" [from *Midas*] was *encored* three times'.[120]

Sinclair is reported to have set out for Paris in 1819 where he studied for a time with Pellegrini, thence travelling to Italy where he had further tuition, including advice from Rossini. He later appeared in some of the leading Italian opera houses, mostly in Rossini's operas, remaining in Italy until 1823. Consequently, it is as the mature and experienced singer that he later became at the time when he made a return visit to Dublin to the Theatre Royal in Hawkins Street in 1824, that he can be most fairly judged.

Following the departure of Sinclair and Catalani, activities mainly involved the engagement of artists and the refurbishing of the theatre for the forthcoming 1814/15 season which commenced on 12 November.[121] The opening had been delayed 'on account of necessary alterations in the Vice-regal Box; amongst others, a fire-place and complete chimney have been erected. The House, and particularly the scenery have been retouched, and, where it was found requisite, repainted. [Not always satisfactorily however. In December it was noted that 'the red paint which has been laid on the inside of the boxes [has] a ruinous effect on such clothes as come in contact with them'.[122]] The seats of the boxes are covered anew with scarlet cloth, and we trust that those (we can hardly call them Gentlemen) who have been in the studied habit of cleaning their boots with such covering, will be so good as to recollect that they thereby defeat its sole object, the accommodation of the public, and particularly of the Fair Sex. So much for what may be deemed the body or external of the drama; and we shall now give some account of what is justly considered the *vivida vis animi* of the scene. Mr Rock, the acting Manager, has just returned from a tour of two months in England, where he has been successful enough to engage several performers of merit'.[123] Among those of particular interest were Miss Griglietti, a soprano from the King's Theatre, and Mr Le Clerc, a dancer from 'the Theatre Royal, Covent Garden'[124] who had also danced with the opera company performing at the Pantheon in 1812. For Dublin he had been engaged 'to preside over the department of the ballet',[125] and would make his first appearance there dancing 'A Grand Pas Seul on 15 November'.[126] It was further announced, 'All boys are in future to be excluded from the orchestra, where none but persons eminent in their profession will be admitted. Mr Blewitt is to preside at the piano forte, which, we are glad to find, is to be restored to its station there'.[127]

Most significant of all was an announcement from the Lord Mayor, John Claudius Beresford, which appeared in the press on 19 November. It reads: 'It is proposed to perform Italian Operas both serious and Comic, with Ballets at the Public Rooms of the Lying-in-Hospital [Rotunda] for the benefit of the Charity, if a sufficient number of persons, in addition to those who have already put down their names shall subscribe to the undertaking by the beginning of December, as it will be necessary to give final decisions to the Performers on the Continent and in London, now on treaty, and to enable them to arrive in time to open the Theatre early in January. The proposals and the Subscription List are to be seen at Mr Delvecchio's in Westmoreland-street, where the plan of the Opera and division of the boxes are to be seen. It is my intention as Lord Mayor, to grant a Licence for this purpose'.[128]

Alas, while the Lord Mayor's intentions were obviously of the best, a sufficient number of people were not forthcoming to endorse their alleged interest with hard cash at Mr Delvecchio's music shop, and so the season never materialised. In its place there was a return visit of Mme and Sig. Ferlendis to the Rotunda 'Under the Patronage of the Right Hon. the Lord Mayor, the Lady Mayoress and Several Persons of Distinction'.[129] They were joined on this visit by Urbani, and later by 'Mademoiselle Gerbini, Professor of the Violin, and Signor Gerbini, Tenor Singer just arrived from Liverpool'.[130] Three concerts were given, on 8 November and on 1 and 13 December, and included works by Mozart, Sacchini, Cherubini, Nasolini, Pucitta, Haydn, Cimarosa, Steibelt, Paisiello, Urbani, Braham, and others. Tickets for the eighth of November (the only concert apparently under the Lord Mayor's patronage) cost half a guinea, for the other two evenings they were 5s 5d. John Mahon was leader of the orchestra on all three occasions, the third was for his benefit, when Dr Philip Cogan presided at the piano.[131] The pianist on 8 November was the controversial Johann Logier,[132] and on 1st December, the versatile Luigia Gerbini.[133]

Tom Cooke had 'an operatic anecdote' produced at Crow Street Theatre on 29 December in which both he and his wife took part. This was *Frederick the Great; or, The Heart of a Soldier*, adapted by S. J. Arnold from Pigault-Lebrun's novel, *Les Barons de Felsheim*. Both music and libretto were well received when the work was first performed some months before at the Lyceum Theatre, London.[134] There may have been an element of chauvinism therefore in a Dublin review which states, 'Its interest is entirely derived from its music, which is, as our readers are aware, the composition of Mr T. Cooke. As to story, it has none...good playing, however, gives some of its incidents no little attraction... Some of the songs were very beautiful'.[135]

A storm which raged throughout Ireland and devastated Dublin on the morning of 16 December causing buildings to collapse, including the houses of Counsellor Campbell and Dr Jolie in York Street, where the chimney stacks crashed through the roofs, while 'in Mr Campbell's the floors all sunk in, and buried three servants in the ruins',[136] presaged the theatrical tempest which almost demolished the Theatre Royal that evening. The event has passed into history as the 'Dog Riots', and so much has been written about it already that here one need only explain that the performance required a dog specially trained to attack the actor playing the villain who has murdered the hero. It was a zoological stage episode in the manner of the renowned goat in Meyerbeer's *Le Pardon de Ploërmel*, that must cross a fallen tree trunk amid flashes of lightning.

The work presented as *The Forest of Bondy; or, The Dog of Montargis** and is at least of peripheral interest to the history of opera in Dublin, since it was a melodrama, and because it would be the cause of sending Frederick Jones yet

---

* The forest of Bondy was formerly a haunt of brigands, and the expression 'forêt de Bondy' has remained in the French language to denote a locale for thieves.

again into exile. The libretto was adapted by the elder William Barrymore from Pixérécourt, and Henry Bishop had composed the music for its recent London performance. In Dublin the overture was by Blewitt, while music to heighten the dramatic action was by Tom Cooke. The star of the performance was a Newfoundland dog, called Dragon, who, from what we read, gave a virtuoso performance, acting 'his part with wonderful address'. He attempts to defend 'his master from the assault of the murderers—he gives the alarm and points to the pit where the dead body has been thrown'. Little wonder that 'the House was crowded in all parts'.[137]

The first performance took place on 8 December, and all went splendidly until the evening of the 16th, when it was commanded by the Lord Lieutenant. The day previous the dog's owner, a Mr O'Connor, who when not engaged in show business was a rope-maker, believing, no doubt, that the viceregal visit had placed him in a puissant position, decided as artist's manager to raise his canine performer's fee. His demand is said to have been a perpetual free admission 'with other terms',[138] reported by another source to have been a *douceur* of about £200.[139] Not unreasonably, Jones rejected this attempted extortion, cancelled *The Forest of Bondy* and put on *The Miller and his Men* in its place. The audience, incensed that this substitution had been made without prior announcement, took prompt revenge. In spite of the Vice-regal party being present, 'every chandelier and lustre with the exception of one or two was broken. The great pier-glass that ornamented the Vice-regal box shared the same fate; and no small injury was done to the *piano-forte*; which was the only instrument that it was not found practicable to remove from the orchestra. Besides all this, the seats, ballustres, and plaistering, suffered a great deal'.[140] One cannot be too surprised to learn that 'about a quarter of an hour after the row began, the Vice-regal party and that of the Commander of the Forces, retired. Neither the Lord Lieutenant nor her Grace the Duchess bowed to the audience'.[141] The rioting continued into the following week, although with the auditorium laid waste, it is difficult to understand why performances were not suspended. So, we read, 'On Tuesday night [20 December] the disorders were at their height. Crow Street Theatre was entirely demolished: not a lamp, chandelier, lustre, foot-light, or seat, was left unbroken in any part of the house. The audience was as numerous as that which originated the tumult on Friday night; but it was much more determined and enthusiastic. It was a miracle indeed that a particle even of the scenery escaped'.[142]

John Sinclair, Scottish tenor (1791-1857)—as Apollo in Midas.

The matter became a contest of will-power between Jones and the audience, allegedly on sectarian grounds, since Jones had been a member of a Grand Jury that had brought in bills of indictment against the Catholic Board then continuing to pursue emancipation. But Jones would not apologise from the stage for altering the programme which the audience demanded of him, and so, in the end, he was forced to withdraw. This he did in a letter 'to the public', published as follows in the Dublin press on 26 December: 'In the present situation of the Theatre I have thought it my duty to take a speedy and decisive step, in order to avoid involving in my ruin the other unoffending Proprietors, and the numerous Performers and Artists, depending upon the establishment for bread. I have been required to apologise personally on *the Stage* (for an offence never intended by me to the Public, and which I have been at all times ready to explain and apologise for any where else) or to retire from the management. I have adopted the latter course, and I now publicly declare that I, from this hour, withdraw from the direction of the Theatre Royal... I trust that I have have one credit at least, that, surrounded as I was by the ruins of my property, and, perhaps personal danger, I have not forfeited my birth right, that of thinking, feeling, and acting as a Gentleman. F. E. Jones'.[143]

The theatre was then temporarily placed in the hands of four trustees, the Earl of Meath, Mr Denis Bowes Daly, Colonel Talbot* and Mr Taylor, and in a further letter to the public, 'the remaining proprietors', John Crampton, E. T. Dalton** and George Gregory, having averred that they 'had had no part in the late transactions' pledged that it 'shall be their study to remove every reasonable cause of complaint which the public may feel'.[144]

That some complaints could be directed with justice against Jones is apparent from a reasoned article in *The Freeman's Journal*, and since some of them apply equally to opera as they do to the drama they may be quoted as follows: 'It is, we think, obvious to the most inexperienced, that some performers have been put forward in high characters who were not fit for low ones, and that many have been obtruded in low ones who should not be seen in any character whatsoever... Greater attention should be given to the general business of the

---

* Richard Wogan Talbot, later first Baron Talbot of Malahide.
** The Earl of Meath, John Crampton and Edward T. Dalton were all members of the Kilkenny Theatrical Society, and the latter two had appeared on the stage there (*Kilkenny Theatrical Society*, 1818, British Library 1890, e5(119))

stage. Within the last month we have seen one performer drunk, another totally ignorant of his part, and two or three out of costume in the same piece*... The public should demand a full and well attended orchestra. The leader should never be absent and we should have some little variety in the music. The musicians, all of them, but especially the leader, should get salaries that would make them less careless than they have been in attending to their professional duty. There is no department of the theatre that stands in greater need of reformation than the orchestra'. Finally, with a nice touch of social propriety, it was recorded: 'Women *who come in unattended* should not be allowed to sit with modest females in the boxes. They should be compelled to resort to a less conspicuous part of the house'.[145]

Following Jones's retirement, John Crampton once again attempted to direct the theatre, and once again after a brief spell, failed. Anthony Rock, the stage-manager, was then placed in charge, but Rock died just as the theatre was about to open for the Autumn season of 1815. In the dilemma, and no doubt in despondency, Jones was forced once more to return.

---

* But see Appendix D

Chapter 8
Two English Nightingales 1815-1816

The year 1815 began with an innovation for Dublin, the arrangement of operas by foreign composers for the English stage, by Henry Bishop. (He also cornered the market with musical adaptations of Sir Walter Scott's novels before Rossini and Donizetti got to them.) The first to be introduced on 18 February was Boieldieu's *Jean de Paris*, the libretto anglicised to *John of Paris*, by Isaac Pocock. For the Covent Garden production which had taken place three months earlier, Bishop had not only arranged the music but had substituted nine[1] to thirteen[2] pieces of his own. For Dublin it was advertised that 'the Overture and Ballet' were composed by Blewitt,[3] and that 'Songs, Trios, Choruses, &c' were by Bishop, Boieldieu and Pucitta.[4]

Although it seems to have been an elaborate production with 'New Scenery, Decorations, &c',[5] and offering in 'Act Second, a Grand Banquet and a Ballet', the latter danced by Le Clerc, St Pierre, and the Misses Dennett,*[6] dramatically it was coldly received by the Dublin press. 'Any thing, indeed, so completely destitute of incident we do not remember',[7] declared *The Freeman's Journal*; 'without plot or humour'.[8] reported *The Dublin Evening Post*. Nevertheless, all were agreed that the work was a complete success 'from the exquisite style of the music',[9] 'as long at least', added *The Freeman's Journal*, 'as it has the powerful assistance of Miss Hughes'.[10]

Miss Hughes, it will be remembered, had arrived in Dublin during the previous August with Catalani, and was now appearing in the principal role of the Princess of Navarre, and receiving eulogies from all sides. 'Many years indeed have elapsed since English Opera obtained so rich a treasure', we read. 'Miss Hughes is what the world would call, a fine figure, and in her carriage and action she is particularly graceful. Her countenance is expressive, and her eye quick and intelligent. Her quality of voice very much resembles that of Mrs Billington; but her style and execution bear a stronger affinity to those of Catalani. Her simple melodies are exquisitely touching, and the most difficult passages she executed with a precision and sweetness, that at once

---

* There appear to have been at least four Misses Dennett, and later their place in the production was taken over by a single Miss Cassidy, who proved to be 'a considerable acquisition'. (*Freeman's Journal*, 8 April 1815)

surprise and delight the hearer. Her shake, her cantabiles, and, as we believe the professors would call it, her *rosalia*, the art of repeating a passage a note higher, must ever secure her the applause of all who have taste and judgment… In dialogue, the greatest improvement of which she stands in need, is a more audible enunciation than inexperience and diffidence yet suffer her to adopt'.[11] With such a review one is not surprised to learn that 'the piece was announced for a second representation this evening amid thunders of applause'.[12] Such is the everlasting supremacy of the star.

Alas, for Dublin, her radiance could not be contained there. She soon travelled to London, making her first appearance at Covent Garden on 22 September 1815. As Mandane in *Artaxerxes*, she was considered by Hazlitt to be 'a very accomplished singer, with a fine and flexible voice, with considerable knowledge and execution'. 'But', he asks, 'where is the sweetness, the simplicity, the melting soul of music?… She lisps and smiles, and bows, and overdoes her part constantly… This lady would do much better at the Opera'.[13] This she did, at the King's Theatre from 1817 until 1821, where her roles included Donna Elvira and Berta in the first performances there of *Don Giovanni* and *Il Barbiere di Siviglia*. Whereupon Hazlitt altered course in his criticism, observing, 'Signora Hughes's Donna Elvira was successful beyond what we could have supposed. This lady at the Italian Opera is respectable: on the English stage she was formidable'.[14] All in all, it explains why *John of Paris* would have 15 performances in Dublin between February and July,* when the season ended.

John of Paris was followed by a second opera adapted from the French. This was Nicolò Isouard's *Le Magicien sans Magie*, with a new libretto by a Mr W. H. Hamilton, and with 'The Overture and Music composed by Mr Blewitt'.[15] From this are we to assume that nothing of Isouard remained? It seems so,

---

* A performance on 3 July (postponed from 29 June 'in consequence of a General Illumination for the late Glorious Victory [at Waterloo] being ordered by the Lord Mayor') was 'By desire of his Grace the Duke of Leinster, Grand Master of Masons in Ireland—for the Benefit of Distressed Free and Accepted Masons and their families. The Members of the different Freemason Lodges of Dublin and its Vicinity, are required by the Most Worshipful the Grand Master, to assemble preparatory to this occasion at six o'clock precisely at the Shakespeare Gallery in Exchequer-street, from whence a Procession will move to the Theatre. The Brethren are requested to appear in Full Clothing: Officers with Jewels. Such members of Provincial Lodges as may be desirous to join the Procession will be placed according to the seniority of their respective Lodges. Arrangements will be made on the Stage (in the centre of which will be placed the Grand Master's Throne) for the accommodation of the whole Body. In the course of the Evening several *Masonic Songs and Choruses* &c by Brothers Fullam, Johnson, Williams, Lees, W. Farren, T. Short &c. (*Saunders' News-Letter*, 28 June 1815)

for a confirmatory announcement reports that 'The admired Overture and Songs…by Blewitt are, we understand, to be published immediately'.[16] *The Magician without Magic* had its first performance at Crow Street on 1 March, and like *John of Paris*, included a 'Grand Ballet'. On the whole the opera 'met with a most favourable reception'. 'The ingenuity and novelty of the plot, the smoothness of the dialogue, the beauty of the music, and the general excellence of the *getting up*'[17] were all noticed, yet, while the piece was said to have received 'the encouragement of a good share of approbation…it also met with some disapprobation'.[18] As a second critic explains: 'It wants curtailment; and, we think it would be materially benefited by the omission of certain phrases, or at least by their less frequent repetition… The hits against *conjuration* were well received, particularly at this time, when two celebrated professors of the Black Art* are performing in Dublin'.[19] It might have been wiser had Blewitt retained some of Isouard's music, but then, he also lacked Miss Hughes among his cast.

It may have been an attempt to replenish diminished box-office receipts which brought forward the next production, for it had been first produced at Drury Lane as early as 1803. Two factors seem to have dictated the choice: a temporary withdrawal from the French scene, and the hope of a return to the proven success of *The Dog of Montargis*, for the chosen work, *The Caravan; or, The Driver and his Dog*, composed by William Reeve to a libretto by Frederick Reynolds, also had a dog in the cast. It was presented in Dublin on 27 March 'with considerable Alterations and Additions…and entire new Scenery, Machinery, Dresses and Decorations. The Scenery painted by, and under the direction of, Signor Mari[nari]. The Scenery and Machinery by Messrs Savage and Dyke. The Properties &c by Mr Kelly and Assistants. The Overture and additional Music by Mr Blewitt'.[20] But that was just customary publicity; a much more sensational announcement appeared at the end of the advertisement where we read: 'The Last Scene will present A Grand Sea View with Light-House, Fire-Ships &c, and for the First Time in this Kingdom a Body of Real Water, into which the Child is thrown by order of *Navaro*, but instantly rescued by the exertions and sagacity of The Driver's Dog!!!'[21] Once again great emphasis is placed not on

---

* One was a Mr Day Francis, who, with his partner Miss Young, a performer on the slack-wire, was then appearing at the theatre in Capel Street, where he was advertising himself as a 'Magician with Magic'. (*Freeman's Journal*, 6 March 1815). The second was a Mr Ingleby, who was performing at the Fishamble Street Theatre. (*Freeman's Journal*, 25 February 1815)

Miss Hughes (b. 1789) as Mandane (1816) in T. A. Arne's *Artaxerxes*.

the actors or musicians, but on the dog, and after the first performance we read, 'The Dog has much to do and he does it well'.[22] With such titillation on stage there were—not surprisingly—16 further performances before the season ended. Present-day film moguls presenting productions featuring Lassie and Rin-Tin-Tin are no more original nor keener psychologists of public response than were Frederick Jones's acting managers. Modern technology merely permits them to be more spectacular.

An English opera of dissimilar antecedents, but of contemporary equivocal arrangement, now followed. This was *Cymon*, which, with music by Michael Arne to David Garrick's libretto, had been first performed in Dublin on 4 March 1771. Garrick's text had now been provided with entirely new music by Sir John Stevenson, which caused some raising of critical eyebrows. Following its first performance on 9 May, one reviewer perorated: 'We have left ourselves room only to say one word about the Music—it was Sir John Stevenson's. But why did the Managers think it necessary to employ the Musical Knight's talents on an Opera which had been previously composed by one of the greatest names in the Musical World, Dr Arne?*... We do not wish to speak harshly of a man, whose talents on various occasions we have had reason to admire; but really it is enough to disturb our constitutional tranquillity of temper, when such a man challenges, as he does most evidently, a comparison with the only Musician who composed an English Opera'.[23] News that it was *Michael* Arne and not his more illustrious father, Thomas Augustine, who had composed the original *Cymon*, and that there were other composers of English opera, Henry Purcell for one, was evidently brought promptly to the attention of the critic. Consequently, some time later he was more indulgent when he stated: 'we do candidly acknowledge that every night we have heard it [*Cymon*] these merits appear more striking... The Overture seems to us an exquisitely wrought, and most gracefully combined, piece of harmony. The songs are not, however, of the highest class, not so high as Sir J. Stevenson would lead us to expect. They are light, airy, and *fluent*, but it is the fluency of mannerism—a sort of key note seems to be given, and a set of chords and symphonies appear to flow from it... Yet there are some exquisite airs'. It would have been as well had the critic ended his elementary exercise in musicology here, but he would persist. 'What

---

* Another 'Musical Knight', Sir Henry Bishop, though yet undubbed, also 'refurbished' *Cymon* later that year.

has been said,' he continues, 'cannot be construed into a palinode of the opinion we formerly expressed on the indecorum of Sir John Stevenson's substituting his Composition for Arne's, even Mathew Arne's [Mathew ?!]… Now, though the son may not be as clever as the father, it is nevertheless true, that the old Music of *Cymon* has been pronounced by adequate judges as excellent, and particularly delightful for the simplicity and sweetness of its melodies'.[24] As criticism, at best a non sequitur, at worst, an example of biased prevarication. Although the cast included Miss Hughes, who 'displayed the happiest talents, and exhibited a compass, sweetness, and variety of voice, with a taste and science, and a force of musical expression that leave her without the shadow of competition in the English Opera',[25] yet she could not save Stevenson's prosaic work. In all, it had only five performances during the season.

Because of Frederick Jones's withdrawal and subsequent unavoidable return, coupled with Anthony Rock's untimely death, 1815 must have been an even more difficult year than usual for Crow Street Theatre. There were other problems, too, such as the orchestra. Concerning this, *The Freeman's Journal* declared it had 'been a long time endeavouring to effect a reformation', but appended hopefully, 'we have at length, we believe, pretty well succeeded. It is at all events certain, that new music has been distributed among the performers, and that the three or four pieces which have maintained so sturdy and obtrusive a station, for a succession of at least seven years, are to be laid on the shelf… We believe, too, there have been some strict rules laid down with regard to practice and regular attendance. Mr Blewitt, we understand, has proved a powerful auxiliary in the effectuation of those most necessary regulations—and the musical portion of the community must therefore hold themselves bound to him in a particular obligation'.[26] Yet when *John of Paris* was produced in the following month there was criticism from another newspaper, that the work would be more popular 'if the Songs and Choruses were given without any retrenchment; several were omitted on Saturday'. Moreover, we read, 'In the arrangement of the Orchestra, too, we perceived some omissions of the Music incidental to the Piece, which, notwithstanding our readiness to pay every tribute to the talents and judgment of Mr Blewitt…we think should not have occurred'.[27]

Then there was a letter to the paper in May, presumably from a justly disgruntled actor, which throws light on some unusual aspects of benefit performances. He wrote: 'Sir: When benefits are frequently seen advertising in our Theatre, not for the Actors or Actresses, but for any purpose or for any person that will pay 86*l*, every considerate being must feel for the Performers of that house, as such coming before their benefits which they are obliged to take, and pay 68*l* 5*s* for each night, operates considerably to their disadvantage. Thus the town is delayed in this way by benefit tickets from John Oaks or Peter Stiles,* and afterwards, when the performer of merit who has a claim upon the citizens of Dublin, and draws money to the theatrical treasury—when he or she comes to endeavour to pass their benefit tickets, by many they are spurned, and told that the town has been so much pestered with such as we allude to, that they cannot afford to be then generous to them. This is a miserable and unjustifiable treatment to our Thespians… Their benefits should precede all others from their connexion with the Theatre. And beside, instead of being obliged to take the least attractive plays for the amusement of their friends, they should have a fair share of those in most estimation… Many instances have occurred where such persons, failing in their expectations on their benefit nights to pay their debts, have afterwards been cast into prison and endured long confinement. A Friend to Justice'.[28]

Charles Incledon once more brought a small group of artists to the Fishamble Street Theatre during August. He was again accompanied by Master Williams, and by a Mr Broadhurst, pianist. On this occasion their entertainment was called 'The Minstrels' and consisted of 'Recitation, Songs, Duets, Trios &c &c'.[29] On the evening of 24 August, they changed their venue, entertaining a summer audience at Seapoint House, Black Rock'.[30] Unhappily, by November we read of 'the late Theatre in Fishamble-st' having been altered to 'an Olympic Pavilion for the performances of Equestrian Exercises'.[31] Where legitimate theatre was concerned, Dublin was truly entering a lean period.

Much of this assuredly was due to the uncertainty of affairs at Crow Street. In August, it had been announced, 'The Dublin Theatre has been let to Mr Rock and Mr Webb, the Lottery Office-Keeper, for 5000*l* for the ensuing season'.[32] But, as has been noted, Rock died not long afterwards leaving Frederick Jones with a theatre which he no longer had any ambition to run. The inevitable press controversy followed, which Jones attempted to answer as follows:

---

* Fictitious names formerly used in law proceedings.

'Complaints have been publicly made that the Theatre has not been long since opened. To this Mr Jones has to reply that it will appear upon record, that the Theatre has seldom been opened until November, the period when those who are most in the habit of frequenting the Drama generally return to Town. He has to add, however, that it would have been opened on the 20th ult. had he not been engaged in a variety of treaties for the Sale of his Property therein, in consequence of having been forcibly expelled from the personal Management of it. His negociations upon this head have terminated ineffectually, which will surprise none who have had to dispose of property of late.

The next complaint is that Mr Jones dismissed the Company. The fact is, that Mr Jones was, from the predicament in which he was placed, compelled to serve notice to such effect upon those of the Performers whose engagements were regularly terminated; a necessity emerging from his situation—for purposing to dispose of the Property of the Theatre, whoever might be the Purchaser would naturally expect to have the privilege of selecting his own Company.

To the third charge, that of having so far interfered in the interdicted Management of his own Property, as to cause the Theatre to be new Painted, Mr Jones must plead guilty. His apology, however, is to be found in the concurrence of circumstances in which he was placed. The ravages committed in the late riots required to be repaired, and there was nobody else to give orders to that effect. If the repairs were to be deferred until the transfer of the Property, by sale, or bargain took place, they must have been deferred still farther, even until the next Season, an inconvenience which the Patentee thought it his duty to spare to the Public...

Mr Jones reiterates his most ardent desire to dispose of the Theatre upon reasonable Terms. The bulk of his property is embarked in that Establishment; and the duty which he owes to himself, and still more to a large Family, imperatively forbids him to dispose of it upon inadequate Terms. The situation of the Theatre is within the recollection of many at the period when Mr Jones obtained the Patent. Its present situation will speak for itself, and constitute a proof to any impartial and calculating mind, how much of his Property the Patentee must have devoted to its present state of Improvements. The actual *Disbursements* at present exceed three times the full amount of the *Receipts* at the period referred to.

Upon this candid Statement Mr Jones wishes for the Indulgence and Support of a Liberal Public to whose judgment and sympathy he confidently Appeals'.[33]

The fact was that Jones was in negotiation with William Farren* at the time, and in discussions as protracted and uncertain as those which presently occupy the Common Market Commissioners in seeking to reach agreement on agricultural subsidies, the matter was finally resolved as follows in mid-November. 'The Dublin Theatre is rented, principally, we hear, by Mr Farren and Mr Shaw, a young Gentleman of fortune. All the Proprietors retire. Mr W. Farren is to have the sole management; and M. Fitzgerald Esq.,** uncle to the late Chancellor of the Irish Exchequer is, we are informed to be the Trustee between the Proprietors, the Renters, and the Public'.[34]

The theatre eventually reopened on Monday, 13 November, with the old favourite opera, *Love in a Village*, which reintroduced an old favourite artist, Thomas Philipps, 'his first appearance here these six years', and brought a newcomer, Mrs Bellchambers (née Singleton) 'from the Hanover-square Concerts, London'.[35] This lady, whose name was so appropriate to her profession, is reported to have made her Dublin debut 'with more qualifications to obtain our partiality than any candidate we recollect for some time. She has youth and very considerable personal attractions—a clear fine voice, full, soft, and flexible, and of extensive compass with a share of good judgment, and a careful restraint to keep within the limitations of its musical exertion, form this Lady's vocal abilities. She seems, as an Actress, free of that redundant motion and gesture, which too often accompany the exertions of youthful Candidates for Stage reputation'. The notice concludes by announcing, 'The Theatre has been a good deal improved and has a very good effect to the eye'.[36]

The first new opera produced by Farren was yet another 'popular Melo-Drame' by Bishop, again taken from the French. What makes it interesting is that it was *The Maid and the Magpie* based by Isaac Pocock on the French melodrame, *La Pie voleuse, ou la servante de Palaiseau* by Caigniez and d'Aubigny, which, shortly, Rossini would use as the story for his opera *La Gazza Ladra*. We learn that three versions had been performed at three respective Theatres

---

* The two Farrens, William and Percy, whose names appear in the cast lists, were brothers. (*Freeman's Journal*, 15 May 1816)

** On 29/30 May 1816, Frederick Jones assigned an interest in the Crow Street Theatre property to Maurice Fitzgerald. (Theatrical Manuscript Records of the Eighteenth Century, Office of the Registry of Deeds, Dublin, MS 702/481724. Reported by La Tourette Stockwell, p 386)

in London with considerable success, which prompted Farren to acquire the three different versions for Dublin, where 'the materials of each [were] so amalgamated into one piece as to render the forthcoming production a species of very superior dramatic excellence'.[37] Bishop contributed 'The Overture and Music…in which will be introduced admired French Airs',[38] to the production. A description of the plot in the press shows it to have been very similar to Gherardini's subsequent libretto for Rossini.

Among the cast, the role of Annette (Rossini's Ninetta) was taken by Mrs Edwin, née Elizabeth Rebecca Richards, who by 1815 had been appearing intermittently on the Crow Street stage for 35 years, having made her first appearance there in 1780 when only seven years old. Her husband, John Edwin junior, comedy actor and singer, is reputed to have died from humiliation through being lampooned in John Wilson Croker's satirical poem reported earlier. The hurtful lines were:

> '…Heaven that dooms to equal fate
> The thespian and the human state,
> With Mrs Edwin blessed our vows,
> But cursed us with her lubbard spouse—
> Yet let us spare him for a *name*
> High on the rolls of comic fame,*
> And on degeneracy take
> Compassion, for the parent's sake.'

We are told that 'in his fevered frenzy, [Edwin's] imprecations on his destroyer…were as horrible as awful'.[39] There is evidence however that his demise** was not entirely due to the shame of having had an honoured name disgraced, but to a longer-standing excess of cognac. A report of the first Dublin performance of *The Maid and the Magpie* records, 'That it is interesting was proved by the attention of the audience during its progress, and the marks of satisfaction with which a second representation of it was announced'.[40] That it was highly successful is confirmed by its having received no fewer than 22 performances between 12 December 1815 and 22 April 1816.

Farren's next new opera was also taken from the French. It was an adaptation of *Lodoiska*, which originally had had two versions produced in Paris in 1791

---

* His father's, John Edwin, senior.
** The *Dictionary of National Biography* reports, 'a tombstone to his memory, erected by his wife in St Werburgh's churchyard, Dublin, attributes his death to the acuteness of his sensibility.'

The three Misses Dennetts of Covent Garden Theatre

within a month of one another, the first with music by Cherubini, the second composed by Kreutzer. Dublin chose a version originally produced at Drury Lane in 1794, with a libretto by John Philip Kemble and music arranged by Stephen Storace. Storace was impartial in his choice of composers, borrowing music from both Cherubini and Kreutzer, and including an aria by Andreozzi for good measure. Michael Kelly thought highly of this hybrid when it was produced at Drury Lane, and in which he played the role of Count Floreski, recording: 'I was in Paris at the first representation of "Lodoiska" at both theatres.* Kreutzer's was performed at the Théâtre des Italiens, and Cherubini's at the Feydeau—both got up with great effect and care; but, partiality apart, the Drury Lane piece surpassed them both. Storace selected the most effective music from either, and enriched the piece with some charming melodies of his own composition'.[41]

In Dublin the work had been in rehearsal as early as April 1815, but the confused situation there had evidently postponed its production. When it was performed eventually on 8 January 1816, once again 'The Scenery [was] painted by, and under the direction of, Sig. Marinari. The Scenery and Machinery [were] by Messrs Savage, Dyke, and Seagrave. The Properties, Banners &c by Mr G. (sic) [John] Henderson and Assistants. The Dresses by Miss Ereby and Mr Lord'. 'In Act I', we are informed, 'Mr Philipps will sing the original Cossack Melody, and in Act II, Beauty in Tears (a Cambrian Air). The Poetry of both written expressly for the Piece. In Act III, A new *Pas de Deux* incidental to the Piece (the Music by Mr Moran) by Master and Miss St Pierre (their first appearance on any stage) and, the Song of Home, Love and Liberty (composed by Bishop) by Mr Hodson'. The scenery was advertised as 'Act I, The Forest of Ostropol, Castle, Drawbridge &c. Act II, The Courtyard of the Castle, Terrace, Towers, &c. Act III, The Grand Hall of Audience; Castle of Lovinski in ruins',[42] all of which helps to justify the announcement that 'The scenery and decorations are truly magnificent. No less indeed, than a sum of *One Thousand Pounds* has been expended'.[43] The dramatic highlights of the production, in all senses of the word, appears to have been the final scene in which the castle goes up in flames, and genuine flames at that. We are told that the tower in *Lodoiska* 'seems to have been the model of all the fortresses subsequently stormed'.[44] The first performance

---

* Dr Roger Fiske has pointed out the unlikelihood of this and propounds that Kelly saw the operas in 1792. He adds, 'Also it was surely Kelly who brought back the libretto of *Lodoiska* for Kemble to translate and the two scores from each of which Storace borrowed a few pieces'. Kelly, p 364

appears to have gone extremely well. 'The House was crowded, and the piece received throughout the loudest and most unqualified approbation'. There was only one reservation—'there was sometimes too great a delay between the acts, which was perhaps not to be avoided, in consequence of the changes of scenery and mechanism'.[45] 'On the whole', concludes another report,'*Lodoiska* is much the most striking exhibition in its way, which has been yet presented to a Dublin audience. In the representation it was cheered all through, and it was given out for this night, amid shouts of applause'.[46] Strangely, for all this superb acclaim it apparently had only 11 performances during the season.

There is little interesting or original to report during the next two months of 1816. As part of William Farren's benefit on 7 February, a new melo drama, *Egbert and Ethelinda; or, The Draw Bridge* had been 'written by a young Gentleman of this City, the Overture and Music composed by Mr Blewitt',[47] (who, 'for the accommodation of his Pupils' had recently 'removed to No. 22 Sackville-street').[48] Tickets were to be had of Mr Farren at 9 Aston's Quay.[49] This was followed on 16 March by an operatic farce, *My Spouse and I*, the music by John Whitaker, once known for his ballads and some comic songs adapted from Irish airs. Mrs Mountain was then engaged 'for 3 nights',[50] making her first appearance on 19 March in *The Siege of Belgrade*, (having retired from the stage less than a year before!).

It is 'a new Musical Entertainment', '*Edwin and Angelina*, taken from Dr Goldsmith's Poem of *The Hermit*',[51] a joint effort by Stevenson and John Charles Clifton, a minor English composer then settled in Dublin, which attracts most interest. This was first performed on 3 April, and a week later, a learned gentleman signing himself 'Philotheat' (perhaps the librettist) was rushing to its defence in *The Freeman's Journal*, coincidentally revealing how defenceless it was. He writes: A contemporary Journal of yours, Sir…has very ill-naturedly observed that a general hiss was given to some parts of the piece, notwithstanding the revision it had undergone. This assertion I know Sir, to be unfounded. Some few parts were hissed by those who came to laugh at it merely as a farce; but the whole of the piece was applauded, by all who came to be entertained and delighted with it, as a production of fancy, taste and science'.[52] Little here of interest admittedly, but among the artists taking part was Miss Griglietti, and she does deserve attention. She had, in fact, been a member of the company since January, and as Clarissa in *Lionel and Clarissa* was described as 'the best…that we have had since Mrs Billington'.⁵³ Throughout the remainder of the season she appeared in *The Lord of the Manor*, *The Beggar's Opera*, *The Haunted Tower* and *Lodoiska*, and for her benefit in May, selected Tom Moore's *M. P.; or, The Blue Stocking*, at which it was understood 'his Excellency the Lord Lieutenant and her Grace the Duchess of Dorset intend honouring the theatre with their presence'.⁵⁴

Very little is known of Elizabeth Augusta Griglietti's early life. A report in *The Times*, following an early appearance by her at the King's Theatre, refers to a 'Signora, or rather Miss Grigletti [*sic*], a young lady from Bath'.⁵⁵ Another source amplifies this, reporting that she was the daughter of a baker from that city.⁵⁶ We know that on her arrival in London she sang at the Hanover Square Vocal Concerts, that she may have been the first in England to sing Beethoven's 'Ah! perfido' (at her benefit concert in the Freemasons' Hall in 1810), and that she undertook the roles of both Donna Anna and Zerlina in a concert performance of *Don Giovanni* when it was first presented in England, mainly by amateurs, (but in which Charles E. Horn 'enacted' Elvira!) in a 'floor-cloth manufactory*…amidst the *mingled* effluvia of canvas, oil, and turpentine'.⁵⁷ She first appeared at the King's Theatre during the 1805/06 season, singing in three operas, including the role of Servilia in *La Clemenza di Tito*. She then continued to be engaged there for nearly every season until 1815, almost invariably singing secondary roles, but very occasionally appearing in more major parts such as The Queen of the Night in *Die Zauberflöte*, Elisetta in *Il Matrimonio Segreto*, and Donna Aristea in *Il Fanatico per la Musica*. On her first appearance as Servilia, it was recorded, 'She has a pretty voice, a pleasing person, and we have no doubt will prove an acquisition'.⁵⁸

During her time at the King's Theatre she took part in at least one English opera at the Lyceum, in June 1810. This was, *Oh! This Love! or, The Masqueraders*, by James Kenney, with music by M. P. King; Leigh Hunt, reviewing the first performance, reported: 'The curiosity of the evening was the appearance of Miss Griglietti from the Opera.** The part allotted to her was sufficiently insipid to favour a want of exertion, and her manner was both indolent and timid, except

* Identified as the firm of Thomas Hayward & Co., Floor-cloth manufacturers, 195 Whitechapel Road, London E.1, in *Musical Pursuits*, Selected Essays by Alec Hyatt King. British Library 1987. p 144)

** A 'curiosity' probably because foreign opera singers very rarely performed on the English stage, This was not due to any language barrier, but because it violated the Opera Agreement of 1792. Exceptions were made for the Lenten oratorios and, occasionally, for benefit performances for privileged artists. (See: *Sheridan to Robertson*, by Ernest B. Watson, Harvard University Press 1926, p 15)

when she was singing, and then she gave us all the pomp and affected emphasis of her Italian stage… At the Opera Miss Griglietti's acting was of course overlooked, and her powers may not have been very effective, but on a little stage like that of the Lyceum, some acting is very requisite and the audience would be content with less powers of singing, or rather with less shew of them'.[59] When her career ended at the King's Theatre in 1815, she apparently set out for Dublin, for it was said of her, 'without the *materiel* necessary to place her amongst the leaders of her profession…want of encouragement here drove her to the sister Isle where…she has attained a station free from the vicissitudes of the boards'.[60] This she did by marrying Mr George Lazenby of Parliament Street (presumably a relative of her fellow-actress, Mrs Lazenby, who lived at No. 23), at St Werburgh's Church, Dublin, on 7 September 1817.[61] By all accounts it was a satisfactory conclusion to her career, for Dublin too remained unconvinced of her merits. Earlier, the *Dublin Examiner* had decided she 'possesses a voice of a full and pleasing tone and displays much natural feeling and expression: but her taste wants elegance, and her execution, neatness'.[62]

One must now return to the adverse life and times of Pietro Urbani. As far as can be traced, he last appeared in public at a 'Grand Concert of Sacred and Miscellaneous Music' at the Rotunda on 6 February 1815, when he sang Handel's 'I verdi prati'.[63] At a later concert for his benefit 'Under the Patronage of his Pupils' again at the Rotunda on 28 May 1816, there is no evidence that he himself took part, nor is it likely that he did, for the concert is advertised to procure 'some assistance to lighten the adversity which a long illness has produced'.[64] His condition must have been truly distressing, for it was further announced that 'The friends of Signor Urbani most humbly hope that a generous public will defer giving parties on the night fixed for his Concert, it being a case of real charity. Many respectable professionable persons who may not have it in their power to assist in the Orchestra, are humbly called on by those who do assist, to send the smallest donation which will be thankfully acknowledged by those appointed to manage the concert'. Even the tickets were not unreasonably priced. Those for gentlemen cost 7s 6d; ladies, 5s.[65] But the end was near for poor Urbani, and on a date between 8 and 15 September 1816* he died at his home or lodgings in No. 1, South Cumberland

---

* Reported without giving a precise date in *The Freeman's Journal* on 16 September, in *Saunders' News-Letter* on 17 September, and in *The Dublin Evening Post* on 21 September.

Street,* aged 67. His obituary notice records, 'Those who had the advantage of his instruction will sincerely regret his loss as a Master—and those who had the pleasure of his acquaintance will deplore the Man of Science, who, alas! as is too often the case, was one to whom fortune refused her smiles in the latter part of his life, when he needed them most'.[66]

He left a widow who had come with him to Dublin as a dancer** when he had first arrived as an opera singer in 1781. In the following May a benefit concert was organised for her. That she was direly in need of it is evident from the following: 'Never was an Appeal to the feelings of the Benevolent made on a more distressing occasion than the present. The Widow of Signor Urbani has been left by his death, totally destitute of, and unable to obtain, any means of subsistence. A foreigner, almost unacquainted with the language of the country she resides in, without money—and, *but a few* excepted, without friends. We understand a concert has been proposed as the best means of relieving her from the present difficulties, and of raising a small fund towards her future support, or to enable her to return to Italy. The patronage and Support of the Pupils of her late Husband—of a humane and generous Public, to whom the Widow and Stranger never appealed in vain, is on this occasion most humbly and earnestly solicited'.[67] The concert held on 11 June had among those taking part, Braham, Philipps, and the Misses Griglietti and Cheese.[68] Such was the end of a musician who even today, over 170 years later, still commands half a column in *Grove's Dictionary of Music and Musicians*.

An observation in the *Dublin Chronicle* in June that 'the Dublin stage presents at this moment the most respectable Operatic Company in the empire',[69] was fully confirmed in August with the arrival of the distinguished English soprano, Catherine Stephens. She was born on 18 December 1791,[70] the daughter of Edward Stephens, a carver and gilder of 85 Park Street, Grosvenor Square'.[71] In 1807 she was articled by her father to the singing teacher, Gesualdo Lanza, who in time brought her before the public, first at concerts in Bath and Bristol, and then on a tour of southern towns and cities, where she occasionally sang under the pseudonym of Miss Young.[72] She also sang at the Pantheon in London in May 1812 'in a burletta entitled *Figaro* (music by Mozart) in which,

---

* The street, in spite of much political and economic upheaval, remains as named, running between Pearse Street and Fenian Street, parallel to Westland Row, but the house close to the railway bridge immediately south of Pearse-Westland Row station, has long since disappeared.

** See *Opera in Dublin 1705-1797*

though her part was trifling, she was always encored in the duet with Mme Bertinotti, the only piece of music that was encored each evening'.[73] Her final appearance as Lanza's pupil was in a concert at Ramsgate on 3 October 1812[74] when, the five years of his contract having expired, much to his chagrin, she was removed from his care by her father and placed under the tutelage of Tom Welsh, who at once launched her at a concert in Manchester on 9 November.[75]

Her first important engagement, announced as Welsh's pupil, was at Covent Garden on 23 September 1813 as Mandane in *Artaxerxes*, when one reviewer declared that she 'displayed talents so captivating and brilliant that we have perfect confidence in her success. Her voice possesses the richest and most affecting power, and she sings and acts with graceful simplicity. There has been no quality of voice equal to her's produced for many years…we have no doubt but we shall find her a worthy successor to Mrs Billington'.[76] A most happy augury for all concerned except for poor Lanza, who some days later was protesting in the press that 'Miss Stephens is my pupil, having studied from my theory the Elements of Singing for five years, and [with my having] prepared her for the stage, affording her the instructions of Mr Wright in elocution out of my own pocket'.[77] In appearance at the time she was said to be 'about the medium height. Her figure was pleasing but not strictly symmetric. It inclined also rather too much towards the *embonpoint*, to agree with Hogarth's "analysis of beauty"… Her hair and eyes were dark, and her countenance agreeable without being handsome'.[78]

She continued to enhance her reputation during the season in performances of *The Beggar's Opera*, *Love in a Village*, and *The Duenna*, although J. E. Cox records: 'Of dramatic impulse or power Miss Stephens had not the slightest notion; and even when she had become accustomed to the stage—a result brought about by her vocal qualifications rather than by any histrionic talent—it could never be said that she had learned how to act. Her manner was cold and unimpressive. She said her "words" accurately, but evidently by rote, and added to the effect of the scenes in which she was engaged, whether in opera or in comedy, in no other way than by her singing talent'. But, he goes on, 'So full and rich however was her voice, so true was her intonation, and so facile her execution, that the supposed absence of imagination, of which complaints were neither few nor far between, was altogether forgotten the moment she opened her mouth to sing… As an English and Scotch ballad-singer she has never been equalled either

John Jackson's portrait (c. 1822) of Catherine Stephens (1794-1882).

before or since her time. I have myself on more than one occasion witnessed the emotion she produced—by simply singing the quaint old strain to which the well-known words of "Auld Robin Gray" are set—upon audiences, amongst whom, young and old, hard and sensitive, there was scarcely a dry eye... In sacred music Miss Stephens likewise manifested neither want of feeling nor unconcern, but threw an amount of pathos into its interpretation... In that most trying of all Handel's sacred songs, "I know that my Redeemer liveth" (*Messiah*) I do not believe Miss Stephens was ever excelled'.[79]

It is only just that the supplanted Lanza should receive some credit for such perfection, and 20 years later it was recorded that he had 'proceeded to form her voice [which eventually could ascend to D in alt[80]] with care, but also with the slow progression of the Italian method... No female singer perhaps ever built so true an English style upon Italian rudiments'.[81] Just over a year after her debut in opera she received the supreme accolade when 'Mrs Billington and Madame Catalani were...in one of the public boxes at Covent Garden Theatre to hear Miss Stephens in Mandane, and, after the performance, paid her their joint tribute of admiration of her vocal powers'.[82]

Understandably, from the foregoing, her appearance in Dublin was eagerly awaited, it was even recommended in July to Farren that 'if the report be true that Miss Stephens is to be at Crow-st Theatre in August [he] should endeavour to engage Incledon [who was just then leaving Dublin following a short engagement], as those operas in which she performs could not be got up effectively without him'.[83] Since Thomas Philipps was a member of the company Farren no doubt considered this an unwarranted extravagance. A star of Stephens' magnitude could fill the theatre on her own; besides, Incledon at age 53 was now passé.

She made her first appearance on 12 August 1816 as Rosetta in *Love in a Village* to tremendous acclaim. 'Her voice,' we read, 'unites power and sweetness, compass and flexibility, in a degree of which we scarcely know an example'.[84] She followed this two evenings later with Clara in *The Duenna*, when it was reported, 'She has the power of Mrs Dickons, with a great superiority over that eminent performer in nature and simplicity. Her notes possess much of the liquid sweetness of Catalani's singular strains... We may be mistaken but we think that Miss Stephens excels Mme Catalani in ear. Her singing has an equability which that lady cannot command'.[85] Next, she 'charmed a crowded audience as Patty in *The Maid of the Mill*, 'though evidently labouring under indisposition',[86] while, for a performance of *The Beggar's Opera*, 'the house overflowed in every part. There were more than two hundred persons who could not find admission to the Pit, and who found it difficult to find places in the Boxes and Gallery'.[87] The season continued with unabated success with a performance of *The Cabinet*, and with further performances of *The Duenna*, which had 'the additional support of Miss Griglietti as Louisa', while for Saturday, 24 August, a performance was commanded by the Lord Lieutenant.[88] Philipps partnered her each evening, appearing 'to peculiar advantage'[89] as Carlos in *The Duenna*, and to 'well deserved…and general applause' as 'the best Lord Aimworth [in *The Maid of the Mill*] the stage has ever produced', while as Macheath in *The Beggar's Opera* he 'ably sustained his unrivalled reputation'.[90]

Shield's *Rosina* was also produced, and on 27 August the first of three performances of 'a new Musical Piece', *Brother and Sister* with text by Dimond (based on Patrat's play, *L'Heureuse Erreur*), lyrics by Charles Dibdin, and music by Bishop and Reeve. Stephens had sung in the original production of this opera at Covent Garden in the previous year, when she 'was very successful in a song by Bishop, in which…she sang an echo to her own voice with great ability'.[91] She then gave 'much satisfaction' when announced to sing at a 'Concert of Sacred Music at the Clarendon St Chapel' (with Griglietti and Philipps) in aid of the General Asylum,[92] on 30 August. It was reported that the ceremony 'was fully attended'. 'Thus have been rescued from want and every attendant vice, between three and four hundred children whose fathers fell in the late war'.[93] She took her benefit on 3 September, including in her programme 'a "Farewell Song" written by a Gentleman of this City, the music by Mr Smith of the Choir',[94] and then on 6 September 'sailed for Holyhead'. In farewell *The Freeman's Journal* reported, 'We do not recollect that any other performer in so short a time became so great a favourite with a Dublin audience as this Lady'.[95]

Unlike so many of her fellow artists, Catherine Stephens was to end a very successful career with an even more successful marriage. In 1838 (when she was 46) 'it pleased a peer of the realm, who had only been a widower three months (his first wife, [from whom he had long been separated] having died on January 16, 1838, to pay her marked attention'. He was George Capel-Coningsby, fifth Earl of Essex, 'a distinguished patron of the drama, and one of the first to notice as a rising star, Edmund Kean…her noble admirer laid his heart and his coronet together with a splendid settlement, at her feet'.[96] He married her on 14 April 1838, and following his death on 23 April 1839, aged 81 years, Catherine continued to live at No. 9, Belgrave Square, the home into which she had married, until her own death on 22 February 1882.

William Farren could indeed be satisfied with the result at the end of his first season as manager of the Theatre Royal, Crow Street. Even shortly after he had obtained the post he was being commended that 'With the lighting of the House, and the general accommodation of the Audience, there is much more reason to be satisfied than hitherto, and we have observed a more dramatic attention to the busines of the Stage, and a more correct adherence to costume, than on former occasions, and without a strict observance of which the illusion of the scene is lost. There is still, however, room for improvement in costume well worthy of consideration… No doubt there [will] be performers who will insist on dressing characters *in their own way*, but when these persons do not chance to be a Kemble or a Philipps, who are always classic in whatever belongs to the personage they represent, it becomes the duty…of the Manager to interfere'.[97] That he should have had his lean periods during the year was as ineluctable as the endurance of opera itself, and so there is no surprise in reading in March 1816: 'though the present season of private festivity has left the Theatre worse attended than it should be, and impaired its revenue to the Proprietorship…we are certain these disadvantages are but of very temporary duration, and that fashion and rank and respectability, will speedily again resort to the Theatre, as to the most rational and the most gratifying source of entertainment they can have access to'.[98] Rank and fashion did speedily resort once again five months later, with the arrival of Catherine Stephens. Farren consequently could approach his 1816/17 season with courage and even a modicum of confidence.

Chapter 9

An Irish Lark and that extraordinary Child, Master Balfe 1817-1818

If London discovered its Catherine Stephens in 1813, three years later Dublin discovered its Mary Byrne. While not to be compared vocally with her English colleague, yet for reasons perhaps social rather than musical, Miss Byrne had a brief career which has carved a permanent niche for her in the history of Dublin opera. To begin with, her grandfather, Edward Byrne of Mullinahack (an area situated between Usher's Quay and Thomas Street West, close to the rear of Bridgefoot Street) had the distinction of being considered the wealthiest Roman Catholic merchant of his time in Ireland. He was a distiller and sugar baker, who reputedly paid £100,000 a year in excise duty, and who had been elected chairman of the Roman Catholic Committee, then seeking emancipation. On his death on 21 December 1804, at North Great Georges Street,[1] his estate was valued at £400,000.[2]

Besides daughters,[3] he also had five sons, of whom Edward junior was Mary's father.[4] We learn of him that he became 'ruined through a series of unsuccessful mercantile operations which he carried on in Liverpool', and in an attempt to retrieve something of his fortune induced Mary 'to appear as a public singer'. In contemporary estimation this was degradation to the lower depths, which prompted 'his respectable relatives' to offer 'to secure him an annuity on condition of his not permitting the young lady to go upon the stage'.[5] Upon the stage Mary went, however, (perhaps she wished to) but first she had some tuition. There is uncertainty concerning her teacher. The *Dublin Evening Post* describes him both as 'one of the first Musical Professors at Naples',[6] and 'the celebrated Pucitta',[7] while Gilbert has placed her 'under the instruction of the eminent Master, Liverati'.[8] She may, in fact, have had lessons from both composers since in 1815 Liverati succeeded Vincenzo Pucitta as composer and music director at the King's Theatre.

Described as 'a Young Lady (of this city), her first appearance on any Stage',[9] she made her debut at Crow Street as Adela in *The Haunted Tower* on 19 December 1816, following it with a second performance some evenings later, when she 'sustained the high reputation she had gained on the night of her debut'.¹⁰ She next played in *Love in a Village*, *The Padlock*, and other popular favourites, all with success. An assessment of her as a singer at this early stage of her career described her as 'of much higher character but inferior in quality of voice' [to Griglietti], and continues, 'she need not blush to acknowledge herself an imitator of Miss Stevens [*sic*], and we think we cannot pay her a higher and more deserved compliment, than to say that she reminds us constantly of the perfections of that lady's style. She has much indeed of the same purity and delicacy of taste—the same truth of expression, and the same affecting pathos'.¹¹ Her social position naturally attracted attention, which drew down a rebuke from *The Correspondent* advocating good manners. 'We should think it would be much more grateful to this young Lady's feelings', it announced, 'if less were publicly said of her private circumstances, and more of her abilities... Her motives for choosing the profession of the Stage are generally known...and there the matter should rest'.¹² Yet all the time her social position kept intruding. At her benefit performance in May, box-office receipts came to over £500, which amount it was expected would be doubled 'by Gold Tickets'.¹³ Then, at a concert which she gave in Waterford in July, while Crow Street Theatre was in summer recess, the attendance included no less a distinguished gathering than the Judges, [it was Assizes week], the Marquis of Waterford and his family, Admiral Beresford, the Lord Bishop of Waterford and family, and many of the Grand Jury and other County Gentlemen.¹⁴

Notwithstanding all this pretence, however, Mary Byrne really did have talent, and this she demonstrated by being engaged for Drury Lane less than a year after her Dublin debut, when she won a 'rave' notice from Leigh Hunt. Once again she chose Adela in *The Haunted Tower* for her first appearance, which took place on 14 October 1817.

'She is young', he wrote, 'and is of a prepossessing appearance, with fine dark eyes and hair, and a little lady-like figure. Indeed, we think we have not seen an actress for a long time so genteel in her air and natural deportment. Her mouth is her least handsome feature, being somewhat too prominent under the lip, though the chin itself recedes; but her smile redeems it in an instant; and this is one of the best things that can be said of a face. She has the ease and unaffectedness of conscious ability, and yet at the same time apparently the true kind of good breeding, a disinclination to make herself over-conspicuous and a consideration for others... We have involuntarily been speaking of Miss Byrne

The Castle of Andalusia (S. Arnold) had its first Dublin performance at Smock Alley Theatre on 13 January 1783; *Lock and Key* (Shield) at Crow Street Theatre on 29 August, 1796.

as an actress; and the truth is she is a very promising one, and decidedly the best we have yet seen among the professed singers. She has a feeling for bye-play as well as for dialogue; and taps her fan on the palm of her hand with a very natural sprightliness. Her voice has one considerable defect in the quality of the louder tones which are reedy; but she has some sweet ones among the softer, both high and low; her ear is excellent, and the general style of her singing is sensible, correct and powerful. Her taste is so good that she will doubtless get over a certain hardness of outline, or over-distinctness in the enunciation of her notes; yet this was most observable in her first song, "Whither my love" and was perhaps summoned up by her in order to hinder her timidity from going to the opposite fault of over-slurring; for she was at first much agitated, though she soon recovered herself. The execution, however, of this song, which is Paesiello's air of *La Rachelina** (one of the sweetest in the whole circle of melody) was upon the whole a very promising announcement of her powers and told us at once that we had no common singer before us'.[15] The review continues in this style, (with certain slight reservations she is even again compared favourably with Catherine Stephens) until one gets the impression that public relations efforts of Miss Byrne's well-connected Irish friends were having effect in London and in the prestigous pages of *The Examiner*.

During her engagement at Crow Street Theatre in January and February 1817, John Braham was also a member of the company. In fact gossip from London whispered that 'a terrible feud has broken out between his admirers and those of Philipps',[16] But while Miss Byrne's sun was now in the ascendant, poor Braham's was beginning to decline into total eclipse. This was noted in several of the Dublin papers, one of which commented: 'Though we do not think Mr Braham's voice by any means so sweet as it was seven years ago, and feel strongly the conviction that the endless embroidery of his songs accords not altogether with the improved and improving taste of the times, yet there is still enough to admire and applaud, and certainly nothing so offensive to a correct ear as to deserve the unqualified censure of a hiss. The principal cause of disappointment…arises, we are convinced, from the great improvement in vocal music within the last seven years. Mr Braham was the first of his school; he introduced the style of singing to which we allude, and while it was *new*, people heard him with astonishment and delight; but Mr

---

* Rachelina's cavatina in his opera, *La Molinara*.

Braham did not keep pace with those who adopted his style; his pupils advanced—he stood still; they acquired his execution, or at least enough of it, but they preserved the chastity of other times, while he continued those embellishments to excess'.[17]

It may have been the same reviewer who summed up the practical consequences of Braham's predicament as follows: 'The spell that bound us seven years ago is broken, and consequently Mr Braham's attraction is not to the extent that had been so confidently anticipated. When the combined talents of Mr Braham, Miss Byrne and Miss Griglietti support an Opera the House is crowded to excess; but when the Proprietors rest upon the individual power of Mr Braham, the Theatre, with the exception of the first night of his appearance, has been comparatively thin'.[18] Nevertheless, he would continue to visit Dublin until 1844, and did not finally retire until 1852 after 65 years on the stage.

Two operas were performed in Dublin for the first time on 29 January 1817, for William Farren's benefit. These were *The Slave*, composed by Bishop to a libretto by Thomas Morton, and 'a Grand Serio-Comic Burletta, *The Queen of Carthage and the Prince of Troy*, with music by William Shield selected by George A. Hodson, to a text by the now aging Leonard MacNally.* Hodson was also a member of the cast. Neither work attracted more than a passing reference in the press. Of *The Slave*, it was reported, 'As a compound of the various materials which constitute Comedy, Opera, and Melo-Drame, it must be estimated without a reference to any of those distinctly', while the afterpiece, *The Queen of Carthage*, was dismissed as 'less laughable than those things usually are'.[19]

The latter had amongst its cast, with Miss Griglietti, a second singing actress of some note. She was Lydia Kelly, younger sister of the more famous Fanny Kelly (who would arrive in Dublin a year later), and daughter of Mark Kelly, Michael Kelly's brother, himself an actor and singer, though of little talent. She was born on 2 June 1795 and is said to have made her first appearance in a joint debut with her sister on the Glasgow stage in 1807. She next joined the Drury Lane company (then playing at the Lyceum following the Drury Lane fire), where she made her first appearance—again with her sister's support—on 11 October 1810. We learn, 'Her person possesses many charms which well

* See *Opera in Dublin 1705-1798*.

qualified her to represent the gentle lovely Rosina [in Shield's opera] and her voice is both powerful and sweet'.[20] She did not succeed there, however, apparently because of 'diffidence and fears',[21] and perhaps her youth; she was only 15. Engagements at Bath and Dublin then followed and ultimately, a trip to America, where she was more successful (though in plays rather than ballad operas), than she had been in London. She is reported to have married a French nobleman but the marriage was not successful and they separated. She seems to have created no better impression in Dublin than she had in London. One reviewer records of her performance as Rosa in *Fontainbleau*, 'It was the first time I had heard this Lady although I had frequently heard her praised as a singer; as Shakespeare says, "I never heard so musical a discord"'.[22]

Some concerts arranged at the Rotunda in July, by Sir George Smart, included among the artists taking part, Mrs Salmon, who we learn, 'was particularly happy in a *scena* from the works of Rossini ['Di tanti palpiti' from *Tancredi*] described as a new Italian composer who is rapidly advancing to celebrity'.[23]

An English composer and singer of minor celebrity, though of some consequence, made his first Dublin appearance in December 1817. He was Charles Edward Horn, born in London on 21 June 1786. He had his first music lessons from his father, Karl Friedrich Horn, a musician who had come to England from Saxony, had later studied briefly with Venanzio Rauzzini at Bath, and had made his first appearance as a singer at the Lyceum Theatre on 26 June 1809. At that time Leigh Hunt thought he had a 'voice and style' that were 'delicate and tasteful, though of small power', and that he 'seemed to want confidence'.[24] Almost a decade later, he refers to his 'very formal and theatrical' acting, adding, 'The best feature in this gentleman…is his good looks. He is what they call a handsome man. Of his compositions we would rather not speak unless we are compelled'.[25] About 1810 Horn took further lessons in singing from Tom Welsh, and reappeared on the Lyceum stage in 1816 as the Seraskier in *The Siege of Belgrade*, winning 'much and deserved applause'.[26] 'His voice and style of singing are good', Hazlitt confirmed, 'and his action spirited and superior to that of singers in general'.[27] Later, Hazlitt considered his Macheath 'much better than what we have lately seen. He sung the songs well with a little too much ornament for the profession of the Captain: and his air and manner, though they did not fall into the common error of vulgarity, were rather too precise and finical. Macheath should be a fine man and a gentleman, but he should be one of God Almighty's gentlemen, not a gentleman of the black rod…something between gusto and slang, like port wine and brandy mixed'.[28]

Horn is described as possessing 'what the Italians denominate *a veiled voice*',[29] and predictably both his voice and style would deteriorate with time. In 1826 it was reported, Mr Horn 'lowers the estimation of his talents by the excessive force and violence of his manner. In a room this gentleman sings very sweetly, and though in a florid style, yet in good taste. His vehemence on the stage is therefore doubly to be regretted'.[30] Ironically, it was about this time that he really established his reputation as an opera singer. This was in the role of Caspar in *Der Freischütz*, (the compass of his voice enabled him to sing both baritone and tenor roles), when the opera was performed at Drury Lane in 1824. Moreover, it was in 1826 that Oxberry described him as 'the best actor for a singer, on the metropolitan boards'.

As a composer it will be recalled he had orchestrated the music for Tom Moore's *M. P.; or, The Blue Stocking*, besides providing the overture. Another opera, performed in Dublin in 1813 in which he had collaborated, was *The Devil's Bridge*. Of his music very little is remembered today (if it is remembered at all) except the ballads, 'Cherry ripe' and 'The banks of Allan Water'. He combined both his vocations of composer and singer when he made his first Dublin appearance on 4 December 1817, as Count Belino in *The Devil's Bridge*. This was followed by performances of Young Meadows in *Love in a Village*, Don Alfonso in *The Castle of Andalusia*, and the Seraskier in *The Siege of Belgrade*. During a performance of *The Castle of Andalusia* on 13 December, in which he sang '"The Summer Gale" with great sweetness and delicacy', there was an interruption at the beginning of the third act, when 'some ill-conducted young men in the Lettices set up a sort of howl…which they continued to repeat at intervals, to the great annoyance of the audience and the performers'. The reviewer goes on to censure the manager for not having 'Peace Officers in attendance to punish, or at least prevent, outrageous violations of decorum'.[31] It was in this performance also that a new singer, Mr O'Callaghan, a pupil of Signor Naldi, made his first appearance on any stage, as Don Caesar. 'He possesses a fine bass voice, improved by science and judgment', we read, 'and promises to fill respectably a line of characters which, for a long period, have found no adequate representative at our Theatre'.[32]

On 8 January 1818, Horn appeared in a work new to Dublin. This was 'the Grand Serious Opera', *The Libertine*, the libretto, based largely on Shadwell's play by Isaac Pocock, according to F. Cordier, 'cruelly adapted' from Mozart's *Don Giovanni* by Bishop, but with some interpolations added in Dublin. When originally produced at Covent Garden in May 1817, Hazlitt declared: 'Almost everything else was against it, but the music triumphed. Still it had but half a triumph, for the songs were not *encored*, and when an attempt was made by some rash over-weening enthusiasts to *encore* the enchanting airs of Mozart…the English, disdaining this insult offered to our native talents, *hissed*… There was some awkwardness necessarily arising from the transposition of the songs, particularly of the duet between Zerlina and Don Giovanni, which was given to Masetto because Charles Kemble is not a singer, and which by this means lost its exquisite appropriateness of expression'.[33]

In Dublin it suffered the indignity of having been performed as an afterpiece, although announced as 'the Grand Serious Opera', and some amends were made 'in point of scenery'. This is reported to have been 'the very best thing we have for a long time witnessed at Crow-street, Marinari having exerted his admirable skill and taste in the preparation of it'.[34] It was also agreed that 'all that part of [the music] which belongs to the original bears the stamp of Mozart's genius, and the selection and additions which have been made from the works of other Composers are not unworthy of association with his production. The piece possesses scarcely any plot, but the charms of the music and the beauty of the scenery must always secure it popularity. In the commencement it appeared rather heavy owing to the imperfection of the performers, especially in the dialogue; but the interest afterwards rose very rapidly till the fall of the curtain. Mr Horn performed Don Juan (The Libertine), and was excellently supported by Miss Hammersley as Zerlino [sic]. Their duet in the first act, ['Come place your hand in mine'] which is unquestionably one of the finest things in the piece, passed off without any striking effect, which a good deal surprised us'. Johnson performed Leporello 'remarkably well', but 'we object to his burlesqueing such songs as that in which he develops the enormities of his master to Elvira. He may either omit the song altogether, or shorten it considerably'.[35] The opera was repeated on 10 January 'with material improvement'. The players were 'more perfect in their parts' and the 'You place your hand' duet 'was sung upon a higher key, and was much bettered by the alteration'. Johnson, however, 'persisted in burlesquing' the Madamina air 'and met some marks of disapprobation. Mr Horn introduced a very fine song in the banquet scene—the words of which have been written by Mr [M. J. O']Sullivan, and the music taken from another opera of Mozart's, *The Magic Flute*. It was very well sung and pleased the audience so much that they loudly encored it. This, we think, ought to conclude the operatic part of the entertainment. There is something extremely ridiculous in the duet between the Ghost and The Libertine. It is also quite inconsistent with the melodramatic and terrific character of the scene that immediately suceeds, in which the Furies seize upon Don Juan'.[36] Apart from the players mentioned above, other artists taking part whose names have been earlier noted were Lydia Kelly as Donna Leonora (Anna), and O'Callaghan as Don Pedro (Commendatore). There was also 'A Dance incidental to the Piece'[37] in Act II, by a Mr St Albin and a Miss Aylett, who were replaced on 14 January by M. and Mlle Simon 'from the King's Theatre, Haymarket, and Opera House in Paris'.[38]

The impending arrival in Dublin of Lydia Kelly's sister, Fanny, was announced on 9 January 1818.[39] Frances Maria Kelly was born in Brighton on 15 October 1790.* At the age of seven she was articled to her Uncle Michael for nine years,[40] and at the age of ten, through Kelly's influence, joined the Drury Lane chorus.[41] At 'that early period she commenced studying the art of acting',[42] making her first adult appearance as already noted at the Glasgow Theatre in 1807. In October of the same year she returned to Drury Lane, and in the summer of 1809, following the destruction of that theatre by fire, she was engaged by George Colman for the Haymarket Theatre. Subsequently, she transferred again to the Drury Lane company, remaining a member—with occasional absences for provincial tours, and brief seasons at the English Opera House—for over three decades. Her engagement in Dublin may have stemmed from Lydia's performing there at that particular time.

Fanny Kelly had many admirers during her lifetime, including Thomas Philipps, and a suitor named George Barnett, who took a pistol shot at her

---

* A discrepancy occurs in her father's Christian name which her baptismal entry records as: '7 November 1790 Frances Mariah, daughter of Luke and Mary Kelly'. To a query, Mr C. R. Davey, County Records Officer, Lewes, East Sussex, kindly replied as follows: 'In the hope of resolving the discrepancy over the christian name of Frances' father, and to find the mother's name, I looked for the marriage in a transcript for the period 1786-1790 but unfortunately found no entry, nor could I see the father listed in the land tax assessments for 1790 but these unfortunately do not list all the occupiers of property in the town'.

Frances Maria (Fanny) Kelly (1790-1882) as Annette in Samuel Arnold's version of *The Maid and the Magpie*. She was a niece of Michael Kelly; Lydia Eliza, her younger sister (b. 1795) also appeared in Dublin.

one evening from the Drury Lane pit. Fortunately he was as inexpert a marksman as he was unsuccessful in love; he was conveyed to Bedlam. Remarkably, a similar attack is said to have been made on her in Dublin in 1818. Even Charles Lamb, 15 years her senior, proposed to her by letter, but she rejected them all, and died, unmarried, aged 92. It was Lamb who, after seeing her in *The Jovial Crew*, wrote, '"What a lass that were", said a stranger who sate beside us, speaking of Miss Kelly in Rachel, "to go a-gipseying through the world with." We confess we longed to drop a tester [sixpence] in her lap, she begged so masterly'.[43] Her acting seems to have outshone her singing, for while Hazlitt comments, 'She is a charming little vixen: has the most agreeable pout in the world and the best-humoured smile; shews all the insolence of lively satisfaction, and when she is in her airs the blood seems to tingle at her fingers' ends'.[44] He also writes, her 'attempts to supply the place of *Prima Donna* of the New English Opera, do great credit to her talents, industry, and good nature, but still have not given her a voice, which is indispensible to a singer, as singing is to an Opera'.[45]

Whatever the reason, there is evidence that her Dublin engagement was not entirely successful. At least some of the blame must be ascribed to the stage management, for we read that in the first performance of *The Innkeeper's Daughter* on 19 January, 'the mechanical blunders were nearly as numerous as in *The Maid and the Magpie* on Saturday night. On both occasions Miss Kelly's best exertions were frequently marred—and it was quite apparent that she had not only to sustain the principal character, but to superintend and direct the minutest details of the stage business'.[46] Both she and Horn took part in the latter's opera, *Rich and Poor* (text by M. G. Lewis) which was performed at Crow Street for the first time on 31 January 1818, and was said to have been dramatically, at least, a complete success. We read, 'there is some good singing and some rather laughable incident[s], and one scene of most excellent acting by Miss Kelly. There was not a dry eye in the house during the interview between her and Rivers, her father'.[47]

On 23 May, Crow Street Theatre endured yet another uproarious evening when John Davy's and Isaac Pocock's opera, *Rob Roy Macgregor; or Auld Lang Syne**

---

* It had had its first Crow Street performance on 11 May 1818, when it was reported, 'Notwithstanding the length of time which this piece has been in preparation, some of the performers were grossly imperfect in the dialogue, and the Prompter did not appear to be very active in his exertions to refresh their memories'. (*Freeman's Journal*, 12 May 1818)

was performed with Horn in the cast. The work, after Sir Walter Scott's novel, enjoyed the felicity of having some of Burns' and Wordsworth's poetry* inserted for three of its songs, while Davy had drawn upon Scottish airs, including 'Corn rigs are bonnie' and 'The White Cockade', for most of the music. It was followed by an entertainment called 'A Minstrel's Summer Ramble', which was modelled on Incledon's entertainments of previous years. This consisted 'of Recitations, Serious and Comic, interspersed with a variety of the most Popular Songs, Duets and Glees, to be sung by Messrs Horn, O'Callaghan, Hodson, D'Arcy, O'Rourke and Mrs Lazenby, to be prefaced with a Poetical Explanation... The whole to conclude with a Poetical Apology. The Recitations written and to be spoken by an Amateur of Fashion'.[48] Our 'Amateur of Fashion' may have been an eccentric named Luke Plunkett, who believed himself to be a great tragedian. The audience indulged him for the mirth and ridicule he provoked, the management engaged him, because—like a theatrical eccentric of more recent times, Florence Foster Jenkins—he attracted an audience. On this particular evening, however, the amusement palled, and the disturbance which followed was fully recorded in the following day's press.

To begin with, *Rob Roy Macgregor* 'passed off with the same indifference as upon its previous Representations. The House was but thinly attended, but there were some fashionable personages in the boxes, assembled no doubt for the purpose of witnessing the exhibition of the "Amateur of Fashion"'.[49] A considerable delay followed the performance of *Rob Roy*, and the audience became impatient. At length the curtain rose again to reveal a stage, bare except for a piano behind which stood an accompanist, 'a Performer unknown', some chairs on which were seated three of the singers and, to one side, a table 'upon which the Amateur of Fashion reclined: before him, a music stand'.[50] The Amateur then came forward and 'proceeded to recite his "Poetical Explanation"'.[51] The audience laughed for some time heartily and good-humouredly; but at length [grew] weary of the *entertainment* which had been provided for them'.[52] Some songs, glees and instrumental items were then given by the other artists, 'but the "Amateur" being so completely unsatisfactory in his delivery, the general opinion of the House seemed to be that the Entertainment should not proceed further, and a *finale* was about to be

---

* 'O my Luve's like a red, red rose' and 'Auld Lang Syne' by Burns, and 'Rob Roy's Grave' by Wordsworth.

attempted when a Gentleman from the Pit called for Mr Horn; this was re-echoed throughout every part of the House and after some time Mr Farren appeared on the Stage, and stated, that he had sent to Mr Horn's dressing room, and received for an answer from his dresser that Mr Horn "had gone home ill". The Gentleman replied that "Mr Horn was not ten minutes before in the Orchestra, apparently *in perfect health*, and also laughing heartily at the performance; his name was in the bills to sing in the Entertainment...the Public must make a stand against these innovations and disrespect cast upon them..." [great shouts of applause].

'A Gentleman from the Pit then inquired from Mr Farren would he promise, that the Proprietor or Patentee would visit with reprehension upon Mr Horn this dereliction of his duty to the Public? Mr Farren retired to the Stage door, and returned, saying, "If it is the wish of the Public, Mr Horn shall be discharged." Several persons vociferated, "He shall never appear again!" Others said, "Do not dismiss him without trial; let him appear and make an apology"... Loud cries ensued, "Is Mr Horn in the House?", to which Mr Farren said, "He was not...he had seen him certainly about half an hour before." The audience generally exclaimed, "Send for him! we will wait! Let him explain his conduct if possible". Mr Farren retired, and the Audience waited for nearly twenty minutes in the most peaceable manner; at length Mr Farren re-appeared, and stated that the Prompter had been to the lodgings of Mr Horn, but he was not at home: "Although he was so ill and obliged to go home?" said a Gentleman from the Boxes; another Gentleman from the Pit said, "Mr Horn shall not appear on Monday next without an explanation..." A Gentleman from the Boxes observed that this was not the only instance of disrespect manifested by Mr Horn towards the Dublin Audience, as he was almost invariably imperfect in the characters he attempted to perform—['True, true', exclaimed several from the Boxes and the Pit]... Mr Farren expressed himself to understand that Mr Horn being apprized of the circumstances should necessarily furnish some statement in extenuation of his conduct; there were loud cries of "dismiss him, dismiss him", others for "apology, apology", the latter seemed to prevail, and after an acknowledgment from the House of the rectitude of Mr Farren's conduct, the Gentleman retired from the Stage; an ineffectual effort was made to conclude the "Summer's Ramble" by 'God Save the King'; but the Audience being put out of temper by the previous exhibition requested that the curtain should drop. The "Amateur of Fashion" remained at the wing and bore his failure with much apparent good temper'.[53]

On the morning of 25 May, Horn published a statement in the newspapers which, however, was more conciliatory explanation than affirmative apology. He was 'extremely sorry', 'he had performed an arduous part in the Opera', the Manager (Farren) 'had arranged it for the purpose of rest that he should not appear until late in the Entertainment. He had gone round to the Orchestra after the commencement of the Recitation to observe its reception, and having seen that it produced strong and general displeasure, he had too hastily concluded that the House would not allow it to proceed and that his services would not at all be called for'. He had also been 'feeling extremely indisposed' at the time. He regretted 'the circumstances exceedingly'.[54]

But a Dublin audience was not to be fobbed off with that tongue-in-cheek response, and on the same evening of 25 May, when *Rob Roy* was due to be repeated, 'the Theatre was visited by almost every person who was there on Saturday night, with a great number of very respectable Citizens who wished to see the end of the affair'.[55] They came away mollified, for 'after the Overture to *Rob Roy* was performed, and before the commencement of the Opera, Mr Horn came forward to address the audience. He had learned, we presume, that the preceding very injudicious apology would not be acceptable, and he therefore, with great propriety, took another course. He found that he had set up a very bad defence, in as much as an Actor has no right to take any step founded upon an opinion as to the probable success or condemnation of the Entertainment in which he has a part—it is his duty to be at his post, there to use his best exertions to save the Piece, and not to promote its total condemnation by his absence… Mr Horn…very properly…admitted the impropriety of which he had been guilty, and stated that he [had come] there to apologise, expressing his sorrow, and at the same time declaring that *intentional* disrespect to the Audience was far from his thoughts. He was proceeding in this strain when a unanimous cry from all parts of the House of "Enough, Enough" relieved him from his unpleasant situation, and on his retiring from the Stage he received a general plaudit of approbation'.[56]

So, acrimony and oratorical drama ended in reconciliation and domestic comedy. Yet Horn must have sensed the warning expressed by Dr Johnson, and declaimed by Garrick 70 years before:

AN IRISH LARK AND MASTER BALFE 1817-1818

> The drama's laws the drama's patrons give,
> For we that live to please, must please to live.

Frederick Jones had always been too proud to take heed of it, and soon his pride would destroy him.

An artist, not previously mentioned, who was announced to take part in 'A Minstrel's Summer Ramble', must rank pre-eminent, for in time he would bring international renown to Irish Opera—he was Michael William Balfe. On this occasion he was advertised to perform a concerto on the violin composed by himself, although in the resulting *brouhaha*, there is no evidence he ever did.

Balfe's biographers, Charles Lamb Kenney and W. A. Barrett, have interlarded the facts of his early life with so much fancy that it is necessary to sift one from the other. He was born on 15 May 1808, at No. 10 Pitt Street (now Balfe Street) in a house not long since demolished. His father, William Balfe, was a dancing master who began to hold classes annually in Wexford, sometime between the autumn of 1810 (in which year a Mr Cunningham, also from Dublin, had opened his dancing school there on 12 February)[57] and the summer of 1814,[58] Balfe senior would arrive in Wexford with his family in June or early July, remain for about six months,[59] and conclude his season with an Annual Ball at the Wexford Assembly Rooms. In an advertisement for the ball on 4 December 1815, he informed 'the Nobility and Gentry of Wexford' that his pupils would 'dance a Ballet, Tambourine and Castanet Dances, Minuets, Waltzes, Hornpipes, Reels, &c &c. The Rooms to open at Half-past 6 o'clock, and the Dances to begin at Half-past 7. Country Dances to commence at 10. Tickets, Five shillings each, to be had at any of the Schools, and of Mr Balfe, at Mr Gainford's, Main Street'*.[60] It is certain that he visited Wexford during the years 1814 to 1817, a period which covers his son's very early formative years.[61]

As recorded in *Grove's Dictionary* (fifth edition), while in Wexford the child Balfe may have had lessons from Joseph Halliday, band master of the Cavan Militia, and inventor of the Key bugle, for during 1813/14 the Cavan Militia was garrisoned in Wexford, where, in November 1813 'with the Yeomanry corps belonging to this town [they] marched to the Bridge, where they fired a

---

* The Gainford family (also spelled Gainfort and Gainsford over the years) were cordwainers or saddlers. It has been impossible to establish the exact location of where they lived during the period of the Balfe family's visits to Wexford.

feu-de-joie in honour of the late glorious news'.*[62] It is almost certain that another Wexford teacher, ascribed to him, a Mr Meadows (*Grove's Dictionary*, fifth and sixth editions) was T. P. Meadows, a music teacher and piano tuner, who in 1833 returned 'his most respectful and unfeigned thanks to the Nobility, Gentry and all his kind friends for the patronage he has experienced for a number of years... Mr Meadows has spent all the property which he earned in this county (amounting to nearly four thousand pounds) amongst its inhabitants... [He] will attend as usual to give instruction in town and country on the Piano-forte, Violin, &c, &c'.[63] *The Wexford Independent* of 26 January 1889 also affirms that Balfe 'was first taken in hand by the band master of the Militia. Then Mr Meadows, a leading professor...took him under his charge'. The tradition that he composed and scored a polacca for a Wexford militia band in 1815 is probably apocryphal,** the story that his very first appearance in public was as fiddler at his father's dancing classes where, because of his extreme youth, he played on a kit instead of a full-sized violin, is more probable.

It is generally accepted that in 1817, or perhaps as early as 1816, Balfe's father placed him for instruction with William O'Rourke, the deputy leader of the Crow Street orchestra, who later anglicised his name to Rooke and gained belated success with his opera, *Amilie; or, The Love Test*. Both Kenney and Barrett assert that Balfe's first public appearance in Dublin was in May 1816, at the Royal Exchange (now, City Hall) where he played a concerto by Mayseder. No press announcement of this concert (nor of any concert about this time at the Royal Exchange) can be found; nor so far has any other record of it been traced. His name is, in fact, first found in the Dublin papers on 30 May 1817, when he appeared at the Rotunda in a benefit performance for his teacher, O'Rourke, and is announced as 'Master Balfe, aged only Seven Years, whose performance on the violin is said to entitle him to the appellation of Phenomenon'.[64] He is reported again for 20 June at Crow Street Theatre, now described as 'that extraordinary Child' and advertised to play 'a Concertante' by James Barton, leader of the orchestra, in a concert for Barton's benefit'.[65] Five evenings later, he was again playing at a Crow Street benefit, this time

---

* Presumably of Wellington's victory over Marshal Soult, and the final expulsion of the French from Spain, in the Peninsular War.

** His first confirmed composition, a ballad, 'The Lover's Mistake', words by Thomas Haynes Bayly, was published when he was 14, on 24 December 1822, by Isaac Willis, at 7 Westmoreland Street, Dublin.

for Henry Willman, the trumpet player.[66] A report in *Saunders' News-Letter* on 30 June that this was the 'third time' he had 'appeared before the public' is forcible evidence that 30 May 1817 marks the date of his first appearance. His success on 25 June was enormous, and *Saunders' News-Letter* continues: 'We really believe a more extraordinary exhibition of musical taste and talent in a child was never witnessed. The greatest amateurs, the most eminent professors, many of whom we perceived were present, and the audience generally [were] alike astonished and enraptured. The amazing execution, the delicacy, grace, and sweetness, with which this infant performer played the beautiful Irish air of "The Minstrel Boy", was strongly testified by the most unbounded plaudits'.

In January 1818, the Balfes were living in Dublin at 'No. 2 Hamilton's Row* (Near Merrion Square)' where William gave dancing lessons on 'Monday and Thursday Mornings from 9 to 12 o'clock, and in the Evenings from 7 to 10'.[67] Between July and December 1817 Michael Balfe's name cannot be traced in the Dublin papers, (was he in Wexford with his parents during this time?), but on 14 January 1818 he and O'Rourke are to be found performing at a concert in the Rotunda 'for the benefit of a Distressed Citizen and a numerous Family',[68] on 2 February he was at the same venue taking part in 'Mr d'Arcy's Concert',[69] and again, same place on 6 March in one of Isaac Willis's (of the music shop) concerts.[70] His next appearance was on 21 March at Crow Street,[71] and again at the theatre on 27 March, when he took part in a benefit for distressed 'free and accepted Masons'. On this occasion, 'the Brethren of the Order were very numerous in attendance—the Duke of Leinster attended, in all his Masonic Decorations as Grand Master—his Grace bestowed much kind attention on the Musical Prodigy, Master Balfe—he had him in his box, previous to his astonishing performance of a most difficult Concerto of Yaniewicz's on the Violin'.[72] On 17 April he was once more at the Rotunda,[73] and on 13 May, at Crow Street.[74] He is now described as 'Mr James Barton's pupil'.[75] Next there followed two further appearances, the first at Crow Street on 28 May, in a performance 'to create a Fund for the Erection of a School-house, for educating the Orphans under the Protection of the Patrician Society, and 200 other Poor Children',[76] the second on 5 June at the Rotunda,[77] a benefit for Horn and Mrs Willis, when Balfe played 'Rode's Air'. With at least nine professional appearances in Dublin to his credit within five months when barely ten years

* Off South Cumberland Street.

old, Master Balfe seemed set fair for stardom, or, were it today, his parents for prosecution by the I.S.P.C.C.

Returning to the opera programme for 1818, we find *Lalla Rookh; or, The Minstrel of Cashmere*, 'The music composed and Dedicated to Lady Morgan' by C. E. Horn;[78] the libretto, an adaptation of Tom Moore's oriental poem* by M. J. O'Sullivan. It was performed for the first time anywhere at Crow Street Theatre on 10 June 1818, and the following day a critic, having launched his review with soothing words—music 'extremely pretty', 'many striking incidents of a Dramatic cast', 'dialogue easy and flowing', etc., etc., then picked up his scalpel: 'However, there is a want of incident', he continues, 'an unpardonable fault with those who cannot admire the simplicity of former times, but give themselves up to the vitiated taste of the present'. Finally, he relents, adding, 'It is well got up as to Spectacle, Scenery and Dresses, and with some judicious alterations, which will, no doubt, occur to the author, *Lalla Rookh* will become a favourite entertainment'.[79] It did not!

Reading of it today, the outstanding feature of the performance seems to have been Tom Moore's presence in the audience. He had travelled to Dublin to attend a dinner in his honour, which took place in 'Morrison's Great Rooms' on 8 June, when 'About two hundred and twenty most respectable persons' were present with Lord Charlemont in the chair.[80] (Perhaps the reason for presenting the work was his intended visit to Dublin) At *Lalla Rookh*, 'He sat in the back part of the Manager's box, but it was quickly observed that he was in the house, and as soon as the discovery was made plaudits resounded on every side, and there was a general call for Mr Moore. After a short hesitation he came to the front of the box, and laying his hand on his heart, bowed frequently. He was greeted with the most enthusiastic and long continued cheers and applause, and these manifestations of public regard were repeated in every interval of the performance, and were each time most gracefully acknowledged'.[81] Knowing Moore's private opinion of his fellow citizens, expressed in a letter to his friend, James Corry, as 'the low, illiberal, puddle-headed, and gross-hearted herd of Dublin (that "palavering, slanderous set", as Curran once so well described them to me)',[82] one wonders what thoughts were concurrently running through his mind.

* Later set by Spontini as *Nurmahal, oder: Das Rosenfest von Caschmir* (1822); by Félicien David (1862); by Anton Rubinstein as *Feramors* (1863); and, as *The Veiled Prophet of Khorassan* (1881) by the Irish composer, Charles Stanford.

Fifth Edition.

THE LOVERS MISTAKE,
A BALLAD, Sung by
MADAME VESTRIS,
with the most Enthusiastic Applause in
Mr Poole's Popular Comedy Paul Pry.

The Words
BY
T. H. BAYLY ESQ^r.

The Music
BY
MICH^l BALFE.

Ent. at Sta. Hall.   Price 2/-

London, Published by I. WILLIS & Co Royal Musical Repository, 55 St James's Street,
and 7 Westmorland Street, Dublin.
Where may be found every variety of
GRAND, CABINET, COTTAGE, SQUARE & CIRCULAR PIANO FORTES, HARPS, &c.
from the most Eminent Makers at the lowest Manufacturers Prices
also Just Published the following admired Songs sung by *Madame Vestris*, viz.

Cherry Ripe,	C.E. Horn, 2 0	White & Red Roses, J. Barnett, 2 0	O hear my sweet Guitar, C.E. Horn, 2 0
Vauxhall,	J.A. Wade, 2 0	The Bonnie wee Wife, Mr Miles, 2 0	Sing on, J.C. Clifton, 2 0

Throughout the season there had been many signs of theatrical unrest among both artists and audiences (the episode with Horn was but one example) to add to Frederick Jones's ever-increasing difficulties. An indication of the former was the announcement as early as January 1818, that William Farren was to leave the company at the end of the season, for Covent Garden, 'after spending *nine* successive Seasons on the Dublin Stage'.[83] Not only was he leaving Crow Street as an actor, he would also be leaving the theatre without a manager. As for Jones, having assigned an interest in the theatre to Captain Maurice Fitzgerald in 1816, he was obliged in 1817, by a deed dated 9 June and registered on 8 March 1818, to convey and make over 'unto James Taylor and Myles O'Reilly* Esqrs. all his right, title, and interest, in and to six and one-half *twelfth* parts of the Theatre Royal, Dresses, Machinery, &c as also of the Letters Patent, granted to him and his Assigns, and all renewals of said Letters Patent *to be granted*; as also his right, title, and interest in and to one-eighth part of the Theatre Royal and Letters Patent, renewals of same, Dresses, Machinery, &c formerly assigned in trust to Maurice Fitzgerald Esq.; as also his right, title, and interest in and to the house and land of Clonliffe'.[84] With even his home mortgaged, Frederick Jones was, in effect, now bankrupt.

Other occurrences, ranging from blatant misrepresentation to the bizarre, continued to reflect the general uncertainty. An example of the latter was the soldier who, one evening in February, 'drew his bayonet on a Gentleman in the Middle Gallery'. It was recalled 'that Lord Whitworth, when Viceroy of Ireland, manifested his displeasure at the introduction of Soldiers into the Theatre'.** and the caution was added, 'One would suppose that there is a wish, by such means, to drive the Audience beyond that line of moderation and decorum they so properly observe, and which is best calculated to insure them success in their efforts to reform the system of this ill-directed and shamefully mismanaged Theatre'.[85] As if Jones was personally to be held responsible for the ill behaviour of presumably drunken soldiers while on guard duty.

More serious, because more vindictive, were attacks and censures made directly against him. Two such appeared in the press during March. The first

---

* ? Jones's attorney.
** On 3 April 1818, it had been 'resolved that it is the unanimous opinion of the Board of Magistrates that the High Constable is the proper officer to have the direction of placing a guard at the Theatre Royal and at all other public places of amusements'. (*Calendar of Ancient Records of Dublin* Vol XVII. Dublin 1916, Appendix IX, p 525)

declared: 'Having looked these ten or twelve years at the management of the Theatre Royal, we have at length come to the conclusion, that the only effectual remedy for the grievance would be the establishment of another Theatre…it would rouse the dormant spirit of Dramatic talent amongst us—we should not be pestered with such trash as the new Pieces lately produced, and which the excellent acting of Miss [Fanny] Kelly could scarcely save from utter damnation'.[86] The second admonished 'Mr Jones [that] were he to value his own peace, and consider the interest of himself and family, [he] should never wish to have anything whatever to do with the Management of the Theatre'.[87] Whatever friends Jones may have had, he certainly did not lack enemies, especially amongst the press.

The year continued in the same vexatious fashion. Catherine Stephens commenced a season on 25 July playing Rosetta in *Love in a Village*, with Charles Horn as Young Meadows. The press reported, her admirers 'will perceive some alteration, and for the better, in her person and countenance, and very considerable improvement in her style of acting. Of her voice, it is sufficient to say that she retains it in its full perfection, and that she continues to manage it with all the sweetness and taste which have gained her so much approbation'.[88] Her season continued with all the favourite operas from previous years, including, on 13 August, *Artaxerxes*, 'the first time for several years',[89] which 'brought one of the fullest and most fashionable houses' seen that season.[90] She had 'dressed the part of Mandane superbly, and with strict attention to propriety: her head and neck ornaments displayed a profusion of diamonds of great value, enriching Cameos, very tastefully disposed of. The whole, combined with the costume, had a most brilliant effect'.[91] A contretemps then arose about her performing in Limerick, where the theatre was controlled by Montague Talbot, the former Crow Street actor, and now Belfast manager, with the statutory exchange of recriminatory letters in the press. The hiatus led to her being re-engaged for a further short season at the Crow Street Theatre, beginning on 28 August. All went well until 7 September when 'after repeated announcements and repeated disappointments'[92] *The Haunted Tower* was performed, and, once again, there was pandemonium.

The bills had stated that Miss Stephens would introduce a song 'Haste idle time', 'rather pompously put forward and said to be "the composition of Mr Hodson". This song, however, was omitted, and another substituted without any explanation or apology'.⁹³ Musically the composition cannot have been of any great loss, but once again the audience deemed that their rights had been infringed by its replacement, and no doubt encouraged by Hodson,* promptly dealt with the implied affront in typical, arbitrary fashion. We learn that 'when the *Finale* commenced, a call for "Haste Idle Time" became general, and the Curtain fell amidst hisses, and demands for—"The Manager—Song—Miss Stephens—Hodson's Song—Apology to Mrs Hodson** &c. &c."—The clamour continued, and Mr Horn presented himself before the Curtain, but he would not be listened to and he retired. The call for "Manager—Song—Apology" was renewed, and again Mr Horn appeared, leading on Miss Stephens, who was greeted on her entrance with a burst of applause. Mr Horn then said that "Miss Stephens was so indisposed in the *Morning* that she was unable to rehearse the Song". Having said so much, he instantly, and in a very ungracious manner, hurried Miss Stephens off the Stage. This seemed to increase the storm, and the call for "Manager—Song—Who is the Manager?" [Farren had by then left for Covent Garden] became louder than ever. Mr Wallace then appeared, but having persisted in his endeavours to obtain a hearing, much beyond the point of prudence and propriety *he was pelted off!*' Wallace had the temerity to return, although this time accompanied by Miss Stephens and 'said that Miss Stephens would sing any song except "Haste Idle Time", but *that* she was unable to sing. Here there was a cry of, "No, no, enough as to Miss Stephens; no Song but Hodson's; your apology will do as to Miss Stephens but send the Manager—off, off". A few silly *Dandies* in the Stage-box on the right of the House, very injudiciously encouraged Miss Stephens to sing. She came forward for the purpose, but the Audience would not hear her, and she was obliged to retire evidently much affected… The tumult, and call for "Manager", now became louder than ever, and the thunder rolled till

---

* Mr Hodson was George Alexander Hodson, a song composer, whose best remembered air is 'Tell me, Mary, how to woo thee'. He was engaged at Crow Street Theatre as a minor singing-actor and occasional music arranger. While in Dublin he also set up as a music teacher. *Saunders' News-Letter* of 10 November 1818 announces: 'Musical Academy. Mr G. A. Hodson respectfully begs leave to inform his Friends and the Public that he has just returned from London with an assortment of the most esteemed and fashionable Music; and at the same time to acquaint his Pupils (and those who may honor him with their commands) that his Academy (for Singing and the Pianoforte) commences on Thursday the 12th instant at No. 5, Dame-street'. Mrs Hodson, his wife, was an actress with the company.

** Frederick Jones, junior, was said to have insulted her.

the Curtain drew up for the farce of *The Review* when, no manager appearing, the Actors were ordered off; and, as they declined to retire, missiles of all description were thrown from *all parts of the House*, and the Stage was soon covered with them. The Actors continued, very imprudently, to come on in dumb show; and at last, most unfortunately (and we mention it with deep regret), Miss [Lydia] Kelly was hit with some hard substance in the face, when she fell on her knees and was carried off the Stage, senseless'.[94] *The Freeman's Journal* adds, 'When we left the Theatre order was not restored and, we learn, it was a very late hour before the house was evacuated'.[95]

Saunders' News-Letter strongly condemned the affair, declaring: 'There was a total want of manhood and generosity in the conduct of those who wantonly disturbed the House on this occasion... We regret exceedingly the disgrace brought upon our character by this very abominable proceeding, and which the imploring softness and the mild and affecting resignation of Miss Stephens could check only during the time she remained on the stage. We hope most sincerely we shall have no recurrence of it'.[96] That hope, alas, would prove to be in vain.

The immediate cause of the uproar seems to have been an altercation which had occurred earlier between Hodson and Frederick Jones, junior, when Jones had ordered Hodson out of the theatre. Whatever the reason, the intention seems quite emphatically to have been an ongoing and concerted attempt by a well-organised group to force Jones senior to relinquish his patent and his theatre. It is obvious that he was now no longer in control there, neither on-stage, back-stage, nor in the auditorium. But if Jones was no diplomat neither was he a coward, and he would stubbornly continue to fight his losing battle for some time yet.

Catherine Stephens' engagement concluded with a farewell benefit on 11 September, when she set off for Worcester,[97] returning to Ireland later in the month to perform in Cork. On her journey to Cork she stopped off at Kilkenny, where she gave two performances on 28 and 29 September.[98] 'A misunderstanding' arose between her and the Cork manager, for she considered she was 'not bound to perform in After-pieces there' as she had done in Dublin, on the plea that in Dublin she had played 'on shares', while in Cork she had only 'a limited sum' of £35 a night.[99] Accompanied by Charles Horn, she next travelled to Limerick where they commenced a short season on 22 October.[100]

She was soon back in Dublin again, but not to perform at Crow Street Theatre. Instead, 'under the immediate Patronage of His Excellency the Lord Lieutenant, and the Countess Talbot' she had returned 'to give Four Concerts in the Ball-Room at the Rotunda, and to appropriate One Fourth Part of the Profits of the Whole to the Funds of the Association for the Suppression of Mendicity'.[101] As in Cork, so also in Dublin there seems to have been 'a misunderstanding' between her and the manager—here, Frederick Jones. 'The rumour [was] that he [had] invited Miss Stephens to perform at Crow-st and as she would not comply, he was determined to do her as much mischief as possible'.[102] The outcome of this was that Charles Horn and James Barton, both members of the Crow Street company, having been engaged to assist at the concerts had been refused permission by Jones to do so. This led to the usual contention in the press. Addressing himself to Edward Stephens, Catherine's brother and apparently her manager, who, very likely, was the instigator of her managerial misunderstandings, Horn wrote, 'In relying on the good understanding between Mr Jones and me, I concluded no obstacle would take place to my singing for Miss Stephens's Concert *during the close of the Theatre*, but I find on my arrival in town that Mr Jones does most positively object to my singing for *that lady in particular*, as upon any other occasion, he says, I should be perfectly free'.[103] Barton also wrote to Edward Stephens, confiding, 'I have made every possible effort to induce Mr Jones to give his consent to my performing at Miss Stephens's Concerts but without success; he is determined that for *her* I shall not play, or if I do, contrary to his wishes, that my engagement at Crow-st Theatre will be cancelled'.[104]

Robert White, Sheriff and Secretary to the Mansion-House Committee, who now tried to intercede, appealed to Jones as follows: 'Sir, The Committee of the Association for the Suppression of Mendicity recently assembled at the Mansion House, deeply interested in the success of the Concerts which have been announced to commence this evening at the Rotunda—a considerable portion of the Receipts of which is destined to support their Funds—having understood that several of the Performers engaged at Crow-st Theatre for the ensuing season, and whose names had appeared in the list of the persons to perform at those Concerts, feel some delicacy in giving their services to the latter, without your express concurrence, requested of the Lord Mayor to apply to you to obviate this difficulty; and his Lordship, therefore, earnestly entreats, in their name and on behalf of the Institution with which they are connected, that you will be pleased to extend to the individuals in question the sanction they require, so that nothing may obstruct the very promising productiveness of the Concerts for a great object of public charity'.[105]

In his reply, Jones (according to his address, still living at Clonliffe House) was polite but adamant. The opening of his letter, 'Sir. Not feeling myself under any compliment to Miss Stephens, I determined upon her Concerts being first mentioned, that those Persons under engagement to me, should not perform in them', sets the tone. He even shows a fine Italian hand at outwitting the opposition in the following summing up. 'I request you may present my respects to the Lord Mayor and Gentlemen of the Committee, and inform them, that although I find myself so circumstanced as to be under the necessity, very reluctantly, to refuse their request, I should hope they will think I make ample amends to the Charity over which they preside, in the offer of the Theatre Royal for one play in every month during the approaching Season, on the terms of giving them *one half of the Profits*, instead of *one-fourth*. The Theatre, which has been newly and most splendidly fitted up at a great expense, will open for the season on Monday next, or as early in the week as the workmen can be got out; and Miss Stephens's Concert, independent of other considerations, must interfere most materially with that Establishment'.[106]

But as ever, where stars are concerned, Catherine Stephens's supporting artists were of little consequence and easily replaceable. Horn by G. A. Hodson, and Barton by A. W. Bowden from Cork, who, although 'generally speaking a good player', after Barton, 'must have appeared to some disadvantage before a Dublin audience'.[107] Stephens' presence, irrespective of who was supporting her, ensured that the 'first concert [on 12 November] was most numerously and fashionably attended', with 'no fewer than from eight to nine hundred persons in the room',[108] and at her last concert on 5 December, 'There were no less than 1700 persons present…the largest number yet assembled at the Great Room of the Rotunda on a similar occasion',[109] and in the course of the evening, 'from one guinea to five were offered for single tickets' originally priced at 8s 4d.[110]

Miss Stephens did have assistance at this concert to help swell the numbers, although not among the artists. Instead, in the audience was the Archduke

Maximilian of Austria,* then on a visit to Dublin. Before the concert, we read, the stewards were 'preparing the accommodation suitable to a personage of his very exalted rank'.[111] This meant enclosing, and otherwise suitably preparing, part of the room 'in front of the audience and but a little distance from the Orchestra' where he and his suite would be seated, and 'within which also seats were placed for the Commander of the forces, and several of the Nobility. On the arrival of his Royal Highness every mark of respect…was observed towards him, and after encountering some little obstruction in his passage through an immense crowd, anxious to see him, he was shewn to a seat placed as conspicuously as propriety required'.[112] During the concert 'Miss Stephens sang the solo part of the Austrian Hymn, "God Save the Emperor" which had been selected in compliment to the Archduke'.[113] For it, appropriate words had been written 'by H. B. Code Esq., a Gentleman not a little celebrated for his poetical compositions'.[114] Each stanza concluded with a fine strain of harmony in which the whole choral party united'.[115] It had, it seems, passed unnoticed that the Austrian hymn or national anthem, 'Gott erhalte Franz den Kaiser', honouring the Archduke's cousin, Franz II, had been composed by Joseph Haydn. It remains as the theme for the second movement of his Emperor Quartet, and as the national anthem of Germany.

---

* Maximilian Joseph (1782-1863), grandson of the Empress Maria Theresa of Austria. It was noted, 'His Imperial Highness seems extremely intelligent and observing, speaks English tolerably well'. (*Freeman's Journal*, 7 December 1818). Indeed, it was quite a year for dukes in Dublin, Arch and Grand, for in August there had been a visit from the 20-year-old Grand Duke Michael, brother of Czar Alexander I of Russia. He, too, was received with much ceremony and stayed at Morrison's Hotel, where 'a splendid suite of apartments, consisting of fourteen rooms, had been engaged for him'. (*Saunders' News-Letter*, 24 August 1818). There was some reason to think he would be 'one of Miss Stephens's audience' attending *The Beggar's Opera* and *No Song, No Supper* on Monday, the 24th, (*Saunders' News-Letter*, 22 August 1818), but ungallantly, on that evening he dined with the Lord Lieutenant instead. (*Saunders' News-Letter*, 24 August 1818)

Chapter 10
The Original Don Giovanni 1819-1820

The 1819 season at Crow Street Theatre—officially, the last ever to take place there—commenced on Monday, 7 December 1818, with a performance of *The Haunted Tower*. The principal singers were Miss Byrne, making an appearance 'after an absence of two years', Miss Rock who 'had that kind and cheering welcome which mark her popularity', and Charles E. Horn, who 'was as usual, imperfect—even the words of "On this cold flinty rock" he did not give correctly'.[1]

Shortly after, *The Freeman's Journal* had announced: 'Every exertion has been made (and it is hoped not without effect) to render the Orchestra greatly superior to what it has been heretofore. Among the new engagements in this department are Mr O'Rourke, one of the most eminent Violin performers in the Kingdom, Mr Samuellini from London, as 2nd violoncello, Messrs Peterson, Meyer, and Kemp, late of his Royal Highness the Duke of Kent's Band… The number of performers in the Orchestra [has] been increased to twenty-seven. The Chorus, which it has hitherto been found difficult to fill respectably, are now put under Mr Hammersley, the first instructor of the famous Lancashire Chorus Singers, so well known and admired in the Sister Kingdom. The Scene painting still remains under the direction of Signor Marinari, with four assistants of eminence. The Gentlemen's Wardrobe has been put under the superintendence of Mr Carty, who was employed seven years in the wardrobe of Covent Garden… In consequence of a recent custom of admitting strangers behind the scenes, greatly to the inconvenience of the performers, the retardment of the business, and dissatisfaction of the audience, the public are most respectfully informed, that for the future, No Person Whatsoever will be admitted behind the curtain who is unconnected with the arrangements of the Theatre'.[2]

Conversely, *The Dublin Evening Post* was anything but cordial, writing: 'Last night our "well regulated Theatre" had its doors opened for the *first time* this season to admit company. This is one of the natural consequences of that monopoly which has left the Citizens of Dublin without a second place of Dramatic Amusement, and subject to the whim, the caprice, or inability of

Mary Byrne as Adela in *The Haunted Tower* (Storace)

whoever may hold an *exclusive Patent*. However, at last, in the middle of the winter the people were admitted into *their* House, which in ridiculous advertisements, and not less foolish letters, was stated to have been fitted up with unusual splendor, and at an enormous expense, for their accommodation'.[3]

Indeed, a review of Dublin newspapers of the period indicates what amounts to a taking up of positions both for and against Frederick Jones. The theatre was being redecorated prior to reopening and *The Freeman's Journal*, predominantly pro-Jones, had reported on 14 November 1818: 'We understand the Theatre Royal is undergoing a total repair; the inside of the house consisting of the audience part will be entirely new, even to the wood work. The fronts of the boxes will consist of richly embossed gold ornaments on a white and pink ground, surrounded by splendid gold mouldings; the hand-rail on the front of each box to be covered with light blue velvet, the ceiling and proscenium are decorated in the best taste of Marinari's classical pencil'. With slight reservations, the same paper agreed on 8 December that the work had been well done, writing then: 'On first entering the house we were much struck with the general effect, and particularly by the manner the ceiling was painted in, which does great credit to Mr Marinari so far as the execution of it, but the part intended for admitting the chandelier is not placed directly in the centre of the circle, and therefore looks extremely awkward…we were greatly pleased with the fanciful distribution of the ornaments on the panels of the boxes, but are of opinion that the colour they are placed on is rather too glaring and fatiguing to the eye; a much lighter shade would be far better, and we are inclined to think that a light blue might be advantageously substituted in preference to either. The proscenium looks extremely well, and is finished in a very tasteful style. A very injudicious alteration has been made in the seats of the pit, which are deprived of their green cloth covering and are stained in imitation of mahogany; besides the inconvenience of being extremely uncomfortable to the feel, they will have a very bad effect to the sight when the pit shall not be entirely full; we would recommend them to be settled in the way they were formerly. Several changes for the better have taken place in the pit passage, particularly the light at the entrance, but we cannot approve of the place to which the money office is removed…as we are convinced it will be attended with serious accidents in case of a rush on a crowded night'.

Saunders' News-Letter, objective in its approach, was of the same opinion, adding the information that 'it having been represented to the Proprietors that fault is found with the present state of the Seats in the Pit, the public are respectfully informed, that it was adopted from the usage of the London Theatres to prevent the accumulation of *Dust*'.[4]

But *The Dublin Evening Post* was highly critical. As early as 11 June 1818 it had recorded: 'We understand that the passage leading from the Street to the Lord Lieutenant's Box was in so offensive a state, that the Countess Talbot felt much indisposed after entering the House; and it was, we learn, intimated to the Person acting for the Patentee, that his Excellency would in future be very sparing of his visits to the Theatre'. Now, six months later, the paper was damning. 'The more we look at what the Patentee calls *ornaments* and *improvements*', it declares, 'the more are we convinced that they are *not* ornaments or improvements, and that splendour is libelled when applied to things so paltry, so vulgar, and devoid of taste... But where the result has been so bad—where taste is violated and the wished-for effect destroyed, it is satisfactory to know that the expenditure has been of a trifling amount—for the whole of the *Ornaments* may be, and we believe are, produced by some Red and White Lead, and Upholsterer's Brasses from Kennedy's Lane, which cost but little. The glare [from the lighting] is the most offensive that can be well imagined. There is nothing to admire—nothing upon which the eye can repose with satisfaction, and such must be the *effect* of this mock *splendour*, that the Ladies in the Boxes though splendidly dressed will be thrown quite into the shade; all show of beauty will be lost, and dress however gaudy must fade before the glass-shop red and foundery trappings of this *ornamented*, *improved* and *well-regulated Theatre*... The vile painted deal seats in the Pit remain uncovered, and the dangerous spikes on each side under the lower Boxes claim the attention of *the Police* as well as of the Public generally'.[5] Having protested for days against the 'vile Pot-house state'[6] of the pit seats, from which 'respectable females' had 'nearly altogether withdrawn',[7] it did however on 29 December concede: 'We are always willing to bestow praise where praise is due. The Seats in the Pit have been covered—and, independent of the mere comfort which this affords to the respectable Persons who visit that part of the Theatre, it greatly improves the general appearance of the House'.

To give *The Dublin Evening Post* its due, it may have had a point concerning the condition of the theatre, for on the following 21 January, 'After the ballet…in which the Indian Warriors* appear conspicuously, the front panel of one of the Stage boxes gave way, and three gentlemen were precipitated into the Orchestra: but we are happy to state that they sustained no injury which was somewhat singular and fortunate, as, from the dangerous, and we may add, objectionable range of spikes beneath, they were in danger of being lacerated or suspended from them. It was a lucky circumstance there were no females in front of the box'.[8]

Between 23 January and 6 March, a series of seven concerts were given in the Rotunda by the Demoiselles Annette and Victorine de Lihu and Giovanni Puzzi. Puzzi is easily identified as the outstanding French-horn player who was born in Parma in 1792. He came to London via Paris early in 1817, making his debut there at a Philharmonic Society Concert, and was to remain in London until his death in 1876. All that can be reported of the de Lihu sisters is that they were of German-French extraction, that Victorine seems to have been a soprano, Annette, a mezzo-soprano, and that the items they performed consisted mainly of duets, including 'Crudel! perchè finora' from *Le Nozze di Figaro*! They, and Puzzi, were accompanied by a small string ensemble with flute and pianoforte consisting of seven players in all. Besides playing the horn, Puzzi also sang in at least one vocal duet with Victorine—'Ai capricci della sorte', from Rossini's *L'Italiana in Algeri*, which was then about to be performed for the first time at the King's Theatre, London.

Socially the Lihu sisters must have been well connected, since for their first concert we learn: 'His Excellency the Lord Lieutenant has been graciously pleased to signify his intention of being present' and 'Ladies of Distinction' who had 'Kindly promised to become Patronesses' included the Lady Mayoress, the Duchess of Leinster, the Countess of Charlemont, Lady Rossmore, Lady Manners and others.[9] We further learn that their third concert was numerously and brilliantly attended—nearly one thousand persons of the first distinction honoured the entertainment with their presence'.[10] It was perhaps significant of their social position that in at least two programmes they included the air, 'Partant pour la Syrie' arranged as a duet with a French horn obbligato played

---

* *The Indian Settlers*. 'In the course of the ballet will be introduced the Native American Indian Warriors'.

by Puzzi. It was said to have been composed by Queen Hortense of Holland, mother of the future Emperor Napoleon III, and in time would become the Marseillaise of the Second Empire. Following the Dublin concerts, the three artists set out to fulfil 'their numerous engagements in England and Scotland'.[11]

In late January 1819, *The Freeman's Journal* critically observed that 'with such an Opera Company as we have, it is astonishing that the Manager should not furnish more musical variety than he had thought proper to afford the public during the whole of this season', but had then added confidentially, 'We are glad however to find that a new Opera by Sir John Stevenson is about to appear in which Miss Byrne will perform the principal character'.[12] Stevenson had had in fact an opera titled *The Out-Post*, with a libretto by 'W. S. Jun.'[13] performed on 11 April 1818. While the music could not 'boast of a great deal of originality or variety', it was none the less 'throughout highly pleasing, and, in some instances, beautiful', but the libretto 'possessed the fault of most new pieces—excessive length—many of the jokes [were] vapid, and the plot [set in Switzerland] not very easily comprehended'.[14] In the event, it quickly disappeared from performance.

His new opera, *Anziko and Coanza; or, Gratitude and Freedom*, had a libretto by 'Edward Fitzsimons, Barrister-at-law, Dublin'.[15] It was based on another of Maria Edgeworth's *Popular Tales* (*The Grateful Negro*) and was set in Jamaica. At the end of the opera there was 'a New Asiatic Divertissement by Mons. Simon Sen., Mons. Simon Jun., Mr St Albin, Master and Miss St Pierre, Miss Aylett and Mrs Green'.[16] It seems to have been well prepared, with new scenery for the three acts painted by Marinari, yet, surprisingly, it was greeted with considerable opposition. The first performance took place on 4 March 1819, when we learn that 'it was represented under unfavourable circumstances for a new Piece, the House being very thinly attended… The hostility of its friends and opponents was carried to such an extent as to produce a regular *milling* match (as the pugilists term it) in the Pit, which involved no small portion of that part of the Auditory'.[17] When an attempt was made to repeat the opera two nights later, it 'was received on the drawing up of the curtain with such violent and determined hostility, that after a trifling effort to try its fate, and though it had its friends in the House strongly supporting it, the necessity for its being withdrawn was imperative, which was complied with, and the Opera of *Love in a Village*, which was in readiness for such a result, substituted in its place'.[18] All in all, neither of his new operas did much to enhance Stevenson's reputation.

On 20 March, Crow Street Theatre was illuminated by gas for the first time. This was a considerable advance, not only in illumination, but in 'reduction of odours and of soot from murky lamps on theatre upholstery and decorations, scenery, and even audiences' clothes. Moreover, lamps had had to be replenished with oil, and wicks to be trimmed, apparently even during performances, for we read: 'A person, judging from costume, of the lowest class, is sent round to trim and revive the box-lights, which office he necessarily fulfils at the partition of the box towards the pit, to which he must obtain a passage no matter how it may be crowded, or what inconvenience arises from suffering his progress. To perform this duty with some decorum, as he possibly presumes, his shoes are taken off and he is exhibited in coarse woollen stockings, dirty and offensive, with an exposure generally far from being decent or delicate. A night or two ago, some ladies who had not been prepared for bad impressions on the olfactory sense, found it necessary to have recourse to snuff to counteract the effect of this person's intrusion into a box, and many others in all probability were as strongly affected, without perhaps the aid of so serviceable a remedy'.[19]

But that had been five years earlier, and now at a performance of *Guy Mannering* we learn: 'The lighting of the Theatre with Gas took place on Saturday night [When 'the first Night's Receipts go by contract to the Gas Company, in order partly to indemnify them against a large sum most liberally advanced for the embellishment and ornaments of the Lustres and Lights'.[20]] but the arrangement is not yet perfect. The lustre suspended over the centre of the Pit is exceedingly splendid and beautiful, but the light it casts does not reach some parts of the House, which it is necessary should not be left in any kind of obscurity. The back part of the lower Boxes was particularly gloomy and will require some local lighting in addition to what is derived from the lustre. It is an imperfection of the system, that in consequence of there being, as we presume, but one directing Key to the principal Gas pipe, the Stage lights and the lustre are necessarily either dimmed or illuminated together. This, however, can be reformed'.[21] It was, at least partly, by 14 April, when we are told that 'the lower Boxes are now well lit by new lamps, which hang in the place of the old ones'.[22]

The now almost predictable, one might say, statutory annual theatre riot occurred this year in April, provoked by, of all people, Mary Byrne. The first intimation of approaching trouble occurred on the evening of the 14th, when Mr Montgomery, the stage-manager, was called upon 'to give an explanation respecting the dismissal of Miss Byrne'. His reply that 'the difference was in a state of arrangement, [and that] he might venture to promise that it would be amicably settled' seemed to satisfy the house and was greeted with loud applause.[23] Two days later, however, a letter, a column long, written from 1 Capel Street, appeared in *The Freeman's Journal* in which Mary Byrne outlined the contretemps and the cause of her grievance. She first quotes a not unfriendly letter received from Montgomery on 10 April, which states: 'I have it in direction from Mr Jones, to inform you that he understands you gave your Vocal assistance at Mr Panormo's Concert on *Saturday evening last*. If 'tis true, he regrets to say he is, in justice to the preservation of his own property, obliged to consider your engagement violated, and to look on you as no longer a Member of the Company'.

She thereupon sets forth her side of the case, explaining that while she was in England the previous October she had made an engagement by letter with Mr Taylor, acting on Jones's behalf, when no mention was made of her being prevented from assisting at Concerts. However, on her arrival in Dublin an 'article' was sent to her to sign which prohibited her from assisting at any musical entertainment without Jones's permission. She immediately objected to the clause, but on being reassured by Taylor that it was merely a matter of form, and that he was convinced Jones would never refuse his permission if applied to, she signed the contract. She then sought permission to assist at concerts for the benefit of Mrs Willis, George Hodson, Percival Panormo and Master Attwood. Assistance to Mrs Willis was refused so she had a gentleman friend of her family wait upon Mr Jones and ask his permission to permit her to oblige the three gentlemen. To this request she was informed, 'that to assist at Mr Hodson's Concert he peremptorily refused, but that he had no objection to [her] singing for the others, provided that their Concerts took place on nights for which no Opera was announced at the Theatre'. Accordingly she informed Panormo that she had received permission from Jones, and her name was announced in all the papers for eight days. In fact it was not until sixteen days later that she received, by his desire, Mr Montgomery's letter. She confirms that no opera had been announced for the night of Panormo's concert, and that the injury Jones now seeks to inflict on her, breaks her engagement, besides depriving her of a benefit in Dublin in the coming June, and a free benefit later in Cork.[24]

On the evening that this press announcement appeared, 16 April, 'The West Indian was announced for representation...but almost immediately on the drawing up of the curtain, strong marks of dissatisfaction were apparent, and before the end of the first act a tumult commenced which continued throughout the whole of the evening... There were frequent cries of "Miss Byrne and an Opera", and repeated calls for the Manager, which Mr Montgomery constantly obeyed; he professed the utmost anxiety to give every explanation and satisfaction in his power, but as he would not state that Miss Byrne was a member of the Crow-street Company, every thing he said was unavailing. The seats in the Galleries were speedily torn up and flung into the pit, which was cleared of its occupants in a moment. The velvet-covered hand-rails of several of the upper boxes and lattices were also torn off and thrown into the pit and upon the stage. Whenever there was any appearance of a partial calm, the performers came on and endeavoured to go through their parts, but they were quickly assailed and compelled to retire. Missiles, of the most dangerous description, flew about; a large sharp stone, we are informed, was aimed at Mrs Broad, who, in all probability, would have lost her life had it struck her'.[25]

Sheriff Wood then came forward and made several unsuccessful efforts to restore peace, which merely led to charges that police* had been placed in various parts of the house, and disapproval at his having his sword drawn. He immediately threw down his sword and formally requested the orderly part of the audience to retire and so, about eleven o'clock the house was cleared.[26]

Nevertheless uproar continued for a further four nights. On the following evening the 'Theatre was much fuller' and 'clamour and confusion reigned

---

* The introduction of police into the theatre (ineffective as it seems to have been) was then a cause of grave dissatisfaction to certain members of the audience. In a letter published in *The Freeman's Journal* on 22 April, Sheriff Wood refutes a report which had accused 'certain gentlemen of being the principal rioters'; he had received from them, he explains, much assistance in his 'endeavours to protect the property of the Proprietors, as well as the peace of the Theatre'. The gentlemen alluded to, he confirms, were in the undress uniform of the Corps of South American Patriots, a voluntary body of solvent citizens organised to keep the civic peace in Dublin. It will be recalled that as early as 1804, the *Hibernian Magazine* had energetically proposed, *'a theatrical police'*.

almost uninterruptedly from the opening to the closing of the doors'. Placards were displayed throughout the house proclaiming, 'Friends of Jones be silent', 'The Friends of Order are requested not to Clap or make any Noise' and

> 'Miss Byrne from London
> Came speedily back,
> For there she could meet with
> No Mullinahack'

but a sort of truce was observed only while the dancers were on stage, and while the orchestra played 'God Save the King' and 'St Patrick's Day'.[27] Of Monday 19 April, it was reported, 'Affairs appear to us to grow worse at the Theatre... The noise and clamour was considerably greater than on Saturday night and there were several warm contests between individuals'.[28] Even on the following night trouble continued, but encouragingly, 'the play experienced a greater share of attention than was expected. Between the first and second acts the chandelier was let down (for the first time since the tumult commenced) amidst the general applause of the House'.[29] As on the earlier evenings, placards were once again displayed, but a truce seemed to be in the offing when it was reported: 'Miss Byrne has made a proposal through one of the Papers which, we should hope, must lead to an accommodation between her and Mr Jones. It is as follows: "If Mr Jones will appoint two friends, she will appoint two, to confer together upon the matter in dispute and she will most cheerfully abide by their decision, whatever it may be; and will submit to any penalty that they may think proper to inflict if she has, in any respect, violated her engagement with Mr Jones" '.[30]

It was a generous, indeed humble approach, and to it Jones replied magnanimously and at once. He wrote: 'Madam. I have this moment learned with great concern from authority that I cannot doubt, that your father lies most alarmingly ill in prison [there, presumably, for debt]. When I consider the melancholy situation that such a circumstance must necessarily reduce you and your family to, God forbid, even after all that has happened, that any act of mine should be the means of encreasing those afflictions, which my own heart tells me must already be so great. If therefore, Madam, you think that your re-instatement in the Theatre will tend to the future prospects of yourself and family, I shall no longer oppose it, nor shall I even look for an apology on my own account, in the full confidence that on cool reflection, you will see the impropriety of your conduct towards me, and that I shall not experience a repetition of it'.[31] To this letter Montgomery received the following reply at six o'clock on the same evening: 'Dear Sir, I have only time to tell you that Mr Jones in a very handsome Gentlemanlike manner, has offered to re-engage me, which re-engagement I *accept*, and all matters now between us are *amicably adjusted*. Yours truly, M. Byrne'.[32]

Although uproar still continued in the theatre on the evening of 21 April,[33] that really ended the affray, and on 23 April it could be reported, 'We have very sincere pleasure in stating that "order" is really once more restored to Crow-street Theatre... Some Ladies were present for the first time for several nights past'.[34] One can merely add that following so much theatre tumult, one is not entirely surprised to learn of the death, in Gardiner Street, on the following 27 July of 'Mr Montgomery, formerly Stage Manager of the Dublin Theatre'[35]—presumably from exhaustion.

Ironic as it may appear, while 1819 was close to the end of Frederick Jones's term as patentee, operatically the time would also mark a pinnacle in his career there, for in 1819 Dublin had its first performances of *Don Giovanni* and *Le Nozze di Figaro*. On 24 and 27 September, advertisements in the Dublin daily papers respectfully informed the public that the Theatre would re-open on the latter night with 'Mozart's celebrated opera of *Il Don Giovanni*, as performed three seasons with the most rapturous applause at the King's Theatre, London, with the following principal performers from the said Theatre: Signora Corri, Signora Mori, Miss R. Corri, Signor Begrez, Signor Romero, Signor Deville and Signor Ambrogetti—the original Don Giovanni'. (He was not, of course; that distinction belonged to Luigi Bassi, but Ambrogetti having created the role at the King's Theatre, and having sung it there since, presumably he was considered to have some claim to it. After all, Prague, where Bassi had first sung it, was a long way from Dublin). Readers were also informed that the leader of the band would be 'Mr Mori (Leader of the Philharmonic Society of London)' who would 'perform a grand solo concerto on the violin between the acts'; and that Mr Corri would be 'the conductor of the pianoforte' and that 'the additions to the band [would] be select and more numerous than on any former occasion'. All complimentary admissions were to be suspended during the Italian Opera Season. Prices of admission were, Boxes, 5s 5d; Pit, 3s 3d; Galleries, 2s 2d; doors to open at seven, and the performances to commence at eight o'clock precisely.[36] Once again, as in previous Italian seasons there was no mention of chorus.

The company was reported to have arrived at Faulkner's Hotel on the 22 September.[37] A day or so later 'the first rehearsal of *Don Giovanni* took place at the Theatre, and as the band was judiciously selected and very ably led by Mr Mori, and the vocal performers all prepared and experienced in the several parts, it was a very delightful thing of that description. Several *Amateurs* who were permitted to be present, spoke very highly even of this first, and in the nature of things, rather imperfect specimen of the talents of the Italian Opera performers'.[38] 'Many', we also learn, were applying 'for the Box-Sheet, among whom are persons of leading rank and fashion and great respectability'.[39]

The cast was certainly exceptional for Dublin. In fact, not since 1811 had there been one capable of being compared with it. Heading the group was the baritone, Giuseppe Ambrogetti. Little is known of his early years except that he is reputed to have been born about 1780. In 1813 he was singing at Verona.[40] In 1805, and again in 1815, his name is to be found among artists at La Scala, Milan, where in 1815 his roles included Count Almaviva in the first performance there of *Le Nozze di Figaro*.[41] In 1816, he is reported at the Théâtre-Italien, Paris* singing Don Giovanni, where he was described as 'the last of that group of newly-engaged singers who—as we have seen—have all more or less either failed to please or disappointed our high expectations… In spite of his pleasant stage presence and his somewhat exaggerated liveliness, which left us cold, his actual performance suggested rather more a grotesque buffoon than a flattering and ingratiating seducer of young girls'.[42] On 1 February 1817, he made his first appearance at the King's Theatre, London, again as the Count in *Le Nozze di Figaro*, and on 12 April as Don Giovanni, when 'it was necessary to throw open to the public disengaged boxes in order to accommodate the crowds demanding admission'.[43] He remained at the King's Theatre for six seasons until 1822, where his fee, at least for the 1821 season, was £600.[44]

---

* About the same period his wife, the soprano Teresa Strinasacchi, was engaged at the same theatre. As late as 1840 a singer with the same surname (Christian name not given) is reported to have been singing at the Teatro alla Fenice, Venice (*Allgemeine Musikalische Zeitung*, 19 August 1840), in 1841 at Ivrea (Piedmonte) (*A.M.Z.*, 23 June 1841) and in 1844—(when, if Ambrogetti's wife she would reputedly have been aged about 76—Berta, in *Il Barbiere di Siviglia* at Vercelli. (*A.M.Z.*, 4 December 1844).

Sir John Andrew Stevenson, composer (1761/2-1833).
Portrait by Charles Robertson who, like Stevenson, was Dublin-born (1760-1821).

London's opinions of his abilities varied, ranging from that of *The Harmonicon*, which was in no doubt that 'genius and enthusiasm, combined with a versatility of histrionic talent seldom equalled on any stage, elevated to the first rank of his profession a man whose voice was neither distinguished for compass, tone, execution, nor any one of the qualities usually looked for in a first singer. In *Don Giovanni* he was the veritable reckless profligate, glorying in his crimes, and whom even the cold grasp of the spectre might appal, but could not awake to repentence… The next night you might see him embodying the sorrows and insanity of Mrs Opie's distracted father [in Paer's *Agnese*]. But here the *vraisemblance* was too harrowing; Ambrogetti had studied the last degrading woe of humanity in the hospitals, where all its affecting varieties were exemplified and had studied it too well. Females turned their backs to the stage to avoid the sight… Kemble, Young and Siddons combined in the confession that madness had never found such a representative'.[45] *The Examiner*, while agreeing that he 'gave considerable life and spirit to the part of Don Giovanni', yet demurred, 'we neither saw the dignified manners of the Spanish Nobleman, nor the insinuating address of the voluptuary. He makes too free and violent a use of his legs and arms. He sung the air "Finch'han dal vino"…with a sort of jovial, turbulent vivacity, but without the least "sense of amorous delight". His only object seemed to be, to sing the words as loud and as fast as possible. Nor do we think he gave to Don Juan's serenade, "Deh vieni alla finestra" any thing like the spirit of fluttering apprehension and tenderness which characterises the original music'.[46] Nevertheless, it was also reported of 'Finch'han dal vino' that 'the gaiety, spirit, and rapidity with which he sung it were quite new to our ears. We were alarmed when he began so rapidly, because we did not think it possible he could have completed the song at the same rate, but he was as articulate, and his voice was as sound, at the close as when he began, and his repetition of the song, immediately upon a unanimous call, with as much vigour and voice was more surprising'.[47]

It was similar with his performance of the Count in *Le Nozze di Figaro*. *The Morning Chronicle* announced, 'His voice is termed in musical language a baritone or a high base; it is smooth, very pleasing, and sufficiently powerful. To these qualities he adds a purity of style and judgment—we might indeed say an elegance—in spite of his person, which inclines to corpulency—in acting, which lead us to expect much from his future exertions',[48] while *The*

London Magazine (admittedly five years later when he was about to retire[49] to his native Italy) commented, his 'acting and singing in the Count are completely at variance; the one is as excellent as the other is execrable. He is certainly the worst singer that ever took the rank that he maintains with so much popularity'.[50]

The tenor, Pierre (Ignace) Begrez, was a Belgian, born in Namur in 1787. He was first a violinist, a member of the orchestra of the Théâtre de l'Impératrice. However, finding that he had a good tenor voice he entered the Conservatoire in 1806, where he became a pupil of Pierre Garat. In 1815 he made his first appearance at the Paris Opéra in Gluck's *Armide*, remaining there until the following year, when he joined the King's Theatre, London. In that season his roles included Guglielmo in *Così fan tutte*, which he continued to perform throughout his time there—Ferrando on this occasion being taken by John Braham. He remained a member of the King's Theatre until 1822, for which season his salary was £400,[51] but returned for further performances in 1825.[52] He is described as 'a fine young man with a gentlemanly taste; but the tones of his voice, as he manages them, are somewhat over-sweet and cloying, and have also a tendency to the lachrymose. He should, moreover, in addition to the fold of the arms and the intertwining of his fingers, get unto himself two or three more attitudes'.[53]

The two other male singers engaged, Deville and Romero, were second-rate. Deville creates a problem of identification, since there were two artists of the same name at this time: one, Paolo, a bass, the other, Giuseppe, a tenor. Paolo is widely reported to have made his debut at the King's Theatre in *La Gazza Ladra* on 10 March 1821,[54] and is said to have died in Venice about 1825 or '26.[55] Another Deville, presumably Giuseppe, arrived at the King's Theatre in 1815, singing minor tenor roles there, such as Guglielmo in *Elisabetta, Regina d'Inghilterra*, Ruggiero in *Tancredi*, and Basilio and Don Curzio in *Le Nozze di Figaro*. *The Times* of 20 December 1819 records, 'the effect of the opera suffers, however, by allowing the two characters of Basilio and Don Curzio to merge in the same performer, and that performer, Signor Deville. His voice is too feeble to produce the least effect in the finale and the concerted pieces'. His engagement at the King's Theatre at this time strongly suggests that he was the artist who travelled to Dublin. He may also have been the artist who, in 1814, sang Licinius in Spontini's *La Vestale* at the Kärntnerthor Theatre, Vienna, when it was reported, 'He lacks everything that one has a right to expect from a singer. He displeased totally',⁵⁶ He was certainly the singer heard by the Dublin barrister and librettist, Edward Fitzsimons in *Otello* at the Théâtre-Italien, Paris, in 1821, whom he describes as 'Deville (*quasi*), the scientific representative of Iago, singing last night for the most part *sotto voce*, "Pursued the *noiseless* tenor of his way".'⁵⁷ Giuseppe Deville is reported to have died in January 1833.⁵⁸

Of Romero* it can at least be said that he was a Spaniard and a bass who had arrived at the King's Theatre for the 1819 season. There, his opening performance of Leporello on 27 February was not encouraging, for it was announced, 'his reception must have convinced him, however potent the suggestions of vanity and self-love, that something better was expected, and is in fact indispensible to secure a passport to public favour. His figure is bulky, and ill-suited to Leporello; a fault, however, to be forgiven, if we could discover in him the requisites of a good musician or a tolerable actor: in the first he is very moderately gifted, and his negligence in acting frequently called forth the displeasure of the audience'.⁵⁹ A second notice confirms that he 'appeared to be a true son of dullness and gave so little satisfaction that he was lustily hissed; a circumstance of rare occurrence here'.⁶⁰ Nor does he seem to have improved, for a year later it was rhetorically observed, 'When Signor Romero first appeared in this country last season, it was considered a total failure: why he is retained we are at a loss to guess; for if his voice have any affinity with music, a knife-grinder's wheel is harmony itself'.⁶¹ He would nevertheless remain one season more when he received a salary of £410.⁶²

There were three women in the company: Frances and Rosalie Corri and Signora Mori. The Corris were sisters, and daughters of Natale Corri who, born in Rome in 1765, came to join his brother Domenico in Edinburgh about 1784. Frances (Fanny) was born in Edinburgh either in 1795 or 1801. Her first singing teacher was her father, but she was soon taken to London where she had lessons, first from John Braham, and then from Catalani, with whom she toured throughout the Continent in 1816 and 1817.

She joined the King's Theatre in 1818, remaining for three seasons. Her first role there was the Countess in *Le Nozze di Figaro*. In this she was reported to have been 'gifted with a rich and powerful voice of considerable compass; its

---

* A 'basso-comico' named Emanuele Romero was singing at Vicenza in 1825. (*Allgemeine Musikalische Zeitung*, 23 February 1825)

chief strength and fulness seem to be in the middle tones, but when the intimidating effects of a first appearance are removed, perhaps the higher notes may be found equal to the rest. Her intonation is perfect'.[63] She was in fact a mezzo-soprano, and possessed one quality 'almost ignored today, one that appears to have gone out of fashion…a beautiful rounded even shake'.[64] This she had probably learned from Catalani, for Leigh Hunt, while dismissing her acting 'at present' as 'a non-entity' due to her 'evident agitation', continues: 'Her voice is sweet and distinct; she has an excellent ear, and displays even through her trepidation a facility and power worthy of her reported instructress, Madame Catalani. Her deficiency seems to be in the intellectual part of her art— in propriety of expression. She throws her lights and shades too indiscriminately, now dropping her voice and now darting it forth like Catalani, but not, like the latter, upon the proper places. She is said, however, to be very young, and appears so… Miss Corri's personal appearance, though not remarkable, is lady-like; and she seems so modest as well as clever, that it is impossible not to wish her success heartily'.[65] According to one report, she 'was driven from her situation in our Italian Opera by a cabal among some of our aristocracy, because she happened not to be handsome'.[66] This, too, may have been the cause of her indifferent success on her tour of the Continent in 1823, in which she was joined by her two sisters, Rosalie and Angelina, although at least one report (from Milan) records that 'she received much applause for the well-known cavatina from *Tancredi*'.[67] About this time she married a bass named Giuseppe Paltoni, and spent the last five years of her career singing in opera throughout Italy, including, in 1828 and '29, at La Scala, Milan. Her last recorded continental performance seems to have been in 1835 at Alessandria in *Norma*, which 'fell flat because Corri-Paltoni seemed at first unequal to her role, but improved in subsequent performances…and received much applause'.[68]

Her younger sister, Rosalie, was also born in Edinburgh, in 1803. Her singing teacher was Tommaso Rovedino, son of Carlo Rovedino, the bass who sang in Dublin in 1808. Her very first public appearance occurred at Drury Lane Theatre on 30 January 1818, in an oratorio conducted by Sir George Smart. It was then reported: 'her voice is melodious, full and flexible; her execution is very considerable, but she is perhaps too fond of exhibiting her uncommon powers in this particular'.[69] As far as can be discovered her next appearance seems to have taken place in Dublin, for in August 1820 we read: 'At the Opera

[King's Theatre] on the 22nd of June appeared in the part of Zerlina in *Il Don Giovanni*, Miss Rosalie Corri, a younger sister of Miss Corri, the joint Prima Donna. Towards the end of last year, a large number of *Corps de l'Opera* embarked for Ireland, and treated the good people of Dublin with an Opera—a real Italian Opera! *Il Don Giovanni* was acted; Miss R. Corri made her debut with her sister and Ambrogetti *cum multis aliis*. Her success was such as to warrant her introduction to the higher tribunal of the English public, and at the benefit of Miss Corri she appeared. This young artist is *petite* but neat; light, and shapely in her figure—full of liveliness, gifted with a sweetly-toned voice of sufficient power and compass, and with great natural facility of execution'.[70] Following her appearance at the King's Theatre, she turned to English opera, making her first appearance at the Haymarket Theatre on 22 July 1820 as Polly in *The Beggar's Opera*,[71] with Madame Vestris in one of her renowned breeches roles as Macheath. By 1826 she had married a Mr W. Geeson, and continued her career, singing as Mrs Geeson.

It is difficult to be biographically precise about the third lady, Signora Mori, since it has been found impossible to be certain of her Christian name. One can merely say that she may have made a first appearance at the King's Theatre on 23 March 1813, when she sang in the second act of Pucitta's *Boadicea* and was reported to be 'a pupil of Naldi and promises to be a singer'.[72] Certainly a singer named Mori joined the company in 1817, remaining there until 1821 when her salary for the year was £300.[73] We read of a Miss A. Mori, a pupil of Lanza,[74] 'a young lady who played Rosetta at Covent Garden Theatre a few weeks since, [who] also made her first appearance as an Italian singer'[75] in Cimarosa's *Penelope* on 11 January 1817. In the autumn of 1821, a Marietta Mori was singing at Lodi,[76] and in the Carnevale season of 1822, a Maria Mori was the seconda donna at Pavia.[77] We are on surer ground by 1823, however, when we read of a 'Mlle Mori, so well known at the King's Theatre for her useful talents [who] is now performing in Paris. She made her debut at the Louvois [Théâtre-Italien] on the 6th of the last month [March—as Rosina in *Il Barbiere di Siviglia*] and was tolerably well received. She is engaged as Seconda Donna at that theatre'.[78] Here her salary was 10,000 francs for the season.[79] Among the new singers engaged for the Teatro São Carlo, Lisbon, in 1824, we again meet Maria Mori,[80] the same singer surely who, two years later, re-appeared in Paris, but now at the Académie Royale de Musique (the

Opéra) where we learn, 'the novelty here' has been her appearance 'in the *Flauto Magico*, and in the principal character in [*La*] *Vestale* of Spontini. This singer, who formerly filled only second-rate characters on your stage, has now risen to comparative eminence in her profession; and, having quarrelled with the directors of the Opéra Italien, passed over to this theatre where she is rising rapidly into favour'.[81] She was still singing there in 1827, when it was announced, 'she certainly improves in the pronunciation of our language'.[82] Finally, she is discovered at a Fétis Concert at the Conservatoire on 8 April 1832 singing a duet from Cavalli's *Xerse* with the bass, Levasseur.[83] Whereupon it could be said that Fétis treated her rather cavalierly in his *Biographie universelle des Musiciens* in confusing her with a Rosina Mori who was singing in Italian theatres in the early 1840s. The voice of the earlier Signora Mori is described as a contralto, 'full-toned, and in its lower and middle parts rich and sweet; its upper notes are somewhat acidulated by the manner of producing them, and her shake is very imperfect'.[84]

Mr Mori, the leader of the band, was her brother Nicolas, who, it will be remembered had previously visited Dublin in 1814 with Catalani. Having been born in 1796 or '97, he was now just 22 or 23. In the same year of 1819, he married the widow of the music publisher, Lewis Lavenu, entering into a business partnership with a stepson some nineteen years later. Following Barthélemon, he had subsequently studied for six years with Viotti, who, at the time, was exiled in London. His subsequent career indicates that, however good or bad the Dublin orchestra may have been, he must have made an excellent leader.

Finally, 'the conductor of the pianoforte' was Haydn Corri, who, as earlier noted, had appeared with the Italian Opera Company in 1811. He was the third son of Domenico Corri, and consequently a first cousin of Frances and Rosalie. His father, who had been a pupil of Porpora at Naples, had come first to Edinburgh in 1771, temporarily visiting London in 1774, where, in the same year he had had his opera, *Alessandro nell'Indie* performed. He then settled in Edinburgh for 15 to 18 years, later returning to London, where again as in Edinburgh, he set up both as music publisher and composer. Haydn Corri was a pianist and organist as well as a composer of some glees and songs, and in 1821, with his wife, née Adami, engaged as second soprano at the new Dublin Theatre Royal in Hawkins Street, he settled at No. 25 Bachelor's Walk as a teacher of singing and the piano. From 1827 to 1848 he was organist and choirmaster at the Pro-Cathedral, Marlborough Street. He died on 18 February 1860 at 13 Queen's Square (now Pearse Square) at the age of 75.

In contemplating the Mozart productions of 1819 one need only think back to the *Così fan tutte* of 1811. Firstly, there were too few members in the company to permit a satisfactory distribution of roles. True, a precedent established at the first performance of *Don Giovanni* in Prague allows the roles of Masetto and the Commendatore to be sung by one artist, but while Masetto was sung by Deville it seems impossible that his feeble tenor voice could have produced any effect in the deep bass role of the Commendatore. In fact, no mention whatsoever is made of the Commendatore in any newspaper report, and although Deville has been reported to have sung 'bass roles deplorably' at Strasbourg in the absence of a 'bass-buffo' there,[85] one gets the impression that this role was omitted. Consequently, with Ambrogetti singing Don Giovanni, the cast consisted of Begrez as Don Ottavio, Romero as Leporello, and Deville as Masetto. The female roles are not identified, but it seems probable that the distribution was: Donna Anna, Frances Corri; Donna Elvira, Mori; and Zerlina, Rosalie Corri. The omissions in *Le Nozze di Figaro* appear to have been as bad, if not worse. The given casting here was: The Countess, Frances Corri; Susanna, Mori; Cherubino, Rosalie Corri; Figaro, Begrez; Count Almaviva, Ambrogetti; while Romero presumably sang Bartolo, and Deville, Basilio and Don Curzio—thus apparently disposing of Marcellina, Barbarina, and Antonio.

A review of the opening performance of *Don Giovanni* records that 'A very genteel and fashionable auditory filled the Theatre last night at the first representation of the Italian Operas. *Don Giovanni* was the piece, and a more exquisite musical performance, or more perfect and delicious harmony, we have never listened to. We must now, at rather a late hour for particular criticism, limit our observations upon the Opera to somewhat general terms. It was performed throughout, acting as well as singing, with incomparable skill and talent. There is no exaggeration whatever in saying that the audience was greatly delighted. Several of the songs, duets and trios were *encored* and every effort was applauded. The finale of the first Act elicited a manifestation of rapture very rarely produced by theatrical exertion. It continued sometime after the dropping of the curtain. The representation of this Opera has convinced us of one fact—that none but Italians, or those bred altogether in the Italian School, can sing, with its national softness and peculiar expression, Italian music;

for though we have heard the music of *Don Giovanni* sung by the first English singers now in public life, it was by no means the same kind of thing we heard last night'.[86] A subsequent review of the performance adds: 'The vocal performers are greatly indebted to Mr Mori, under whose skilful superintendence and direction the Orchestra has been rendered perfectly qualified to the difficult effort of doing justice to Mozart's music. The gentleman performed a Concerto on Monday night [between the acts], in which he displayed very singular powers of execution and consummate science and taste. He met with loud applause as a tribute to his abilities. The Harmony of the Orchestra owes no small portion of its excellence to Mr Corri, who presides at the Piano-forte. This gentleman's reputation needs no particular notice, it is long established in the first class of the musical profession'.[87]

Withal, this was Ambrogetti's performance and we read that 'independent of the singing, it was as complete a piece of acting in that line as we ever remember to have seen; we know not which to admire most, the graceful easiness of his deportment, or the arch gaiety of expression which he threw into his features. He was rapturously encored in the "Là ci darem la mano" and was greeted with frequent bursts of applause during the evening. His voice does not possess much compass, but he supplies the deficiency by the exquisiteness of his taste. In the last scene, when surrounded by the fiends, the vehemence of his actions, his exclamations, and his appropriate attitudes, confirmed him in our opinion as an accomplished actor'.[88]

The reference to 'fiends' makes this an appropriate place to introduce the bizarre story which circulated between 1826 and 1839, and had Ambrogetti retire to a Trappist monastery. Reports, such as 'he had been living in a monastery in France; when that country no longer welcomed these monks on their soil, he travelled with them to Ireland where he is living now',[89] are contradicted by assertions such as 'That he became very religious is true, but that he [has] never entered into any monastery…is equally certain: he merely lived in a quiet, self-denying manner, in a small house in a retired quarter in Rome. His fit is now said to be over'.[90] To add verisimilitude to the story, a report from Dublin attributed his condition to his having seen one more than the correct number of six 'fiends' surrounding him each evening in the final scene of *Don Giovanni*.[91] That he was a recurrent depressive is more or less confirmed by John Ebers, who writes: 'With an overwhelming humour, the

Giuseppe Ambrogetti (right) as Don Giovanni and
Giuseppe Naldi as Leporello in Mozart's opera.
Portrait by John Partridge (1790-1872)

outgushings of which never failed in its effect on others, Ambrogetti was himself the most wretched of men, a prey to the horrors of hypochondria'.[92] That he never entered a Cistercian monastery is definitely confirmed from Mount Melleray Abbey in Ireland, and the Abbey of La Grande-Trappe in France.[93]

Don Giovanni was repeated on 29 September 'to a very respectable, and apparently very critical auditory',[94] and on the following evening there was announced a performance of the old Dublin favourite, Mayr's *Il Fanatico per la Musica*, preceded by a 'Grand Dramatic Selection from Mozart's celebrated operas of *Il Flauto Magico*, *Così fan tutte*, and *Il Barbiere di Siviglia* by Rossini'.[95] This performance was postponed however until 2 October, when *Il Fanatico* was preceded by the overture and quintette, 'Hm! Hm! Hm!' from *Il Flauto Magico*, a quintette (presumably 'Sento oh Dio!') from *Così fan tutte*, the finale of *Il Barbiere di Siviglia*, and arias and duets by Paisiello, Pucitta, Guglielmi and Bishop.[96] *Il Fanatico* was almost certainly produced in the one-act version as performed in 1808. Ambrogetti had already appeared in one-act performances at the King's Theatre in July 1819. It was noted of him as Don Febeo that 'perhaps neither fancy nor anticipation could reach the humorous, laughable, and admirable personation of the musical fanatic… His voice imitated the various instruments of the Orchestra with ludicrous precision, and the extravagance of his gesticulation denoted the despotism of his favourite propensity, to a degree that was almost painfully pleasing. The applause was vociferous'.[97] *Don Giovanni* was performed again on 4 October, and *Il Fanatico* on the fifth, when Ambrogetti 'raised the shout of applause and laughter as on the first night. We have never witnessed an audience in better humour towards the entire performance, than on last night'.[98]

Next, on 7 October, *Le Nozze di Figaro* 'was performed to a very fashionable audience. The cast of the piece was such as to place the talents of the several performers in the most advantageous light, and their exertions were regarded by the most enthusiastic applause. The duet of "Crudel! perchè finora" between Signor Ambrogetti and Signora Mori was *encored*. The finale to the first Act, which is amongst Mozart's finest compositions, was executed with great effect, and was loudly applauded'. It was advertised that 'between the Acts Signor Mori will play a Concerto in which he will introduce the favourite Air of "Robin Adair". In Act Second, a Pas Seul by Mlle Simon'.[99] *Figaro* was repeated on 9 October, this time for the benefit of Frances and Rosalie Corri.

Then, on the 10th, a Sunday, at 'St Michael's and St John's Chapel' (where Smock Alley Theatre had once stood) there was 'a Grand Concert of Vocal and Instrumental Music…at 2 p.m., for liquidating the Debts incurred for building the Organ and improving the Chapel'. Begrez, Ambrogetti and Romero sang, Mori led the orchestra, and Mr Panormo was conductor and organist. The concert ended with the grand chorus, 'Halleluja' from Beethoven's *The Mount of Olives*, and there were also trios, quartettes, etc. by local singers and a violoncello solo by Mr Pigott. Admission tickets cost five shillings.[100]

On the following evening, there was a joint performance of *Figaro* and *Il Fanatico*, a benefit for the two Moris, which was said to afford 'the best of two of the most popular pieces of that class, and as music is the main attraction of Italian opera, nothing material is sacrificed in this instance by the resumption to the Dramatic progress of either *Figaro* or *Il Fanatico*. The finest music of both is preserved in the arrangement'.[101] This was followed by a final performance of *Don Giovanni* on the twelfth, when Ambrogetti and Begrez took their benefit. The season then ended on Wednesday, 13 October, when the Italian Company gave a morning concert at the Rotunda for the benefit of the widow of Pietro Urbani.[102]

But time was now fast running out for Crow Street Theatre. On 15 November it was announced: 'We understand that until the arrangement of the patent is decided upon, the Theatre will be opened for the season by permission of his Excellency the Lord Lieutenant'.[103] The question of Jones's patent had in fact been in the public domain for some years. In February 1817, it had been declared, 'We agree fully with the writers of five or six communications we have before us on the subject that "the *Theatre Royal*, its *situation* and *management* should be closely examined into by the Citizens of Dublin, and particularly now, when the Patent is about to expire".'[104] Late in 1818, a writer to *The Patriot*, obviously well intentioned towards Jones, though in the circumstances lacking wisdom, spoke 'of the Theatre Royal as the private property of Mr Jones, and that the Press and Public have no right to interfere with the Engagements, the Expenditure, or with any of the proceedings of the Patentee behind the curtain'.[105] For his pains he was sharply informed by *The Dublin Evening Post* (continually hostile) that 'we have already said more than once, that we could only consider Mr Jones as a trustee placed for certain purposes between the Government and the Public. The Legislature of Ireland held it to be useful and necessary that there should be one, *or more*, well-regulated Theatres in the Capital of Ireland, and therefore, by the 26 Geo. III c.57 [1786] it was enacted, in as much as such Establishments would "be productive of public advantage, and tend to improve the morals of the people" that "it shall and may be lawful for his Majesty, his heirs and successors to grant under the great seal of this Kingdom, for such term not exceeding twenty-one years, and under such restrictions and limitations, as to him or them shall seem meet from time to time, and when and as often as he or they shall think fit, one or more letters patent to one or more person or persons for establishing and keeping one or more well regulated theatre or theatres, play-house or play-houses in the city of Dublin, and in the liberties, suburbs and county thereof, and in the county of Dublin". Now, on this Act, is the Patent to Mr Jones founded… It is clear that it was granted for specific *public purposes*, and not as *private property* to Mr Jones'.[106]

Jones, for his part, had in 1817 presented through Sir Robert Peel, then Secretary for Ireland, a memorial to the Lord Lieutenant, Viscount Whitworth, seeking the usual renewal, an application which he considered had been favourably received. Having presented a second petition in February 1818 to Earl Talbot, (who had meanwhile—on 9 October 1817—succeeded to the Viceroyalty), he, as has been noted, refurbished his theatre and installed gas at a reputed cost of £1800.[107] But a group calling themselves 'Friends of the Drama' was organised about this time to oppose his being granted a patent. They convened a public meeting, and issued a report setting forth their objections to the manner in which the theatre had been managed.[108] Jones fortunately had friends too, one declaring, 'there is no stimulus for principal actors to engage in Dublin, where the public would rather encourage a company of Indian jugglers, an Italian puppet-show, or the burlesque pantomime of a company of monkey rope-dancers and French dogs'.[109]

Then, on Wednesday 25 November 1818, a meeting of bond-holders of the theatre was held. A Captain Fawcett who 'was called to the Chair' informed the meeting that, for his own part, he had not received one shilling of interest upon his bond since the time he purchased it, now four years; when he applied to the Treasury of the Theatre there was always an indifference, amounting in fact to rudeness, expressed. He had written several polite letters to Mr Jones on the subject, but that gentleman had not even had the courtesy to return an answer. Under all these circumstances, and the fact of Mr Jones's patent being nearly expired, it was imperative for the Bond-holders to come to some decision in order to protect their property... Mr Myles O'Reilly (Mr Jones's Attorney) came forward and said that Mr Jones was extremely anxious to meet a deputation from the Bond-holders, to whom he would submit an arrangement:

> *Mr King*: Has Mr Jones obtained a renewal, or promise of renewal, of the patent?
> *Mr O'Reilly*: I cannot answer that question at present.

The meeting ended inconclusively following the appointment of a committee of five to meet Frederick Jones at the theatre at two o'clock on the following Sunday.[110] But the committee was no more successful in resolving the problem. Indeed, it was asserted that 'for some time previous to the expiration of the theatrical Patent [on 15 December 1819] Jones's partners, Crampton and Dalton, together with his trustee [? Myles O'Reilly] privately combined to prevent his obtaining a renewal, being desirous of effecting new arrangements to exclude him totally, and to acquire for themselves the entire control and emoluments of the Dublin stage'.[111]

In February 1819, Jones enquired from William Gregory, the Under-Secretary, whether it was necessary to present another petition to Earl Talbot, but was assured that this was not needed since his former application was under consideration. However, in May of the same year, Crampton and Dalton presented a petition to the Lord Lieutenant containing complaints against Jones's management, to which Jones replied, declaring the charges to be unfounded. So the matter continued to drag on until 20 October 1819, when Jones received the following letter from the Chief Secretary, Charles Grant (later Lord Glenelg):

> Dublin Castle... Sir, In reply to the letter with which you have favoured me, it is my duty to inform you that the Lord Lieutenant, on the fullest consideration, has determined that you should not be sole or joint patentee in any patent to be granted for conducting a theatre in the city of Dublin. But, desirous of attending to the interests of the several parties having rights or claims on the Theatre Royal, his Excellency is disposed to grant a joint Patent to the present proprietors of the Theatre Royal, as trustees for the several parties interested therein, according to their respective rights and proportions, and for the creditors having claims thereon, or to two or more trustees to be named by them, and to whom all the parties interested in the Theatre Royal must previously assign over their interests therein as trustees for the same, for the parties interested as aforesaid. Should the parties interested in the Theatre Royal not be able to agree to what is so generously offered, his Excellency will in such case grant a patent to such person or persons as he shall think fit.[112]

Portrait of Giuseppe Ambrogetti by Abraham Wivell (1786-1849)

Crampton and Dalton, apparently seeing an opportunity here of deposing Jones, refused to agree to the nomination of trustees, which gave the government the right to grant the patent to any person approved by the Crown, but it appears that some legal obstacle prevented either of them being appointed the new Patentee. Ultimately, Henry Harris of Covent Garden was decided upon. He arrived in Dublin on 11 March 1820, accompanied by his brother, Captain Harris of the Royal Navy,[113] 'with a patent in his pocket' which he had obtained through personal interest with the Duke of York.[114] On 30 March it was reported, 'We understand that Mr Harris has finally concluded his treaty for the Dublin Theatre. He is to pay an annuity of £3,600,* of which £600 is to be enjoyed by Mr Jones and the remainder by the other Proprietors and Creditors of the Concern',[115] which was immediately contradicted by the news: 'Mr H. Harris returned last night to town from Dublin—The business of the Theatre had been arranged with Mr Jones in the manner we stated yesterday, but we now learn that it has been defeated by the interference of a principal Creditor who objects to the arrangement. The erection of a new Theatre in that city will probably be the result'.[116]

And so it was, at first a temporary one at the Rotunda which opened on 19 June 1820, and then the new Theatre Royal in Hawkins Street. An article in *The Dublin Evening Post* of 18 May expresses the annoyance of the public and appears to identify the 'principal Creditor' as Maurice Fitzgerald. It states: 'The Benefits, which Government permitted some of the Performers to take at Crow-street, are over, but there do not appear to be any preparations in progress for the opening of the Theatre under the new Patent—A London Paper indeed tells us that Mr Harris intends opening it in *October*!—a month ago we were told it would open in May, now October is mentioned as the period when the second City in the Empire, after a lapse of twelve months, is to be indulged with a place of national public amusement. When October arrives we shall perhaps be requested to wait patiently till Christmas, or probably till a *Chancery Suit* between Mr Jones and Mr Fitzgerald shall be finally determined. This trifling with the Citizens of Dublin and the Irish Public ill becomes those who have power to put a stop to it'.

---

* The proposed annuity was earlier said to be £3,000: £500 for Jones and £2,500 for the creditors. (*The Globe*, 17 March 1820). The *D.N.B.* tells us: 'Jones lost heavily by this arrangement and was imprisoned for debt'.

Meanwhile, the unhappy Bondholders had held what seems to have been another futile meeting on 2 May at Morrison's Tavern. 'Mr Beaumont', we learn, '(one of the Gentlemen appointed to communicate with the Government and Mr Jones on the subject of the rights of the Bond-Ticket Holders) stated, that the Commissioners (Mr Sergeant Lefroy and Mr Townsend) to whom the consideration of the affairs of the Theatre had been referred by Government, had reported that the rights of the Bond-Ticket Holders ought to be respected, and that they had good *liens* on the Theatre; that Mr Jones had declared his determination to support their rights; but notwithstanding all this, an agreement was now concluded between Mr Harris and Mr Jones in which the claims of the Bond-holders were left wholly out of view. It therefore would be necessary for them to take some immediate and decisive steps to protect their lawful interests'. But 'Mr Abbott, the professional Agent of Mr Harris', explained 'that proportionate shares of the rent offered were to be paid to the Minor Proprietors—that the annuities to Mr Daly's family and the admission tickets issued by Barry, Daly, and Mr Jones, to Mr Latouche, were to be protected but that the Government and Mr Jones appeared to consider the Bond-Ticket Holders as Mr Jones's private creditors… After a good deal of desultory conversation, three Gentlemen were appointed to wait upon Mr Jones',[117] which undoubtedly they did, with an equally desultory outcome.

Earlier, the benefit performances mentioned by *The Dublin Evening Post* had taken place from 18 April to 13 May. One, of *The Haunted Tower* performed for James Barton on 8 May, included 'A concerto on the violin by Master Balfe—his first appearance in public this Season'.[118] The theatre ultimately closed it doors—as a theatre—on 13 May 1820. (Later, we read, 'it has been fitted up for equestrian performances by a Mr Ord from Edinburgh'.[119]) The final programme—'Miss Curtis's Night and the last of performing'—was *Richard the Third* and *Rosina*. 'Richard, Duke of Gloster, by an amateur—his first appearance. Richmond, by a young gentleman, his second appearance. Duke of Buckingham, by a young gentleman, his first appearance. In *Rosina*, Mr Mayhew played Belville, Miss Ford, Rosina. 'With the interlude of *Three Weeks after Marriage*, Sir Charles Rackett, by the gentleman who performs Richmond. In the course of the evening, a Young Gentleman, his first appearance, will recite [William] Colli[n]'s ode on the Passions'.[120]—an ode for music.

So, did the Theatre Royal, Crow Street, under Jones's tenure, finally end as it had begun, with a play by Shakespeare. It had commenced with the comedy, *The Merchant of Venice*, and now it was to close with *The Tragedy of King Richard III*, while an appropriate requiem was intoned with the closing couplet of Collins's Ode:

> O *Music*, Sphere-descended Maid,
> Friend of Pleasure, *Wisdom's* Aid,
> Why, Goddess, why to us deny'd?
> Lay'st Thou thy antient Lyre Aside?

Sadly, the theatre was then 'finally pulled to pieces by instalments'. John William Calcraft, subsequently manager of the new Hawkins Street theatre, records: 'I stood on the stage soon after my arrival in Ireland, in October 1824. The scenery was gone, and there were sundry rents and chasms in the roof. The audience part of the house was still tolerably perfect, but many detachments of unlicensed plunderers were busily employed in all directions (apparently with no one to interfere), knocking out the panels of the boxes, tearing up the benches for firewood, and carrying all off bodily, for such other purposes as pleased their fancy'.[121] Gilbert adds: 'The scene-room, erected by Jones at a cost of £3,000, was converted into a hat manufactory, and the other portions of the vacant premises subsequently became a receptacle for the rubbish of the neighbourhood… In 1836, portion of the site of Crow-street Theatre was purchased by the Company of the Apothecaries' Hall of Dublin, who erected on it a building with spacious lecture-rooms, and a laboratory for their Medical School. These premises were sold in 1852…to the Catholic University of Ireland [for their] Medical School'*.[122]

And what became of Frederick Jones? In various petitions to the government he sought compensation 'and represented the stringent manner in which the State had obliged him to carry out his agreement with Daly, the previous Patentee, to whom and to whose family he had, to the then present time, paid annuities exceeding in the aggregate of £25,500; while his own expenditure on the Theatre, exclusive of salaries to actors or other ordinary disbursements, had been upwards of £30,000. To these applications Jones received the following official reply from Charles Grant:

---

* In time it housed the Anatomy Department of University College, where the writer was a student among the last class to attend in the academic year 1930-31. Today it is converted to a warehouse and offices.

Poney Races, at the Theatre Royal, Crow-Street.

> Dublin Castle. 13th December 1821. Sir, I have submitted to the Lord Lieutenant the application that you have preferred for some provision for yourself and your family, in consideration of the loss you have sustained by the expiration of your Patent for the Theatre, and of its renewal in Mr Harris's favour. His Excellency desires me to observe, that much time and pains were taken in the investigation of the business by the two great, eminent Barristers who were selected by the Government for that purpose; that, in consequence, liberal offers were made to you, which you thought fit to decline; and that it is now wholly out of his power to afford you any relief; nor can he think you entitled to the consideration you claim at the hand of the Government.[123]

In parenthesis it should be reported that the bond and ticket holders were also still being importunate, going so far as to present a petition—without success—to King George IV during his visit to Dublin in 1821.[124] Then, in 1829, the government granted in trust to Jones's sons, Richard Talbot and Charles Horatio,* a patent for a second theatre in Dublin under which the house in Abbey Street was opened,[125] in February 1837.

Of Jones himself it was written, 'no man has been more misrepresented. Jones's puffers and flatterers have bedaubed him with the most fulsome praise; his enemies (and some who were not his enemies, but who did it from a motive of wanton mischief) have denied him every title to virtue and common sense… That he can do many good, many generous things, which no law, no obligation, but his own will, could possible enforce, no one will attempt to deny; and that he frequently says good things, is as true, if we may rely on the report of those who shovel his turbot down their throats, and drink his claret by pailfuls. Few men…are better qualified to govern a theatre: but the misfortune is, that he has ever been in the habit of delegating the office to others; in the choice of whom he certainly has not displayed any great position of intellect'.[126]

He died on Wednesday, 5 November 1834, 'of cholera after a few hours illness',[127] alas, no longer at Clonliffe House, then, retired to Portland Place.

---

* A third son, Frederick, was a sometime actor who played Iago at the new Theatre Royal in Hawkins Street on 2 February 1821. (*Dublin Evening Post*, 1 February 1821)

Appendix A

Cast lists of first performances of representative operas and operatic pieces performed at the Theatre Royal, Crow Street, between 1798 and 1819

THE ADOPTED CHILD (Birch)   Attwood

Sir Bertrand, Rawling; *Record*, Callan; *Le Sage*, Bellamy; *Michael*, Williams; *Boy*, Master Blanchard; *Clara*, Mrs Addison

Original production 1 May 1795, Drury Lane
First Dublin performance 17 Dec 1798

LOVE IN A BLAZE (Atkinson)   Stevenson

Prince of the Island, Hamerton; *Merville* , Bellamy; *Flambeau*, Stewart; *Gentooba*, Byrne; *High Priest*, Gaudry; *Jack Gangway*, Williams; *Captain of the Guard*, Newenham; *Theresa*, Mrs Addison; *Maryanne*, Mrs Creswell; *Elora*, Mrs Williams; *Priestess*, Mrs Blanchard

First Dublin performance 29 May 1799

THE MOUTH OF THE NILE; or,
THE GLORIOUS FIRST OF AUGUST (T. J. Dibdin)   Attwood

Jack Oakum, Johnson; *William*, Williams; *Pat*, Callan; *Michael*, Bellamy; *Susan*, Miss Webb

Original production 25 Oct 1798, Covent Garden
First Dublin performance 18 Dec 1799

THE TURNPIKE GATE (Knight)   Mazzinghi and Reeve

Sir Edward, Byrne; *Smart*, Galindo; *Henry Blunt*, Bellamy; *Crack*, Munden; *Robert Maythorn*, Johnston; *Old Maythorn*, Coyne; *Joe Standfast*, Williams; *Peggy*,

Mrs Creswell; *Mary*, Mrs Addison

First Dublin performance 4 July 1800
Original production 14 Nov 1799 Covent Garden

OF AGE TO-MORROW (T. J. Dibdin)　　　　　　　　　　　　Kelly

Frederick, Baron Willinhurst, Bannister; *Baron Piffleberg*, Williams; *Hans Molkus*, Galindo; *Lady Brumback*, Mrs Dawson; *Sophia*, Miss Webb; *Maria*, Mrs Addison

Original production 1 Feb 1800, Drury Lane
First Dublin performance 19 July 1800

THE BEDOUINS; or,
THE ARABS OF THE DESERT (Irwin)　　　　　　　　　Stevenson

Gusto, Williams; *Volatile*, Philipps; *Steerage*, Johnson; *Hamet*, Stewart; *Abdallah*, Galindo; *Perim*, Byrne; *Cadiga*, Mrs Creswell; *Zeleika*, Mrs Addison; *Shireen*, Miss Webb; *Bedouins and Arabs*, Mesdames Davidson, Dawson, Sparks, King, Jocasto; Messrs Coyne, Kelly, Pitman, Ratchford, Madden, Downes, Dunn, Goodwin, Burton, Hughes &c

First Dublin performance 1 May 1801

PAUL AND VIRGINIA (Cobb)　　　　　　　　Mazzinghi and Reeve

Paul, Incledon; *Alambra*, Davis; *Captain Tropic*, Williams; *Don Antonio*, Wilton; *Diego*, Norman; *Sebastian*, King; *Dominique*, Stewart; *Sailor*, Seymour; *Officer*, Gaudry; *Virginia*, Mrs Creswell; *Jacintha*, Miss Webb; *Mary*, Miss Davidson; *Villagers*, Mesdames King, Jocasto, Dyke, Kelly, G. King; *Slaves*, Messrs Dyke, Pitman, Kelly, Smith

Original production 1 May 1800, Covent Garden
First Dublin performance 4 July 1801

APPENDIX A: CAST LISTS OF FIRST PERFORMANCES

RAMAH DROOG; or,
WINE DOES WONDERS (Cobb)   Mazzinghi and Reeve

Liffey, Johnstone; *Chellingoe*, Munden…

Original production 12 Nov 1798, Covent Garden
First Dublin performance 11 July 1803

THE WIFE OF TWO HUSBANDS (Cobb)   Mazzinghi and Cooke

Count Belfior, Talbot; *Maurice*, (*Baron Werner*), Hargrave; *Armagh*, Lindsay; *Carronade*, Williams; *Montenero* (*Captain of the Banditti*), Philipps; *Eugenia*, Miss Howells; *Countess Belfior*, Mrs Galindo; *Ninetta*, Mrs Stewart

Original production 1 Nov 1803, Drury Lane
First Dublin performance 21 Jan 1804

LOVE LAUGHS AT LOCKSMITHS (Colman the Younger)   Kelly

Captain Beldare, Philipps; *Risk*, Stewart; *Totterton*, Fullam; *Solomon Lob*, Johnson or Warren; *Vigil*, Williams; *Lydia*, Miss Howells

Original production 25 July 1803, His Majesty's
First Dublin performance 27 Feb 1804

MATRIMONY (Kenney)   King

Baron de Limburg, Fullam; *Delaval*, R. Jones; *O'Clogherty*, Lee; *Lisetta*, Miss Howells; *Clara*, Mrs Edwin

Original production 20 Nov 1804, Drury Lane
First Dublin performance 24 Jan 1805

THE FIVE LOVERS (T. Swift)   Cooke

The Mufti, Williams; *Grand Bashaw*, Weston; *Killarney* (*An Irish Soldier*), Lee; R. Jones, Fullam, Mansell, Johnson, Radcliffe, Wilton, Byrne, Carroll. *Zobeide*, Mrs Stewart; *Sophia*, Mrs MacNamara; *Zarinda*, Mrs Nunn

First Dublin performance 22 Feb 1806

THE HUNTER OF THE ALPS (Dimond)   Kelly

Felix, Elliston; *Rosalvi*, Putnam; *Jeronymo*, Williams; *Julio*, Master Grant; *Helena di Rosalvi*, Miss Macauley; *Genevieve*, Mrs Cooke; *Florio*, Miss Grant

Original production 3 July 1804, His Majesty's
First Dublin performance 17 June 1806

THE CABINET (T. J. Dibdin)   Braham, Davy, Moorehead and Reeve

Count Curvoso, Norman; *Lorenzo*, Incledon; *Prince Orlando*, Philipps; *Whimsiculo*, Johnson; *Peter*, Grant; *Marquis de Grand Château*, Fullam; *Constantia*, Mrs Nunn; *Floretta*, Mrs Cooke; *Leonora*, Mrs Johnson; *Bianca*, Mrs Hitchcock; *Curiosa*, Mrs Davis

Original production 9 Feb 1802, Covent Garden
First Dublin performance 14 Aug 1806

THE FIRST ATTEMPT; or,
THE WHIM OF THE MOMENT
(Sydney Owenson)   Cooke and Sydney Owenson

Marquis de los Cisternas, Fullam; *Alonzo* Philipps; *O'Driscoll*, Owenson; *Orlando*, R. Jones; *Benedetto*, Moore; *Pedro*, Weston; *Elvira*, Mrs Nunn; *Flora*, Mrs Stewart; *Nicholette*, Mrs Cooke

First Dublin performance 4 Mar 1807

THE INVISIBLE GIRL (T. E. Hook)   Hook

Captain Allclack, R. Jones; *Sir Christopher Chatter*, N. Jones; *Moses*, Moore; *Lord Flutter*, Carroll; *Tom*, Pitman; *Fac-Simile Moses*, R. Jones; *Fac-Simile Lord Flutter*, R. Jones; *Fac-Simile Mrs Allclack*, R. Jones; *Mrs Allclack*, Mrs McCulloch; *Harriet (The Invisible Girl)*, Mrs Cooke

Original production 28 Apr 1806, Drury Lane
First Dublin performance 28 Apr 1807

APPENDIX A: CAST LISTS OF FIRST PERFORMANCES

THE ENGLISH FLEET IN 1342 (T. J. Dibdin)   Braham

De Mountfort, Bartley; *De Montaubon*, Putnam; *Philip*, Fullam; *Valentine*, Philipps; *Fitzwater*, Incledon; *Mainmast*, Williams; *Countess of Brittany*, Mrs W. Johnston; *Jeannetta*, Mrs Hitchcock; *Isabel*, Mrs Cooke; *Katherine*, Mrs Nunn

Original production 13 Dec 1803, Covent Garden
First Dublin performance 3 Aug 1807

LA MORTE DI SEMIRAMIDE (Caravita)   Portugal

Semiramide, Catalani; *Azema*, Siga. Miarteni; *Arsace*, Siboni; *Seleuco*, Rovedino; *Mitrane*, Miarteni

Original production 27 Sept 1801, S. Carlos, Lisbon
First Dublin performance 20 Aug 1808

IL FANATICO PER LA MUSICA (G. Rossi)   Mayr

Donna Aristea, Catalani; *Celestina*, Siga. Miarteni; *Don Carolino*, Siboni; *Don Febeo*, Miarteni; *Biscroma*, Rovedino

Original production 18 October 1798, San Benedetto, Venice
First Dublin performance 25 Aug 1808

LA DIDONE (Metastasio)   Paisiello

Didone, Catalani; *Selene*, Siga. Miarteni; *Jarba*, Siboni; *Araspe*, Rovedino; *Osmida*, Miarteni; *Enea*, Kelly

Original production 4 Nov 1794, San Carlo, Naples
First Dublin performance 27 Aug 1808

IL FURBO CONTRO IL FURBO (Tottola/Buonaiuti)*   Fioravanti/Ferrari*

Rosina, Catalani; *Olimpia*, Siga. Miarteni; *Federigo*, Siboni; *Melibeo*, Miarteni; *Sciabecchino*, Rovedino; *Gaspero*, Kelly

Original production 29 Dec 1796, San Samuele, Venice
First Dublin performance 1 Sep 1808

* But see p 78.

LA MORTE DI MITRIDATE (Sografi) — Portugal

Vonima, Catalani; *Fedima*, Siga. Miarteni; *Mitridate*, Siboni; *Farnace*, Rovedino; *Arbate*, Miarteni; *Zifare*, Kelly

Original production Carnevale 1806, S. Carlos, Lisbon
First Dublin performance 8 Sept 1808

LA FRASCATANA (Livigni) — Paisiello

Donna Violante, Catalani; *Donna Stella*, Siga. Miarteni; *Il Cavaliere Giocondo*, Siboni; *Don Fabrizio*, Miarteni; *Nardone*, Rovedino

Original production Nov 1774, San Samuele, Venice
First Dublin performance 17 Sep 1808 (3 December 1777, Smock Alley)

THE TRAVELLERS; or, MUSIC'S FASCINATION (Cherry) — D. Corri

Prince Zaphimiri, Huddart; *Delvo*, Fullam; *Admiral Lord Hawser*, Williams; *O'Gallagher*, Lee; *The Emperor of China*, Younger; *Koyan*, Hill; *Duke Posilipo*, Simpson; *Ben Buntline*, Johnson; *Celinda*, Miss Walstein; *Mindora*, Mrs Mason; *Marchioness Merida*, Mrs Stewart; *Safie*, Mrs Cooke; *Dancers*: Mr and the three Misses Giroux, Miss A. Dyke*

Original production 22 Jan 1806, Drury Lane
First Dublin performance 29 Nov 1808

THE EXILE; or,
THE DESERTS OF SIBERIA (Reynolds) — Mazzinghi and Bishop

Count Ulrick (The Exile), Huddart; *Governor of Siberia*, Fullam; *Baron Altradoff*, Johnson; *Count Calmar*, Philipps; *Daran*, Rae; *Yermack* Younger; *Servitz*, Williams; *Welzien*, Duff; *Rimski*, N. Jones; *Empress Elizabeth*, Miss Locke; *Sedona*, Mrs Mason; *Alexina*, Miss Smith; *Catharine*, Mrs Stewart; *Anna*, Mrs McCulloch; *Villager*, Mrs Cooke

Original production 10 Nov 1808, King's Theatre (presented by the Covent Garden Company)
First Dublin performance 20 Mar 1809

* Tom Moore's sister-in-law

APPENDIX A: CAST LISTS OF FIRST PERFORMANCES

THIRTY-THOUSAND; or,
WHO'S THE RICHEST? (T. J. Dibdin)   Braham, Davy and Reeve

Captain Foresail, Braham; *Lawyer Plainly*, Younger; *Mr Dubious*, Fullam; *Arable*, Payne; *Windmill*, Lewis; *Clump*, Johnson; *Gangway*, Williams; *Teddy*, Lee; *Rosanna*, Mrs Nunn; *Mrs Arable*, Mrs Davis; *Henrica*, Miss Dyke; *Mrs Notable*, Mrs McCulloch

Original production 10 Dec 1804, Covent Garden
First Dublin performance 13 Dec 1809

FALSE ALARMS; or, MY COUSIN (Kenney)   Braham and King

Edgar Gayland, Braham; *Sir Damon Gayland*, Younger; *Tom Surfeit*, Farren; *Lt. McLary*, Lee; *Plod*, W. Farren; *Gabriel*, Sloman; *Grinvelt*, N. Jones; *Lady Gayland*, Mrs Duff (late Miss Dyke); *Caroline Sedley*, Mrs Stewart; *Emily*, Mrs Cooke; *Miss Umbrage*, Miss McCulloch; *Susan*, Mrs Nunn

Original production 12 Jan 1807, Drury Lane
First Dublin performance 29 Dec 1809

THE FORTY THIEVES (Ward and Colman the Younger)   Kelly

Cassim Baba, Younger; *Ali Baba*, Williams; *Ganem*, Payne; *Mustapha*, Johnson; *Hassarac*, Gomery; *Hymen*, St Pierre; *Zaide*, Mrs Davis; *Cogia*, Mrs Cooke; *Morgiana*, Mrs Stewart

Original production 8 Apr 1806, Drury Lane
First Dublin performance 4 Jan 1810

THE ESCAPES; or,
THE WATER CARRIER (Holcroft and Dibdin)   Cherubini and Attwood

Count Armand, Kent; *Francisco*, N. Jones; *Antonio*, Sloman; *Michelli* (*The Water Carrier*), Williams; *Constantia*, Mrs I. Williams; *Angelina*, Mrs Cooke

Original production 14 Oct 1801, Covent Garden
First Dublin performance 11 Jan 1811

THE GAY DECEIVERS (Colman the Younger) — Kelly

Mr Candy, W. Farren; *Sir Harry*, Farren; *Nehemiah Flam*, Johnson; *Trap*, Sloman; *Pegasus Puncheon*, Fullam; *Emily*, Miss Sharpe; *Mrs Flaw*, Mrs Cooke; *Jenny Stumps*, I. Williams

Original production 22 Aug 1804, His Majesty's
First Dublin performance 20 Feb 1811

BORDER FEUDS; or, THE LADY OF BUCCLEUCH (?) — Stevenson

Lord Cranstoun, C. Connor; *Sir William of Deloraine*, N. Jones; *Lord Dacre*, Farren; *Monk*, W. Farren; *Lady of Buccleuch*, Miss Smith; *Margaret*, Miss Walstein; *Spirits*: Mesdames Cooke and I. Williams

First Dublin performance 2 May 1811

GUSTAVUS VASA, THE HERO OF THE NORTH (Dimond) — Kelly

Gustavus Vasa, Conway; *Sigismund of Calmar*, Incledon; *Casimir Rubenski*, Younger; *Carlowitz*, C. Connor; *Gabriel*, W. Farren; *Princess Gunilda*, Miss Walstein; *Alexa*, Mrs I. Williams; *Frederica Rubenski*, Mrs Dickons; *Ulrica*, ——

Original production 29 Nov 1810, Covent Garden
First Dublin performance 30 July 1811

DUE NOZZE ED UN SOL MARITO (?) — P. C. Guglielmi

Gioconda, Bertinotti; *Olivetta*, Siga. Cauvini; *Enrico*, Cauvini; *Beltrame*, Naldi; *Farfalla*, Manni; *Giacinto*, Balassi

Original production Autumn 1800, Cocomero, Florence
First Dublin performance 17 Aug 1811

Repeat performances of IL FANATICO PER LA MUSICA
Aug/Sep 1811

Donna Aristea, Bertinotti; *Celestina*, Siga. Cauvini; *Don Carolino*, Cauvini; *Don Febeo*, Naldi; *Biscroma*, Manni; *Carluccio*, Balassi

APPENDIX A: CAST LISTS OF FIRST PERFORMANCES   241

COSÌ FAN TUTTE (Da Ponte)   Mozart

Fiordiligi, Bertinotti; *Dorabella*, Siga. Cauvini; *Vespino* (sic), Balassi; *Ferrando*, Manni; *Guglielmo*, Cauvini; *Don Alfonso*, Naldi

Original production 26 Jan 1790, Burgtheater, Vienna
First Dublin performance 31 Aug 1811

ZAIRA (Botturini)   F. Federici

Zaria, Bertinotti; *Nerestan*, Siga. Cauvini; *Orosman*, Cauvini; *Lusignan*, Naldi; *Chatillon*, Manni; *Corasmin*, Balassi

Original production 1799, S. Cecilia, Palermo
First Dublin performance 12 Sep 1811

ONE O'CLOCK; or,
THE KNIGHT AND THE WOOD DEMON (Lewis)   Kelly and King

Hardyknute, Count of Holstein, Hill; *Sangrida* (*The Wood Demon*), N. Jones; *Guelpho*, Williams; *Willikind*, W. Farren; *Clotilda*, Mrs I. Williams; *Leolyn*, Miss Rock; *Oswy* Miss Lyons; *Una*, Miss O'Neill

Original production 1 Aug 1811, Lyceum (presented by the Drury Lane company)
First Dublin performance 2 Apr 1812

M.P.; or,
THE BLUE-STOCKING (Moore)   Moore. Overture and orch. C. E. Horn

Sir Charles Canvass, —; *Captain Canvass*, —; *Henry de Rosier*, —; *Mr Hartington*, —; *Leatherhead*, —; *Davy*, —; *La Fosse*, —; *Lady Bab Blue*, —; *Madame de Rosier*, —; *Miss Hartington*, —

Original production 9 Sep 1811, Lyceum
First Dublin performance 19 May 1812

THE DEVIL'S BRIDGE; or,
THE PIEDMONTESE ALPS (Arnold)   Braham, Horn and D. Corri

Baron Toraldi, Neville; *Count Belino*, T. Cooke; *Paolo*, Thompson; *Marcelli*, Johnson; *Fabricio*, C. Connor; *Florian*, Nichols; *Petro*, W. Farren; *Julio*, Miss E. Moore; *Countess Rosalvina*, Mrs Cooke; *Claudine*, Mrs Stewart; *Lauretta*, Mrs Fulton

Original production 6 May 1812, Lyceum, (presented by the Drury Lane company)
First Dublin performance 12 Nov 1813

THE MILLER AND HIS MEN (Pocock)   Bishop

Count Frederick Friberg, Thompson; *Lothair*, C. Connor; *Kelmar*, Williams; *Karl*, W. Farren; *Grindoff (The Miller)*, Cooke;* *Claudine*, Miss S. Norton; *Ravina*, Mrs Fulton

Original production 21 Oct 1813, Covent Garden
First Dublin performance 4 Jan 1814

THE CORSAIR; or, THE PIRATES' ISLE (O'Sullivan)   Blewitt

Conrad, Talbot; *Pedro*, McKeon; *Juan*, Neville; *The Pasha Seyd*, Thompson; *Medora*, Miss O'Neill; *Rosa*, Miss Rock; *Gulnare*, Miss Walstein

First Dublin performance 22 Apr 1814

THE FOREST OF BONDY; or,
THE DOG OF MONTARGIS (Barrymore)   Blewitt, Cooke and Corri

Col Gontran, Younger; *Captain Aubry*, T. Cooke; *Lt. Macaire*, Huntley; *Lt. Landry*, N. Jones; *The Seneschal of Bondy*, Thompson; *Blaise*, W. Farren; *Dame Gertrude*, Mrs Burgess; *Lucille*, Miss Rock; *Florio*, Miss S. Norton

Original production 30 Sep 1814, Covent Garden (music by Bishop)
First Dublin performance 8 Dec 1814

* Not Tom Cooke! *The Freeman's Journal* of 18 March 1814 reports a Cooke as 'clown of the Comic Pantomimes', while the same paper of 15 January 1816 refers to 'Mr W. Cooke, Clown and Tailor'.

APPENDIX A: CAST LISTS OF FIRST PERFORMANCES

FREDERICK THE GREAT; or,
THE HEART OF A SOLDIER (Arnold)   Cooke

Frederick II, King of Prussia, Huntley; *Charles, Baron of Felsheim*, T. Cooke; *Theodore d'Hartiman*, Short; *Governor of Schwiednitz*, Fullam; *Brandt*, Williams; *Count d'Herleim*, C. Connor; *Stock*, W. Farren; *Flank*, Johnson; *Matilda*, Mrs Mardyn; *Charlotte*, Mrs Cooke

Original production 4 Aug 1814, Lyceum
First Dublin performance 29 Dec 1814

JOHN OF PARIS (Pocock)   Bishop, Boieldieu, Blewitt and Pucitta

John of Paris, Short;* *The Grand Chamberlain*, Johnson; *Vincent*, T. Short;* *Pedrigo Potts*, W. Farren; *Princess of Navarre*, Miss Hughes; *Rosa*, Miss Johnston

Original production 12 Nov 1814, Covent Garden
First Dublin performance 18 Feb 1815

THE MAGICIAN WITHOUT MAGIC (Hamilton)   Blewitt

Marquis Aliprandi, Short; *Tartine*, W. Farren; *Paolo*, Fullam; *Ricardi*, Williams; *Ordicaldo*, Farren; *Hortensia*, Mrs Cooke; *Lucinda*, Mrs McCulloch; *Fanchette*, Mrs Charles McCulloch

First Dublin performance 1 Mar 1815

THE CARAVAN; or, THE DRIVER AND HIS DOG (Reynolds)   Reeve

Marquis of Calatrava, C. Connor; *Don Gomez* (*Governor of Barcelona*), Fullam; *Count Navarro*, Thompson; *Blabbo*, (*Driver of a Caravan*), Johnson; *Morillo*, W. Farren; *Arabbo*, T. Short; *Julio*, Master Bedford; *Marchioness of Calatrava*, Mrs Cooke; *Rosa*, Mrs C. McCulloch

Original production 5 Dec 1803, Drury Lane
First Dublin performance 27 Mar 1815

* Two brothers named Short 'from Dublin' were members of the company at the New English Opera House, London, in June 1816. T. Short appeared for the first time with John Isaacs, when the two were described by Hazlitt as: 'Mr Short and Mr Isaacs are singers, and we fear not good ones. Mr Short has white teeth, and Mr Isaacs black eyes'. (*The Examiner*, 23 June 1816)

CYMON (Garrick)   Stevenson

Merlin, Younger; *Cymon*, T. Cooke; *Dorus*, W. Farren; *Linco*, Johnson; *Damon*, T. Short; *Dorilas*, Gaven; *Urganda*, Mrs C. Connor; *Sylvia*, Miss Hughes; *Fatima*, Mrs C. McCulloch; *Amynta*, Mrs Cooke; *Delia*, Miss Johnston; *Dorcas*, Mrs McCulloch

First Dublin performance 9 May 1815

THE MAID AND THE MAGPIE (Pocock)   Bishop

Gerald, Williams; *Henry*, C. Connor; *Martin*, W. Farren; *Granville*, Huntley; *Benjamin*, Johnson; *Malcour,* Fullam; *George*, Burgess; *Captain of Gendarmes*, Rowswell; *Bertrand*, N. Jones; *Annette*, Mrs Edwin; *Dame Gerald*, Mrs Burgess

Original production 15 Sep 1815, Covent Garden
First Dublin performance 12 Dec 1815

LODOISKA (Kemble)   Storace, Cherubini, Kreutzer and Andreozzi

Polanders: *Prince Lupauski*, Younger; *Count Floreski*, Philipps; *Baron Lovinski*, Huntley; *Varbel*, Johnston; *Adolphus*, N. Jones; *Gustavus*, Rowswell; *Princess Lodoiska*, Mrs Bellchambers. *Tartars*: *Kerah Khan*, T. Cooke; *Ithorak*, Thompson *or* Hodson; *Khor*, Reid; *Japhis*, Burgess; *Camazin*, Gaven

Original production 9 Jun 1794, Drury Lane
First Dublin performance 8 Jan 1816

EGBERT AND ETHELINDA; or, THE DRAW BRIDGE (?)   Blewitt

Egbert, T. Cooke; *Gondibert*, Huntley; *Oswald*, C. Connor; *Rowland*, Thompson; *Gilbert*, W. Farren; *Edgar*, N. Jones; *Hugo*, Rowswell; *Kenrick*, Bedford *or* Burgess; *Offa*, I. Jones; *Ethelinda*, Mrs Hodson; *Bertha*, Miss Rock

First Dublin performance 7 Feb 1816

APPENDIX A: CAST LISTS OF FIRST PERFORMANCES 245

MY SPOUSE AND I (C. Dibdin jun.)   Whitaker

Frisk, Farren; *Paddock*, Johnson; *Dick*, Williams; *Ned*, Burgess; *Wilton*, Smythson; *Harriet*, Mrs Bellchambers; *Dame Paddock*, Mrs Burgess; *Janet*, Miss Johnstone

Original production 7 Dec 1815, Drury Lane
First Dublin performance 16 Mar 1816

EDWIN AND ANGELINA (?)   Stevenson and Clifton

Baron Mowbray, Younger; *Edwin*, C. Connor; *Sir Jeffrey Knowles*, Smythson; *Davy*, Johnson; *Angelina*, Miss Griglietti; *Beatrice*, Miss Johnstone; *Jenny*, Miss S. Norton

First Dublin performance 3 Apr 1816

BROTHER AND SISTER (Dimond and C. Dibdin jun.)   Bishop and Reeve

Don Sylvio de Flores, Philipps; *Don Christoval de Tormes*, Fullam; *Pacheco*, Johnson; *Bartolo*, N. Jones; *Rosanthe*, Miss Griglietti; *Donna Camilla*, Miss S. Norton; *Agatha*, Miss Rock; *Donna Isidora*, Miss Stephens

Original production 1 Feb 1815, Covent Garden
First Dublin performance 27 Aug 1816

THE SLAVE (Morton)   Bishop

Gambia, Conway; *Governor of Surinam*, Younger; *Colonel Lindenburg*, Montgomery; *Captain Malcolm*, Hewett; *Captain Clifton*, Hodson; *Matthew Sharpset*, Lacy; *Sam Sharpset*, Johnson; *Mr Fogrum*, W. Farren; *Zelinda*, Miss Griglietti; *Stella Clifton*, Miss Rock; *Miss Von Frump*, Mrs Burgess; *Mrs Lindenburg*, Mrs Smythson

Original production 12 Nov 1816, Covent Garden
First Dublin performance 29 Jan 1817

THE QUEEN OF CARTHAGE
AND THE PRINCE OF TROY (MacNally) — Shield and Hodson

Æneas, Fullam; *Achates*, Williams; *Iarbas*, W. Farren; *Plenipotentiary*, Johnson; *Dido*, Miss Griglietti; *Anna*, Mrs Lazenby; *Venus*, Miss L. Kelly

First Dublin performance 29 Jan 1817

GUY MANNERING; or,
THE GIPSY'S PROPHECY (Terry and Scott) Attwood, Bishop and Whitaker

Henry Bertram, Braham; *Col. Mannering*, Montgomery; *Dandie Dinmont*, Williams; *Capt. Dirk Hatterick*, Yates; *Gilbert Glossin*, Fawcett; *Bailie Mucklethrift*, Ailiffe; *Dominie Sampson*, W. Farren; *Julia Mannering*, Miss Byrne; *Lucy Bertram*, Miss Griglietti; *Meg Merrilies*, Mrs Yates; *Mrs McCandlish*, Mrs Burgess; *Flora*, Mrs Lazenby; *Gipsy Girl*, Miss McDonnell

Original production 12 Mar 1816, Covent Garden
First Dublin performance 4 Feb 1817

THE LIBERTINE (Pocock) — Mozart arr. Bishop

Don Pedro, O'Callaghan; *Don Juan*, Horn; *Don Octavio*, Smythson; *Leporello*, Johnson; *Masetto*, Burgess; *Lopez*, Fry; *Peasant*, Rowswell; *Donna Elvira*, Mrs Smythson; *Donna Leonora*, Miss L. Kelly; *Maria*, Miss Johnstone; *Zerlina*, Miss Hammersley; *Alguazils*, Misses Grey and Sutcliffe

Original production 20 May 1817, Covent Garden
First Dublin performance 8 Jan 1818

THE INNKEEPER'S DAUGHTER (Soane) — Cooke and Barton

Richard, Power; *Monkton*, Thompson; *Mary*, Miss F. Kelly; *Marian*, Mrs Smythson

Original production 7 Apr 1817, Drury Lane
First Dublin performance 19 Jan 1818

APPENDIX A: CAST LISTS OF FIRST PERFORMANCES   247

RICH AND POOR (Lewis)   Horn

Lord Listless, Power; *Walsingham*, Younger; *Beauchamp*, Horn; *Rivers*, Williams; *Frank*, W. Farren; *Zorayda*, Miss F. Kelly; *Lady Clara Modish*, Miss L. Kelly; *Mrs Ormond*, Mrs Lazenby

Original production 22 July 1812, Lyceum
First Dublin performance 31 Jan 1818

THE OUTPOST (W. S. jun.)   Stevenson

La Roche (*an old cottager*), Cobham; *Albert* (*his son*), Horn; *Herman* (*son of the Seneschal*), Thompson; *De Vrize* (*an old naturalist*), Fullam; *Edric* (*a young soldier*), Power; *Kieran Molloy*, Ward; *Jero*, Burgess; *Alvina* (*daughter of Count de Morville*), Miss Hammersley; *Mira* (*daughter of De Vrize*), Miss Ford

First Dublin performance 11 Apr 1818

ROB ROY MACGREGOR; or,
AULD LANG SYNE! (Pocock)   Davy, Bishop and Hodson

Sir Frederick Vernon, Younger; *Bailie Nichol Jarvie*, W. Farren; *Rashleigh Osbaldiston*, Thompson; *Major Galbraith*, O'Callaghan; *Francis Osbaldistone*, Horn; *Captain Thornton*, Wallace; *Rob Roy MacGregor Campbell*, Cobham; *Mr Owen*, Burgess; *Dougal*, Ward; *Diana Vernon*, Mrs Lazenby; *Helen MacGregor*, Mrs Yates; *Mattie*, Miss Johnstone

Original production 12 Mar 1818, Covent Garden
First Dublin performance 11 May 1818

LALLA ROOKH; or,
THE MINSTREL OF CASHMERE (O'Sullivan)   Horn

The Emperor Aurungzebe, Younger; *Feramorz*, Horn; *Sadi*, Smythson; *Ibrahim*, Williams; *Fadladeen*, W. Farren; *Mesrour*, Johnson; *Mustapha*, O'Callaghan; *Princess Lalla Rookh*, Miss L. Kelly; *Selima*, Miss Rock

First Dublin performance 10 June 1818

ANZICO AND COANZA; or,
GRATITUDE AND FREEDOM (Fitzsimons) — Stevenson

Molasses (*a benevolent planter*), Williams; *Crout* (*an avaricious planter*), Fullam; *Captain Gayland* (*a naval officer*), Horn; *Anzico* and *Matamba* (*Koromantyn slaves*), Cobham and Thompson; *Balaam Budget* (*a Cockney speculatist*), Johnson; *Lieutenant Capstan* Reid; *Miss Sydney* (*ward to Crout*), Miss Byrne; *Andora* (*an Obi woman*), Miss L. Kelly; *Coanza* and *Miamba* (*Koromantyn slaves*), Misses Whitaker and Hammersley

First Dublin performance 4 Mar 1819

DON GIOVANNI (Da Ponte) — Mozart

Donna Anna, Frances Corri; *Donna Elvira*, Siga. Mori; *Zerlina*, Rosalie Corri; *Don Giovanni*, Ambrogetti; *Don Ottavio*, Begrez; *Leporello*, Romero; *Masetto*, G. Deville

Original production 29 Oct 1787, Prague Nationaltheater, now Tyl Theatre
First Dublin performance 27 Sep 1819

Repeat performances of IL FANATICO PER LA MUSICA, Mayr Sep/Oct 1819

Donna Aristea, Frances Corri; *Donna Rosina*, Rosalie Corri; *Celestina*, Siga. Mori; *Don Carolino*, Begrez; *Don Febeo*, Ambrogetti; *Biscroma*, Deville

LE NOZZE DI FIGARO (Da Ponte) — Mozart

La Contessa, Frances Corri; *Susanna*, Siga. Mori; *Cherubino*, Rosalie Corri; *Figaro*, Begrez; *Il Conte Almaviva*, Ambrogetti; *Il Dottor Bartolo*, Romero; ?*Don Basilio* and *Don Curzio*, G. Deville

Original production 1 May 1786, Burgtheater, Vienna
First Dublin performance 7 Oct 1819

Appendix B

List of performances of Italian Operas given in Dublin between 1808 and 1819

		1808	1811	1819	Total
Paisiello, Giovanni (1740-1816)	*La Didone*	27 Aug 6, 22 Sept	- -	- -	3
	La Frascatana	17 Sept 15 Oct	- -	- -	2
Mozart, Wolfgang Amadeus (1756-1791)	*Così fan tutte*	- - -	31 Aug 2, 5 Sept 21, 28 Sept	- - -	5
	Don Giovanni	- -	- -	27, 29 Sept 4, 12 Oct	4
	Le Nozze di Figaro	-	-	7, 9 Oct	2
Portugal, Marcos Antonio (1762-1830)	*La Morte di Semiramide*	20 Aug 15 Sept	- -	- -	2
	La Morte di Mitridate	8 Sept	-	-	1
Guglielmi, Pietro Carlo (c.1763-1817)	*Due Nozze ed un sol Marito*	-	17, 19, 22 Aug 7, 9, 19, 24 Sept	- -	7
Mayr, Johann Simon (1763-1845)	*Il Fanatico per la Musica*	25, 30 Aug 13, 20, 22 Sept 15 Oct	24, 29 Aug 16, 25 Sept -	2, 5, 8, 11 Oct - -	14
Fioravanti, Valentino (1764-1837)	*Il Furbo contro il Furbo*	1, 3, 10 Sept	-	-	3
Federici, Francesco (17xx-18xx)	*Zaira*	- 17	12, 14 Sept 18	- 10	2 45

Appendix C
Programmes of Catalani's Dublin concerts in 1814

MONDAY, 8 AUGUST. MORNING.
ST WERBURGH'S CHURCH

Messiah by Handel

'In which Mme Catalani will sing: "Comfort ye my People", "Every Valley", "There were Shepherds", "Rejoice greatly", "I know that my Redeemer liveth", and the Grand Sacred Bravura, "Gratias agimus tibi".[1] [P. A. Guglielmi[2]]'.

TUESDAY 9 AUGUST. EVENING.
THE ROTUNDA

'Part I. Haydn's Symphony: The Surprise

Glee: "When winds breathe soft along the silent deep", Webbe [senr.]
    Master Stansbury, Mr Garbett, Mr Spray, Mr Tett & Mr Tinney

Scena ed Aria— Miss Hughes— (composed expressly for her):
    Recit: "Barbara! deh fermate"…
    Aria: "Perchè, perchè dividermi', Pucitta

Cavatina— Signor Chiodi— (his first performance in Dublin):
    "Tutti i gusti sono gusti" [*Pamela Nubile*, Generali]

Recit and Aria— Mrs Bianchi Lacy:
    "Accenni vostri suro…" "A quei lumi, a quel sembiante", Mayr

Concerto— Violin: Mme Gerbini

Duet— Mrs Bianchi and Mr Spray!
    "Prenderò quel Brunettino" [*Così fan tutte*, Mozart]

Grand Scena and Air— Mme Catalani:
    Recit: "Eccomi giunta al fine d'una misera vita…"
    Aria: "Deh frenate oh Dio le lacrime…", Pucitta

Part II. Overture: *Die Zauberflöte,* Mozart

APPENDIX C: CATALANI'S DUBLIN CONCERTS IN 1814 251

Ballad— Mrs Bianchi Lacy:
"Sweet Helen! say not with a sigh, that time at last will bid us part", Mrs Bianchi Lacy

Duet (Buffo)— Mme Catalani and Signor Chiodi:
"Con pazienza sopportiamo, [Fioravanti]

Ballad— Miss Hughes. Words by T. Moore Esq.
"Oh! remember the time in La Mancha's shades" Spanish Air

Concerto— Violoncello. Mr Lindley, Lindley

Air-with variations, Mme Catalani:
"Sul margine d'un rio", Pucitta

Ballad— Mr Spray
"My dark-eyed Maid!", Bishop[3]

The Freeman's Journal of 11 August 1814 reports: 'The Concert concluded with "Rule Britannia" which was given in fine style by Mme Catalani. The Viceregal party did not leave the room before twelve o'clock'. Admission tickets cost 10s 10d[4] and evening performances commenced at 8 o'clock.[5]

WEDNESDAY 10 AUGUST. MORNING.
ST WERBURGH'S CHURCH

A Sacred Performance

'From the most esteemed works of Handel, Haydn, Mozart, Pergolese, &c'.[6] Catalani sang: 'Angels ever bright and fair', 'Gratias agimus tibi', 'Holy, Holy, Lord God Almighty', 'Sing ye to the Lord', 'The Horse and his Rider hath he thrown into the sea', and 'Ombra adorata aspetta' by Crescentini.[7]

THURSDAY 11 AUGUST. EVENING.
THE ROTUNDA

Part I. A New Overture, Wranitsky

Glee: "See our oars with feather'd spray" [*The Patriot*, Stevenson]

Mrs Lacy, Messrs Garbett, Spray, and Lacy

Aria— Signor Chiodi

Scena and Aria— Miss Hughes: "Vittima sventurata" [*La Vestale*, Pucitta]

Concerto— Violin— Mme Gerbini

Song— Mr Lacy: "Angel of Life"— accompanied on the violoncello by Mr Lindley [Callcott]

Duo— Mrs Bianchi Lacy and Miss Hughes, Fioravanti

Bravura— Mme Catalani: "Della tromba" Pucitta

Part II. Concerto: Trumpet, Mr Willman

Ballad: Mrs [Bianchi] Lacy, "O Nanny", Carter

Duo-buffo: Mme Catalani and Signor Chiodi: ["Con pazienza"] from the celebrated Opera of *Il Fanatico per la musica*, [Fioravanti]

Concerto— Violoncello: Mr Lindley

Cavatina— Mme Catalani: "Più non ho", [Sacchini]

Duet— Mrs Bianchi Lacy and Mr Lacy: "Haste my Nanette", Travers

Aria, with variations— Mme Catalani: "O dolce concento", [Mozart]

Finale: "God Save the King". Verse: Mme Catalani'.[8]

SATURDAY 13 AUGUST. MORNING,.
ST WERBURGH'S CHURCH

'A selection from Haydn's celebrated sacred Oratorio of [*The*] *Creation* in which Mme Catalani will sing, "With verdure clad", "The Marvellous Work", "Gloria Patri" [Cimarosa]. And (by particular desire), "Sing ye to the Lord", &c &c'.[9]

APPENDIX C: CATALANI'S DUBLIN CONCERTS IN 1814

MONDAY 15 AUGUST. EVENING.
THE ROTUNDA

Part I. New Sinfonia, Romberg

Ballad: "Fly swift ye zephyrs". Master Stansbury

Aria— Mme Ferlendis

Duet: "Vaga fravola odorosa" [Cimarosa. Occasionally introduced into Paer's opera of *Il Principe di Taranto*]. Miss Hughes and Signor Chiodi

Concerto— oboe: Signor Ferlendis

Terzetto: "O dolce e caro istante" [*Gli Orazi e i Curiazi*, Cimarosa] Mme Ferlendis, Miss Hughes, and Signor Chiodi

Bravura: "Sù, Griselda, coraggio!" [*Griselda*, Paer] Mme Catalani, (accompanied on the violin by Mr Loder)

Part II. Overture. *Die Zauberflöte*, Mozart

Duet: "Ah perdona" [*La Clemenza di Tito*, Mozart] Mme Catalani and Miss Hughes

Aria: Mme Ferlendis

Aria: Signor Chiodi

Concerto: Violoncello: Mr Lindley

Grand Recit. from *Semiramide*: "Lasciami per pietà" and the celebrated cavatina, "O quanto l'anima" [introduced into both of Portugal's operas, *La Morte di Semiramide* and *Il Ritorno di Serse*] by Mme Catalani.

To conclude (by general desire) with "Rule Britannia"—Verses: Mme Catalani'.[10]

WEDNESDAY 7 SEPTEMBER. EVENING.
THEATRE ROYAL, CROW STREET

'A Grand Concert of Vocal and Instrumental Music. The Concert to be concluded by "Rule Britannia" by Mme Catalani. After which, a comic Scene from a favourite Italian Opera [*Il calzolaio o La bacchetta portentosa* by M. A. Portugal] by Signor Chiodi and Mme Ferlendis... To conclude with the much-admired Scene from the Serious Opera of *Semiramide* by Madame Catalani in which she will sing the favourite Bravura Song of "Son Regina"'.[11]

THURSDAY 8 SEPTEMBER. EVENING.
THEATRE ROYAL, CROW STREET

'A Grand Concert of Vocal and Instrumental Music; particulars will be given in hand bills. The concert to be concluded by "God Save the King" by Mme Catalani— After which Signor Chiodi and Mme Ferlendis will perform a Scene from the Comic Opera of *Il calzolaio*... To conclude with the much admired scene from the serious opera of *Mitridate* by Mme Catalani'.[12]

FRIDAY 9 SEPTEMBER.
THEATRE ROYAL, CROW STREET

'Mme Catalani last night...concluded with "God Save the King" which the audience rather clamorously called for, and we do not recollect that she ever exhibited greater power and harmony of voice than in the first verse of the song which she gave in deference to the general desire of the house. The bewitching archness and grace with which she performed a scene in the Comic Opera of *Le Due Nozze e un sol Marito* [with Mme Ferlendis] and the celebrated song of "Papà non dite di no" [included in a comic scena with Chiodi], produced applause more in acclamation than in the ordinary manner of bestowing it. An "Air", with variations was also given with all the effects of vocal excellence and skill'.[13]

APPENDIX C: CATALANI'S DUBLIN CONCERTS IN 1814

SATURDAY 10 SEPTEMBER. EVENING.
THEATRE ROYAL, CROW STREET

'This Present Evening will be performed A Grand Concert of Vocal and Instrumental Music. After which a Scene from a favourite Comic Opera by Madame Catalani and Signor Chiodi. To conclude with a Comic Scena from *Il Fanatico* by Signor Chiodi and Mme Catalani'.[14]

MONDAY, 12 SEPTEMBER. EVENING.
THEATRE ROYAL, CROW STREET

'A Grand Concert of Vocal and Instrumental Music. After which a scene from the Comic Opera of *Il Tutore Accorto* by Mme Ferlendis (in male attire) and Signor Chiodi. And, by desire, the celebrated scene from *Mitridate*. To conclude with *Il Fanatico per la Musica* in which Mme Catalani will sing the much admired duet of "Con pazienza" with Signor Chiodi, and the air of "Nel cor più non mi sento"'.[15]

THURSDAY 13 SEPTEMBER. MORNING (AT 1 P.M.).
DENMARK STREET CATHOLIC CHURCH

'The public will be glad to hear that Mme Catalani has liberally consented to perform…(tomorrow) in aid of the funds towards completing the new parish Chapel of St Michans, North Anne-street… No apprehensions respecting too great a crowd or want of room need be entertained, as the [Denmark Street] Chapel can easily contain *three thousand* persons'.[16]

'Part I. Overture: *The Creation*, Haydn

Recit: "Comfort ye". Air: "Every Valley" [Handel], Mme Catalani

Chorus: "And the Glory of the Lord"

Solo: "Qui sedes ad dextrum Patris", Signor Chiodi

Concert: Clarionet. Mr Willman

Chorus: "Hallelujah"

Part II Occasional Overture

Air: "Ave Regina", Mme Catalani

—: "Grant" by Cemeranic [?]

Chorus: "Lift up your heads", [Handel]

Grand Bravura: "Gratias agimus tibi", Mme Catalani, accompanied on the clarionet by Mr [T. L.] Willman

Concerto: Violin. Signora Gerbini

Air: "Let the bright Seraphim" [Handel], Miss Hughes, accompanied by Mr H. Willman on the Trumpet

Air: Signor Chiodi

Grand Finale: "God Save the King". Verses— Mme Catalani

Tickets to be had of the Reverend Gentlemen of [St] Mary's Lane Chapel, the Sub-Committee, and at the Principal Music Shops'.[17]

WEDNESDAY 14 SEPTEMBER. EVENING. THEATRE ROYAL, CROW STREET

'A Grand Concert of Vocal and Instrumental Music in which Mme Catalani will sing five of her most celebrated Scenes and Songs, assisted by Mme Ferlendis and Signor Chiodi and Miss Cheese…[and] will have the honour of taking leave of the Audience in the National Air of "God Save the King"'.[18] It transpired that 'she was rather indisposed, but her voice could not fail of its usual attractions'.[19]

Appendix D
Rules and Regulations

To be observed by the several Performers and Persons engaged and employed in the Theatre-Royal, Crow-street, and in the Theatres of Cork and Limerick, and in all other Theatres in which FREDERICK EDWARD JONES Esq. may be interested or concerned.

1. No persons at the time of rehearsal or performance are to wear their hat in the green-room, neither are they to talk vociferously, or enter into such altercations there as may tend to dispute or quarrel. The green-room is a place appropriated for the quiet and regular meeting of the company, to be called from thence and from *thence* only, by the call-boy, to attend their business on the stage. The manager is not to be applied to in *that place*, about any matter of personal complaint. A breach of any of the above articles will subject the person to a penalty of *one Guinea*.

2. Improper or ungentleman-like language or behaviour or quarrelling in any part of the theatre, the offending parties shall forfeit *one Guinea* each— if to the interruption of a rehearsal *two Guineas*— but if a blow be given the first striker shall pay *five Pounds* and be discharged for the remainder of the Season at the option of the manager.

3. No excuse can be given for any gentleman appearing in liquor on the stage; but should any performer be capable of the crime they shall forfeit a *week's salary*; or if such flagitious conduct is repeated, it shall be at the manager's option to retain or discharge such performer for the remainder of the Season.

4. Making the stage stand or not being at the proper entrance after being summoned by the caller, *half-a-Guinea*.

5. Every Lady and Gentleman after being properly warned to attend all rehearsals punctually; the green-room Clock is to regulate the time, ten minutes will be allowed for difference of Clocks— if absent the first scene *half-a-crown*, every scene afterwards *one shilling*— the whole rehearsal *half-a-guinea*— if a new or a revived play the forfeit will be double.

6. Any Performer rehearsing with a book or part in their hand the last rehearsal will forfeit a *Crown*; if a new or revived play a *Guinea*.

7. Any Performer who walks across the stage or stands on it during rehearsal (without the business required their attendance there) shall forfeit *half-a-Guinea*.

8. Any Performer introducing improper Jokes not in the author shall forfeit *one Guinea*.

9. Any representation or advice on the stage so as to disturb the performance or alarm the Audience, to forfeit *half-a-Guinea*.

10. Any Performer obtruding behind the scenes when not called or wanted in the representation, to forfeit *five Shillings*. Performers during the representation are requested not to be at the entrance till called upon. Any Performer violating this regulation to forfeit *half-a-Crown*. Any Performer who suffers his hair-dresser or servant to stand behind the scenes will be also forfeited *five Shillings*. Such persons are only to attend in the performer's dressing-room.

11. Any Performer opening the stage-door except required so to do by the actual business of the representation, to forfeit *one Guinea*.

12. Any Person conversing with the Prompter during representation, or talking loud behind the scenes, to the interruption of the performance, to forfeit *five Shillings*.

13. Every Performer concerned in the first Act of the Play, to be ready dressed in the green-room five minutes before the hour of beginning as expressed in the bills, or to forfeit *one Guinea*. Those performers wanted in the second act to be ready as the first finishes. In like manner, every other act. Those performers who are not in the two last acts of the Play, to be ready to begin the Farce, or be subject to the same penalties. When a change of dress is necessary, proper time to be allowed.

14. All Dresses will be regulated and settled on the morning of the performance, when approved by the manager. Any Performers making any alteration of such dresses, or refusing to wear them, shall forfeit *two Guineas*.

15. Any Performer who neglects or refuses to give out a play, when called upon by the Prompter by the Manager's direction, to forfeit *one Guinea*.

16. Any Performer giving out a Play without authority from the prompter by the Manager's direction he will forfeit *five Pounds*.

17. Any Person advertising or causing to be advertised, or printing or causing to be printed, any Play-bill or Entertainment for the Theatre Royal, without the consent of the manager previously obtained in writing, shall forfeit the sum of *twenty pounds*.

18. After the Manager has allotted any part or parts to a performer, and a reasonable time given for the study— that is at the rate of a length a day— should they be refused, or through neglect not attended to, so as not to be ready when called upon, such performance shall forfeit *ten pounds*; and it shall besides be in the option of the manager to retain or discharge such performer for the remainder of the season.

19. Any Performer performing imperfect after a sufficient time allowed— that is at the rate of a length a day in an old Play or Opera, but in a new play after three rehearsals— shall forfeit *ten pounds* and be discharged for the remainder of the Season at the option of the manager.

20. Any Performer who shall through neglect or illness be incapable of attending the business of the Theatre, though it may be no recompence to the manager for the disappointment, shall not receive any salary during their absence.

21. Should any Performer on any pretence whatever (real illness excepted which must be ascertained on the report of the established Physician or Chirurgeon of the Theatre) absent themselves from the theatre, or refuse doing the business or parts allotted them by the manager for the time being, such performer shall forfeit *ten pounds*: and it shall besides be in the option of the manager to retain or discharge such performer for the remainder of the season.

22. Any Performer, or person or persons belonging to the Theatre Royal, disposing of his or their benefit, benefits, or any part or parts thereof, or permitting tickets to be issued in any name or names but his or theirs, or

allowing any part of the receipts or emoluments arising from such benefit, benefits, or any part or parts thereof, to be for the use of any public or private measure, or for any person or persons whatsoever, without the consent of the manager in writing, first had and obtained, shall forfeit *one hundred pounds Sterling*.

23. The Prompter's boy or caller, must attend to summon the Ladies and Gentlemen at all representations or rehearsals, or forfeit *half-a-Crown*.

24. The Prompter will take care to provide the performer with a book or part of a new Piece, or forfeit *half-a-Guinea*; but the Performer must be careful for all old or revived Pieces to provide their own, or forfeit *half-a-Guinea*.

25. If the Prompter converses with any person during representation, or is guilty of any neglect to the prejudice of the performance, he shall forfeit *two Guineas*.

26. If the Prompter should be guilty of any neglect in his office, or not forfeit such performers as incur the penalties annexed to the non-observance of the Rules and Regulations of the Theatre, he shall forfeit for each office or omission, *two Guineas*.

27. If the Prompter should remit any forfeit that has been imposed, or omit returning the same to the treasurer, he shall forfeit *one Guinea*.

28. Any Performer not leaving notice on a Play day where they may be immediately found, in case there should be a change of Entertainments, shall forfeit *five Shillings*.

29. As there sometimes is an absolute necessity for a sudden change of Play or Farce, any Lady or Gentleman refusing to perform any character which they may have formerly played, shall forfeit *two Guineas*.

30. Any Performer…[the remainder has 'fallen off' the page]

31. Should additional Regulations be judged necessary for the good of the Community, such Regulations being hung up in the Green-room must be observed by the Performers under the penalty annexed to them.

(*The Irish Dramatic Censor*, No. III, Dublin 1811/12, pp 49-54)

References

Chapter I—Interval Music 1798-1799
1. Gilbert, John T., *A History of the City of Dublin*, Dublin 1861, 3 vols, II, p. 254.
2. ibid., II, p. 214.
3. J. D. Herbert [Dowling], *Irish Varieties for the last fifty years*, London 1836, p. 267.
4. Gilbert, II, p. 215.
5. ibid.
6. ibid., II, p. 216.
7. ibid., II, p. 217.
8. ibid., Appendix V, ix.
9. ibid., II, p. 218.
10. ibid., Appendix V, ix.
11. ibid., II, p. 218.
12. ibid., II, pp. 218, 219.
13. *Saunders' News-Letter*, 29 January 1798.
14. ibid., 30 January 1798.
15. Gilbert, II, p. 219.
16. ibid., II, 235, 247.
17. Warburton, J., Whitelaw, Rev. J. and Walsh, Rev. R., *The History of the City of Dublin*, London 1818, p. 1121.
18. ibid., p. 1120.
19. Gilbert, II, pp. 316, 321.
20. *Saunders' News-Letter*, 20 December 1798.
21. ibid., 4 July 1798.
22. ibid., 5 July 1798.
23. *The Monthly Mirror*, July 1798.
24. Gilbert, II, pp. 219, 220.
25. Microfilm of Minute Book of the Incorporated Musical Fund Society, in the National Library, Dublin.
26. *The Globe*, 17 October 1811.
27. *Saunders' News-Letter*, 9 January 1799.
28. ibid., 13 February 1799.
29. *Freeman's Journal*, 21 March 1799.
30. ibid., 28 May 1799.
31. ibid., 1 June 1799.
32. ibid., 30 May 1799.
33. ibid., 6 June 1799.
34. ibid., 14 December 1799.
35. ibid., 17 December 1799.

Chapter 2 Curtain Raiser 1800-1807
1. *Freeman's Journal*, 12 June 1800.
2. Kelly, Michael, (ed. Roger Fiske), *Reminiscences* London 1975, pp. 256-7.
3. *The Monthly Mirror*, February 1800.
4. ibid.
5. *Freeman's Journal*, 6 November 1800.
6. ibid., 28 April 1801.
7. *Saunders' News-Letter*, 14 November 1803.
8. *Freeman's Journal*, 2 May 1801.
9. ibid., 5 May 1801.
10. *Cyclopaedian Magazine*, April 1807, p. 209.
11. *Dramatic Censor*, 4v II, p. 122.
12. *Saunders' News-Letter*, 4 July 1801.
13. ibid., 7 July 1801.
14. ibid., 13 July 1801.
15. *Hibernian Journal*, 23 October 1801.
16. *Saunders' News-Letter*, 31 October 1801.
17. *Freeman's Journal*, 6 November 1802.
18. Walker's *Hibernian Magazine*, April 1805.
19. ibid.
20. *Freeman's Journal*, 3 March 1803.
21. *The Examiner*, 22 December 1816.
22. *Freeman's Journal*, 12 July 1803.
23. Walker's *Hibernian Magazine*, September 1803.
24. ibid.
25. *Familiar Epistles to Frederick E. J- s l sq. on the Present State of the Irish Stage*, 4th ed., Dublin 1805, p. xix.
26. ibid., p. 31.
27. ibid., pp. 69-70.
28. ibid., p. 69.
29. ibid., pp. 118-9.
30. ibid., p. 156.
31. ibid., p. 131.
32. ibid., pp. 132-3.
33. ibid., pp. 155-6.
34. *Monthly Panorama*, February 1810.
35. *Freeman's Journal*, 21 January 1804.
36. Kelly, p. 270.
37. *Freeman's Journal*, 29 January 1805.
38. ibid., 22 February 1806.
39. ibid.
40. *The Monthly Mirror*, March 1806.

41. *Dublin Evening Post*, 16 April 1807
42. ibid., 25 February 1806
43. *Saunders' News-Letter*, 21 February 1814
44. ibid., 17 February 1806
45. *The Monthly Mirror*, February 1806
46. Hazlitt, William, (ed. W. Spencer Jackson), *A View of the English Stage*, London 1906, p. 76
47. *The Musical World*, 4 November 1841
48. *Dublin Evening Post*, 21 June 1806
49. ibid.
50. *Freeman's Journal*, 15 August 1806
51. Walker's *Hibernian Magazine*, July 1804
52. *Faulkner's Dublin Journal*, 19 August 1806
53. *Saunders' News-Letter*, 29 August 1806
54. ibid., 3 September 1806
55. ibid., 13 September 1804
56. *Freeman's Journal*, 29 December 1804
57. *Saunders' News-Letter*, 28 March 1805
58. ibid., 18 March 1805
59. ibid., 4 and 10 April 1805
60. ibid., 15 January 1806
61. ibid., 31 January 1806
62. ibid., 6 March 1806
63. ibid., 3 April 1806
64. ibid., 12 June 1806
65. ibid.
66. *Freeman's Journal*, 24 March 1807
67. *Hibernian Journal*, 17 June 1807
68. Read, Charles A. and Katherine Tynan Hinkson, (eds), *The Cabinet of Irish Literature*, London 1902, 4 v, II, p. 177
69. *Dublin Evening Post*, 5 March 1807
70. ibid., 28 February 1807
71. *Cyclopaedian Magazine*, March 1807, p. 186
72. *New Monthly Magazine*, September 1835
73. *Cyclopaedian Magazine*, January 1807
74. *Freeman's Journal*, 29 May 1807
75. *Dublin Evening Post*, 2 June 1807
76. ibid., 21 July 1807
77. ibid., 29 March 1808
78. ibid., 22 April 1809
79. *Cyclopaedian Magazine*, August 1807, p. 508

Chapter 3: Curtain up on Catalani. 1807
1. Radiciotti, Giuseppe , *Teatro Musica e Musicisti in Sinigaglia*, Milan 1893, p. 149
2. ibid., p. 131
3. Wiel, Taddeo, *I Teatri Musicali Veneziani del Settecento*, Venice 1897, pp. 366, 372, 373
4. Radiciotti, p. 149
5. ibid.
6. ibid., p. 151
7. Escudier, Marie and Léon, *Vie et Aventures des Cantatrices Célèbres*, Paris 1856, p. 240
8. Wiel, p. 478
9. ibid., pp. 478, 479
10. Sartori, C. (compiler), *Catalogo Unico dei Libretti Italiani a stampa fino all'anno 1800*, Milan 1973-85, 19 V, IX
11. Rinaldi, Mario, *Due Secoli di Musica al Teatro Argentina*, Florence, 1978, vol I, p. 327
12. Curiel, Carlo L., *Il Teatro S. Pietro di Trieste, 1690-1801*, Milan 1937, pp. 360-1
13. Wiel, p. 513
14. Gatti, Carlo, *Il Teatro alla Scala*, Milan 1964, 2 vols, II, p. 18
15. ibid.
16. Ruders, C. I., *Portugisisk Resa-Beskrifven i Bref till Vänner*, Stockholm 1805-9, 3 Parts, III, p. 110
17. Compiled from Pereíra Peíxoto d'Almeida Carvalhaes, Manoel, *Marcos Portugal—na sua musica dramatica*, Lisbon 1910, and Fonseca Benevides, Francisco da, *O Real Theatro de S. Carlos*, Lisbon 1883
18. Ruders, III, p. 129
19. Radiciotti, pp. 151, 152
20. ibid., p. 152
21. Pereíra Peíxoto, p. 150 *and* Fonseca Benevides, p. 88
22. Carmena y Millan, Luis, *Crónica de la Ópera Italiana en Madrid*, Madrid 1878, p. 43
23. *L'Ambigu*, 20 December 1806
24. Detcheverry, Arnaud, *Histoire des Théâtres de Bordeaux*, Bordeaux 1860, p. 201
25. Radiciotti, p. 152
26. ibid., pp. 152, 153
27. Ferrari, Giacomo Gotifredo, *Aneddoti piacevoli e interessanti occorsi nella vita*, London 1830, 2 vols., II, pp. 131, 321
28. Gardiner, William, *Sights in Italy*, London 1847, p. 325
29. Smith, William C., *The Italian Opera and Contemporary Ballet in London, 1789-1820*, London 1955, p. 84
30. Kelly, p. 297
31. ibid., p. 56

32. Boaden, James, *Memoirs of the Life of John Philip Kemble Esq.*, London 1825, 2 vols, I, p. 449
33. Burney, Charles, *A General History of Music*, London 1776-1789, 4 vols, IV, p. 528
34. *The Universal Magazine*, July 1807
35. Mount Edgcumbe, Earl of, *Musical Reminiscences*, London 1834, p. 16
36. Ferrari, I, p. 171
37. Curiel, p. 151
38. Kelly, p. 56
39. *Journal de Paris*, 27 January 1789
40. *The Examiner*, 9 May 1813
41. *The Harmonicon*, January 1830
42. *The Gentleman's Magazine*, November 1822
43. Smith, William C., p. 36
44. ibid., p. 89
45. Fenner, Theodore, *Leigh Hunt and Opera Criticism. The "Examiner" Years 1808-1821*, Kansas 1972, p. 286 fn
46. *The Quarterly Musical Magazine*, 1825, VII, p. 475
47. Dutton, Thomas, *The Dramatic Censor; or, Weekly Theatrical Report*, London 1800, I, p. 348
48. Houtchens, L. H. and C. W. (eds), *Leigh Hunt's Dramatic Criticism 1808-1831*, New York 1949, p. 151
49. *Freeman's Journal*, 2 September 1807
50. *Louis Spohr's Autobiography*, Trans., London 1878 [1865], 2 vols, II p. 26.
51. Pleasants, Henry, (trans. and ed), *The Musical Journey of Louis Spohr*, Oklahoma 1961, p. 187
52. *L'Ambigu*, 20 December 1806
53. *Kurze Lebensbeschreibung der Madame Angelica Catalani*, Berlin 1816, p. 8
54. W[endt], A[madeus], *I Giudizj dell'Europa intorno alla Signora Catalani*, Milan 1816, p. 35
55. *The Cabinet*, March 1807
56. Greig, James (ed), *The Farington Diary by Joseph Farington R. A.*, London 1924, 8 vols, IV, p. 237
57. *The Quarterly Musical Magazine*, 1818, I, p. 181
58. *The Daily Advertiser*, 23 January 1807
59. ibid.
60. Spohr, p. 187
61. Curiel, p. 360
62. *I Giudizj dell'Europa*, p. 35
63. *The Cabinet*, April 1807
64. *Il Monitore di Roma*, 8 February 1799
65. *The Quarterly Musical Magazine*, 1818, I, p. 183
66. Thurner, A., *Les Reines du Chant*, Paris 1883, p. 95
67. *The Cabinet*, March 1807
68. *The Quarterly Musical Magazine*, 1818, I, p. 183
69. *L'Ambigu*, 20 December 1806
70. Radicotti, p. 158
71. Thurner, p. 95
72. *The Quarterly Musical Magazine*, 1818, I, p. 184
73. Mount Edgcumbe, p. 98
74. *Knight's Quarterly Magazine*, January-April 1824, p. 88
75. *The Quarterly Musical Magazine*, 1818, I, p. 184
76. *The Examiner*, 2 February 1812
77. *The Morning Post*, 6 August 1807
78. *The Daily Advertiser*, 13 August 1807
79. *The Morning Post*, 19 August 1807
80. ibid., 18 August 1807
81. *The British Press*, 15 June 1807
82. *The Morning Post*, 19 August 1807
83. *The British Press*, 19 August 1807
84. Kelly, p. 298
85. Aspinall, A. (ed), *Mrs Jordan and her Family*, London 1951, p. 88
86. ibid., pp. 88, 89
87. ibid., p. 96
88. ibid., p. 97
89. ibid., pp. 95, 96
90. *Freeman's Journal*, 9 September 1808
91. *The Daily Advertiser*, 10 June 1807
92. Radiciotti, p. 152
93. Mount Edgcumbe, fn 107
94. Lewis, Lady Theresa (ed), *Extracts of the Journals and Correspondence of Miss Berry*, London 1865, 3 vols, II, p. 339
95. Farington, VII, p. 204
96. ibid., 35 fn
97. Castle. Egerton (ed), *The Jerningham Letters (1780-1843)*, London 1896, 2 vols, I, p. 283
98. *The Morning Chronicle*, 13 January 1807
99. Farington, VII, p. 204
100. Westminster, the Dean of, (ed), *The Remains of the late Mrs Richard Trench*, London 1862, p. 232
101. Farington, IV, p. 86
102. *Cyclopaedian Magazine*, May 1807
103. Ibid., June 1807

104. *Freeman's Journal*, 21 August 1807
105. *The British Press*, 29 March 1808
106. *Freeman's Journal*, 21 August 1807
107. *The Morning Post*, 21 November 1807
108. *Saunders' News-Letter*, 20 August 1807
109. *Cyclopaedian Magazine*, August 1807
110. *The Morning Post*, 25 August 1807
111. *Freeman's Journal*, 19 August 1807
112. *Dublin Evening Post*, 25 August 1807
113. *Freeman's Journal*, 21 August 1807
114. *Dublin Evening Post*, 25 August 1807
115. *Freeman's Journal*, 20 August 1807
116. *Dublin Evening Post*, 25 August 1807
117. ibid.
118. *Freeman's Journal*, 20 August 1807
119. *Dublin Evening Post*, 25 August 1807
120. *Freeman's Journal*, 25 August 1807
121. ibid., 28 August 1807
122. ibid., 2 September 1807
123. ibid., 28 August 1807
124. *The Morning Herald*, 7 September 1807
125. *Hibernian Journal*, 28 August 1807
126. *Saunders' News-Letter*, 20 August 1807
127. Carse, A., *The Orchestra from Beethoven to Berlioz*, Cambridge 1948, pp. 199, 200 (trans from Fétis, F. J., *Curiosités Historiques de la Musique*, Paris 1830, p. 250)
128. Kelly, p. 298
129. ibid., p. 297
130. *The Morning Post*, 5 September 1807
131. *The Morning Herald*, 12 September 1807
132. *Cyclopaedian Magazine*, April 1807
133. *The Morning Herald*, 9 July 1806
134. ibid.
135. *Freeman's Journal*, 2 September 1807
136. *The British Press*, 29 August 1807
137. *Hibernian Journal*, 12 October 1807
138. *The Morning Herald*, 7 September 1807
139. *Saunders' News-Letter*, 2 September 1807
140. ibid.
141. *Freeman's Journal*, 1 September 1807
142. *Saunders' News-Letter*, 2 September 1807
143. *The Morning Herald*, 12 December 1808
144. *Freeman's Journal*, 2 September 1807
145. *The Quarterly Musical Magazine*, 1818, I, pp. 182, 183
146. *Freeman's Journal*, 9 September 1807
147. ibid., 11 September 1807
148. ibid., 14 September 1807
149. ibid., 7 September 1807
150. *Hibernian Journal*, 4 September 1807
151. *Freeman's Journal*, 7 September 1807
152. *The Morning Chronicle*, 16 July 1807
153. *The Morning Herald*, 15 September 1807
154. *The Morning Advertiser*, 12 September 1807
155. *The Morning Herald*, 25 September 1807
156. *Faulkner's Dublin Journal*, 12 September 1807
157. *Freeman's Journal*, 14 September 1807
158. ibid., 15 September 1807
159. ibid., 21 September 1807
160. ibid., 17 September 1807
161. *The Morning Advertiser*, 5 August 1807
162. *The British Press*, 21 September 1807
163. *Dublin Evening Post*, 24 September 1807
164. *Freeman's Journal*, 18 September 1807

Chapter 4 Catalani Encore. 1808
1. Kelly, p. 303
2. *The British Press*, 2 August 1808
3. *The Morning Post*, 20 August 1808
4. Kelly, pp. 303, 304
5. ibid., p. 304
6. ibid., p. 306
7. *Saunders' News-Letter*, 29 August 1808
8. *Limerick Gazette*, 7, 11 October 1808
9. Kelly, p. 307
10. ibid.
11. Hogan, Charles Beecher (ed.), *The London Stage, 1660-1800*, Carbondale, Illinois, 1960 [1968], Part 5, p. 2189
12. *New Monthly Magazine*, September 1822
13. *The Courier*, 1 June 1807
14. *The Morning Chronicle*, 30 May 1807
15. ibid., 26 February 1812
16. ibid.
17. ibid., 17 February 1812
18. Speyer, Edward, *Wilhelm Speyer, der Liederkomponist (1790-1878)*, Munich 1925, pp. 51, 52
19. Moscheles, Charlotte, *Life of Moscheles*, London 1873, 2 vols, I, p. 119
20. ibid., p. 120
21. Radiciotti, p. 159
22. *The Morning Chronicle*, 4 November 1808
23. Radiciotti, p. 159

24. Bottenheim, S. A. M., *De Opera in Nederland*, Amsterdam 1946, pp. 107, 108
25. *The Morning Post*, 24 February 1808
26. ibid., 9 March 1808
27. ibid., 30 March 1808
28. *The Morning Chronicle*, 13 February 1808
29. *The Morning Post*, 6 April 1808
30. ibid., 24 February 1808
31. *The Morning Chronicle*, 13 February 1808
32. *The Morning Herald*, 6 April 1808
33. *The Morning Post*, 6 April 1808
34. ibid., 9 March 1808
35. *The Morning Chronicle*, 24 December 1806
36. *The Harmonicon*, February 1830
37. Mount Edgcumbe, p. 102
38. *The Courier*, 29 December 1806
39. *Journal des Luxus und der Moden*, November 1812, p. 747
40. *The Monthly Mirror*, May 1807
41. *The Morning Chronicle*, 25 February 1807
42. *The Morning Post*, 23 October 1806
43. ibid., 15 December 1806
44. *The Morning Chronicle*, 5 January 1808
45. *Freeman's Journal*, 7 September 1808
46. ibid., 23 December 1808
47. *The Morning Post*, 15 April 1807
48. *The Morning Herald*, 15 April 1807
49. *The Morning Chronicle*, 14 September 1808
50. *Freeman's Journal*, 21 January 1804
51. ibid., 23 February 1811
52. ibid., 3 March 1803
53. *Dublin Evening Post*, 27 January 1807
54. *Saunders' News-Letter*, 22 November 1804
55. *Freeman's Journal*, 19 November 1814
56. ibid., 6 March 1815
57. *Dublin Evening Post*, 24 January 1807
58. *Saunders' News-Letter*, 16 March 1803
59. ibid., 20 January 1801
60. *Freeman's Journal*, 14 November 1807
61. ibid., 25 August 1808
62. *The British Press*, 11 January 1808
63. *Dublin Evening Post*, 27 August 1808
64. *Faulkner's Dublin Journal*, 23 August 1808
65. ibid., 20 August 1808
66. *Saunders' News-Letter*, 23 August 1808
67. *Dublin Evening Post* and *Freeman's Journal*, 23 August 1808
68. *Freeman's Journal*, 24 August 1808
69. *Faulkner's Dublin Journal*, 27 August 1808
70. *The Harmonicon*, February 1830
71. *Faulkner's Dublin Journal*, 27 August 1808
72. *The Morning Chronicle*, 14 September 1808
73. *The Daily Advertiser*, 27 August 1808
74. *The Morning Herald*, 12 December 1808
75. *The Daily Advertiser*, 27 August 1808
76. *Faulkner's Dublin Journal*, 8 September 1808
77. *The Examiner*, 31 January 1808
78. ibid.
79. Kelly, p. 304
80. *Freeman's Journal*, 30 August 1808
81. ibid., 27 August 1808
82. *The British Press*, 10 October 1806
83. ibid., 15 December 1806
84. *The Times*, 5 March 1812; {*The Morning Post*, 15 December 1806]
85. ibid; [*The Morning Post*], 9 June 1808
86. *The Examiner*, 1 April 1810
87. *Le Beau Monde*, January 1807
88. *The National Register*, 21 February 1808
89. ibid., 1 February 1808
90. ibid., 21 February 1808
91. *The Monthly Mirror*, February 1807
92. *The Daily Advertiser*, 20 February 1807
93. Barrington, Sir Jonah, *Personal Sketches of His Own Times*, London 1827, 3 vols, II, p. 197
94. *Freeman's Journal*, 3 July 1800
95. ibid., 31 August 1808
96. *L'Ambigu*, 30 June 1808
97. *The Morning Chronicle*, 2 March 1808
98. *Freeman's Journal*, 2 September 1808
99. ibid.
100. ibid., 12 September 1808
101. Kelly, p. 299
102. *Dublin Evening Post*, 10 September 1808
103. *The Morning Post*, 17 April 1807
104. *Freeman's Journal*, 16 September 1808
105. *The Morning Chronicle*, 24 June 1808
106. *The Morning Post*, 25 June 1808
107. Parke, W. T., *Musical Memoirs*, London 1830, 2 vols, II, pp. 11, 13
108. *Hibernian Journal*, 19 September 1808
109. *Freeman's Journal*, 28 September 1808
110. *Faulkner's Dublin Journal*, 11, 13 October 1808; *Hibernian Journal* and *Freeman's Journal*,

12 October 1808
111. *Dublin Evening Post*, 13 October 1808
112. *The British Press*, 3 October 1808
113. *Freeman's Journal*, 27 October 1808

Chapter 5 The Great God Braham 1808-1811

1. Gilbert II, p. 230
2. *Cyclopaedian Magazine*, March, April, June 1807.
3. ibid., March 1807
4. Gilbert II, p. 231
5. ibid., pp. 232, 233
6. ibid., p. 235
7. *Cyclopaedian Magazine*, November 1808
8. Gilbert II, p. 235
9. *The Pilot*, 12 April 1811
10. Parke II, p. 5
11. Clarke, William Smith, *The Irish Stage in the County Towns, 1720 to 1800*, Oxford 1965, p. 192
12. *Freeman's Journal*, 30 November 1808
13. *Cyclopaedian Magazine*, November 1808
14. ibid.
15. ibid.
16. ibid.
17. *Freeman's Journal*, 2, 5, 8, 20 December 1808; 2,. 19, 27 January 1809
18. *Cyclopaedian Magazine*, March 1809
19. ibid., November 1808
20. ibid., March 1809
21. Nicoll, Allardyce, *A History of English Drama 1660-1900*, Cambridge 1955, 6v., IV, p. 391; White, Eric Walter, *The Rise of English Opera*, London 1951, p. 241
22. *Cyclopaedian Magazine*, March 1809
23. ibid.
24. ibid.
25. *Saunders' News-Letter*, 4 November 1809
26. *Faulkner's Dublin Journal*, 24 October 1809
27. *Freeman's Journal*, 6 November 1809
28. ibid., 4 November 1809
29. *Faulkner's Dublin Journal*, 24 October 1809
30. Gilbert II, pp. 236, 237
31. *Freeman's Journal*, 6 November 1809
32. ibid., 4 October 1809
33. ibid.
34. *Faulkner's Dublin Journal*, 31 October 1809
35. Lamb, Charles, [Essays of] 'Imperfect Sympathies' in *Elia* (First Series), London 1823, p. 142
36. Edwards, H. Sutherland, *The Lyrical Drama*, London 1881, 2 v, II, p. 157
37. *The Cornhill Magazine*, December 1865, p. 692
38. *The Morning Herald*, 2 May 1796
39. *Allgemeine Musikalische Zeitung*, 4 September 1801
40. *Quarterly Musical Magazine*, 1818 I, pp. 88-93
41. Lamb (*Elia*), p. 142
42. *Quarterly Musical Magazine*, 1818 I, pp. 93, 94
43. ibid., p. 93
44. *New Monthly Magazine*, April 1835, p. 446
45. *The Examiner*, 12 February 1815
46. [Cox, J. E.], *Musical Recollections of the Last Half-Century*, London 1872, 2v, I, p. 66
47. Mount Edgcumbe, p. 95
48. [Mackintosh, Matthew], *Stage Reminiscences ... by an old Stager*, Glasgow 1866, p. 25
49. *Monthly Panorama*, January 1810, p. 59
50. ibid.
51. *The Correspondent*, 7 November 1809
52. *Saunders' News-Letter*, 6 November 1809
53. Certificate at Westminster City Archives Department
54. *Dictionary of National Biography*, London 1894, xxxix (Wilson, C. H.), p. 209
55. *The Examiner*, 12 February 1815
56. Fenner, *Leigh Hunt and Opera Criticism*, p. 139
57. *Freeman's Journal*, [16], 21, 22, [24], 26, 28, 30, 31 July; 1, 2 August, 1806
58. *Monthly Panorama*, January 1810, p. 62
59. *European Magazine and London Review*, May 1815
60. Robson, William, *The Old Play-goer*, London 1846, p. 198
61. *Monthly Panorama*, January 1810, p. 62
62. *Saunders' News-Letter*, 13 November 1809
63. *Dublin Evening Post*, 11 November 1809
64. ibid., 16 November 1809
65. *Saunders' News-Letter*, 21 November 1809
66. *Dublin Evening Post*, 30 November 1809
67. ibid., 11 November 1809
68. *Saunders' News-Letter*, 20 November 1809
69. *Dublin Evening Post*, 12 December 1809
70. *The Correspondent*, 30 December 1809
71. *Dublin Evening Post*, 14 December 1809
72. ibid.

73. ibid.
74. *The Correspondent*, 14 December 1809
75. ibid., 30 December 1809
76. *Freeman's Journal*, 22 December 1809.
77. *The Correspondent*, 23 December 1809
78. ibid., 26 December 1809
79. ibid., 3 January 1810
80. *Freeman's Journal*, 3 January 1810
81. [Genest, John], *Some Account of the English Stage from the Restoration in 1660 to 1830*, Bath, 1832, 10v, VII, pp. 706, 707
82. *Monthly Panorama*, February 1810
83. *The Monthly Mirror*, January 1810
84. Stieger, F., *Opernlexikon*, Tutzing 1975
85. Nicoll, Allardyce, IV, p. 326
86. *Hibernia Magazine*, July- December [August] 1810, p. 122
87. *Freeman's Journal*, 11 January 1811
88. *The Correspondent*, 19 February 1811
89. Kelly, p. 282
90. ibid.
91. *Freeman's Journal*, 2 May 1811
92. Preface to libretto. Printed 1811 by Graisberry and Campbell, 10 Back Lane, Dublin, p. xi
93. *Freeman's Journal*, 13 August 1811
94. ibid., 30 July 1811
95. *Allgemeine Musikalische Zeitung*, 12 February 1817, pp. 121, 128, 129
96. *The Examiner*, 12 July 1812
97. *Saunders' News-Letter*, 1 August 1811
98. *St James's Chronicle*, 20/23 April 1811.

Chapter 6—Mozart Adapted 1811

1. *Freeman's Journal*, 9 October 1811
2. *The Globe* , 9 August 1811
3. *Allgemeine Musikalische Zeitung*, 27 October 1802, p. 87
4. ibid., 28 August 1805, p. 766
5. ibid., 15 October 1806, p. 41
6. *Berlinische Musikalische Zeitung*, January-June 1805, p. 72
7. Bottenheim, p. 109
8. Gordon, Pryse Lockhart, *Personal Memoirs; or, Reminiscences of Men and Manners*, London 1830, 2v, II p. 94
9. *The Examiner*, 30 December 1810
10. ibid.
11. Gordon, II, pp. 96, 97
12. *Freeman's Journal*, 21 August 1811
13. *The Morning Chronicle*, 5 December 1810
14. ibid., 23 May 1811
15. *The Monthly Mirror*, December 1805, p. 407
16. ibid., April 1806, p. 266
17. *La Belle Assemblée* April 1806, p. 169
18. Parke, II, p. 4
19. Mount Edgcumbe, pp. 113, 114
20. *Quarterly Musical Magazine*, 1818, I, p. 403
21. *Caledonian Mercury*, 19 October 1811
22. *The Morning Chronicle*, 24 December 1806
23. *Johnson's Sunday Monitor*, 20 April 1806
24. *Revue et Gazette Musicale*, 14 January 1877
25. *The Morning Chronicle*, 16 February 1814
26. ibid.
27. Aspinall, A. (Ed.), *Letters of the Princess Charlotte 1811-1817*, London 1949, p. 4
28. *Allgemeine Musikalische Zeitung*, 29 November 1820, p. 813
29. ibid., 4 September 1822, p. 593
30. *La Quotidienne*, 25 March 1822
31. Giazotto, Remo, *Giovan Battista Viotti*, Milan 1956, Appendix, p. 282
32. *The Court Journal*, 24 March 1838
33. *Allgemeine Musikalische Zeitung*, 23 March 1808, pp. 403, 404
34. *The Courier*, 23 January 1811
35. *The Times*, 3 June 1811
36. *The Courier*, 23 January 1811
37. *The Morning Chronicle*, 20 February 1811
38. *The Globe*, 4 May 1812
39. *The Times*, 3 June 1811
40. Mount Edgcumbe, p. 104
41. Rogge, H. C., 'De Opera te Amsterdam', in *Oud-Holland*, Amsterdam 1887, p. 256
42. Programme, 6 June 1810, George Smart Collection at British Library, C.61.g.18
43. *The Morning Post*, 6 June 1810
44. Programme, 6 June 1810, George Smart Collection at British Library, C.61.g.18
45. *The Morning Chronicle*, 26 June 1812
46. ibid., 24 February 1812
47. *Saunders' News-Letter*, 17 August 1811
48. ibid., 14 [19] August 1811
49. ibid. 14 August 1811
50. Kelly, p. 319
51. *Hibernian Magazine*, July 1811, p. 374

52. *Saunders' News-Letter*, 19 August 1811
53. ibid., 28 August 1811
54. *The Morning Post*, 20 June 1806
55. *The Sun*, 13 May 1811
56. *Freeman's Journal*, 12 February 1805
57. ibid., 19 February 1805
58. *Quarterly Musical Magazine*, 1825, VII, pp. 189, 190
59. *The Globe*, 4 May 1812
60. *The Morning Chronicle*, 13 and 16 March 1813
61. ibid., 26 April 1813
62. *The Harmonicon*, March 1830, p. 114
63. ibid.
64. *The Examiner*, 19 May 1811
65. *Hibernian Journal*, 31 August 1811
66. *Freeman's Journal*, 20 September 1811
67. *Edinburgh Advertiser*, 1 November 1811
68. *The Examiner*, 19 May 1811
69. ibid., 4 August 1816
70. *Freeman's Journal*, 12 September 1811
71. Gutierrez, Beniamino, *Il Teatro Carcano*, (1803-1914), Milan 1914, p. 17
72. ibid., p. 17fn
73. ibid., p. 23
74. Sorge, Giuseppe, *I Teatri di Palermo nei secoli XVI-XVII-XVIII*, Palermo 1926, p. 412
75. Stieger, *Opernlexikon*, Tutzing 1975
76. Manferrari, Umberto, *Dizionario Universale delle Opere Melodrammatiche*, Florence, 1954-55, 3v, I p. 371
77. ibid.
78. *Allgemeine Musikalische Zeitung*, 13 August 1817, p. 555
79. *Saunders' News-Letter*, 12 September 1811
80. ibid., 30 March 1799
81. ibid., 16 September 1811
82. *The Globe*, 2 March 1812
83. *Saunders' News-Letter*, 18 September 1811
84. *Hibernian Journal*, 25 September 1811
85. ibid., [*Saunders' News-Letter*], 1 October 1811
86. Kelly, p. 319
87. *Saunders' News-Letter*, 14 August 1811
88. ibid., 22 August 1811
89. *Freeman's Journal*, 24 September 1811
90. *Saunders' News-Letter*, 23 September 1811
91. *Freeman's Journal*, 25 September 1811
92. Kelly, p. 319
93. *Freeman's Journal*, 7 October 1811
94. ibid., 11 February 1812

Chapter 7—Catalani Ritorna 1812-1814

1. *Leinster Journal*, 7 October 1809, (Reported *The Private Theatre of Kilkenny*, ? Kilkenny 1825, p. 48
2. *Freeman's Journal*, 13 May 1812
3. *The Examiner*, 15 September 1811
4. *The Alfred*, 10 September 1811
5. Vocal score at British Library, G 176
6. Russell, Lord John, (Ed.), *Memoirs, Journal, and Correspondence of Thomas Moore*, London 1853-56, 8 vols, II pp. 55, 56
7. *Freeman's Journal*, 2 April 1812
8. ibid., 5 February 1812
9. ibid., 21 November 1812
10. *The Morning Chronicle*, 25 November 1812
11. *Freeman's Journal*, 13 November 1813
12. ibid., 12 November 1813
13. ibid., 21 June 1811
14. *The European Magazine and London Review*, August 1813, p. 144
15. ibid., September 1820, p. 259
16. *The Morning Chronicle*, 21 August 1815
17. *New Monthly Magazine*, November 1815
18. *Monthly Panorama*, February 1810
19. *Freeman's Journal*, 14 April 1813
20. ibid., 6 November 1817
21. ibid., 13 May 1814
22. ibid., 20 March 1815
23. ibid., 1 December 1812
24. Martorelli, G. C. (Ed.), *Indice o sia Catalogo dei Teatrali Spettacoli Musicali*, Rome 1822, p. 79
25. *Freeman's Journal*, 1 December 1812
26. ibid., 29 June 1813
27. ibid., 9 August 1813
28. ibid., 23 November 1813
29. ibid., 5 January 1813
30. ibid., 6 January 1814
31. *Cyclopaedian Magazine*, January 1808, p. 56
32. *Freeman's Journal*, 12-24 January 1814
33. *Saunders' News-Letter*, 25 February 1814
34. *Irish Dramatic Censor*, 1812, p. 141
35. *Saunders' News-Letter*, 22 April 1814
36. *Freeman's Journal*, 26 April 1814
37. O'Sullivan, M. J., *A Fasciculus of Lyric Verses*, Cork 1846 (Preface), pp. 7-12

REFERENCES TO PAGES 132-147

38. *Saunders' News-Letter*, 26 April 1814
39. ibid., 22 April 1814
40. *The Private Theatre of Kilkenny*, ? Kilkenny 1825, pp. 116, 117
41. *The British Stage*, January 1820, p. 85
42. *Saunders' News-Letter*, 29 June 1814
43. ibid., 2 August, 1814
44. ibid.
45. ibid.
46. *Oxberry's Dramatic Biography and Histrionic Anecdotes*, 7 vols, London 1825-27, I, p. 160
47. Ruders, II p. 291
48. Pereíra Peíxoto, pp. 205, 206 *and* Fonseca Benevides, p. 69
49. *Journal de Paris*, 6 September 1805
50. *Journal des Luxus und der Moden*, July 1812, p. 451
51. ibid.
52. *The Morning Chronicle*, 18 May 1813
53. Mount Edgcumbe, p. 114
54. *Allgemeine Musikalische Zeitung*, 10 June 1812, p. 399
55. ibid., 22 June 1814, p. 417
56. *The Statesman*, 11 May 1814
57. *Allgemeine Musikalische Zeitung*, 22 June 1814, p. 417
58. ibid., 10 June 1812, p. 399
59. ibid., 29 January 1817, pp. 84, 85
60. *The Morning Chronicle*, 13 May 1814
61. *Freeman's Journal*, 2 August 1814
62. *The Universal Magazine*, April 1807, p. 356
63. *Freeman's Journal*, 2 October 1815
64. Fonseca Benevides, pp. 96, 97; Beard, Harry R., 'Figaro in England', in: *Maske und Kothurn*, Universität Wien, X, 1964, p. 503
65. Fonseca Benevides, p. 97
66. *The Morning Chronicle*, 6 May 1813
67. Mount Edgcumbe, p. 111
68. *Journal de Paris*, 15 November 1790
69. *Gazette Nationale, ou le Moniteur Universel*, 16 November 1790
70. Carmena y Millan, Luis, *Crónica de la Ópera Italiana en Madrid*, Madrid 1878, pp. 38, 39
71. Ruders, I pp. 182-184; II pp. 183, 186, 187
72. Fonseca Benevides, p. 60
73. Smith, William C., pp. 68, 69
74. *The Morning Post*, 6 December 1802
75. *Allgemeine Musikalische Zeitung*, 11 January 1804, p. 247
76. ibid., 30 October 1811, p. 735
77. ibid., 10 June 1812, p. 401
78. ibid., 30 December 1812, p. 868
79. *Freeman's Journal*, 9 August 1814
80. ibid., 2 August 1814
81. ibid., 5 August 1814
82. ibid., 8 August 1814
83. ibid., 11 August 1814
84. *Saunders' News-Letter*, 11 August 1814
85. ibid., 12 August 1814
86. *Freeman's Journal*, 29 August 1814
87. ibid., 10 September 1814
88. *Saunders' News-Letter*, 6 September 1814
89. ibid., 10 September 1814
90. *Freeman's Journal* 8 September 1814
91. ibid., 6 September 1814
92. ibid., 8 September 1814
93. ibid., 9 September 1814
94. ibid., 10 September 1814
95. ibid., 12 September 1814
96. ibid., 13 September 1814
97. ibid., 10 September 1814
98. ibid., 12 September 1814
99. ibid., 13 September 1814
100. ibid.
101. ibid., 14 September 1814
102. ibid., 16 September 1814
103. ibid.
104. *The Morning Post*, 1 November 1814
105. ibid., 18 November 1814
106. *Freeman's Journal*, 16 September 1814
107. ibid., 11 August 1814
108. ibid., 10 September 1814
109. *Saunders' News-Letter*, 12 August 1814
110. ibid., 10 September 1814
111. *Allgemeine Musikalische Zeitung*, 10 July 1816, p. 473
112. ibid., 21 August 1816, pp. 583, 585
113. *The Pilot*, 11 February 1811
114. *The Courier*, 21 September 1811
115. *The Globe*, 25 October 1811
116. *Saunders' News-Letter*, 12 August 1814
117. ibid., 25 August 1814
118. ibid., 1 September 1814
119. ibid., 26 August 1814

120. *Freeman's Journal*, 26 August 1814
121. ibid., 12 November 1814
122. *Saunders' News-Letter*, 9 December 1814
123. *Freeman's Journal*, 5 November 1814
124. *Saunders' News-Letter*, 15 November 1814
125. *Freeman's Journal*, 5 November 1814
126. *Saunders' News-Letter*, 15 November 1814
127. *Freeman's Journal*, 5 November 1814
128. ibid., 19 November 1814
129. ibid., 8 November 1814
130. ibid., 26 November 1814
131. ibid., 7 December 1814
132. ibid., 8 November 1814
133. *Saunders' News-Letter*, 1 December 1814
134. *Freeman's Journal*, 29 December 1814
135. ibid., 30 December 1814
136. *Saunders' News-Letter*, 17 December 1814
137. *Freeman's Journal*, 9 December 1814
138. Gilbert, II p. 240
139. *Freeman's Journal*, 17 December 1814
140. ibid., 19 December 1814
141. *The Times*, 23 December 1814
142. ibid., 31 December 1814
143. *Freeman's Journal*, 26 December 1814
144. ibid., 28 December 1814
145. ibid., 27 December 1814

Chapter 8 Two English Nightingales 1815-1816
1. Northcott, Richard, *The Life of Sir Henry R. Bishop*, London 1920, p. 19
2. Corder, F., 'The Works of Sir Henry Bishop', in *The Musical Quarterly*, IV, January 1918, p. 81
3. *Freeman's Journal*, 18 February 1815
4. ibid.
5. ibid.
6. ibid., 20 February 1815
7. ibid.
8. *Dublin Evening Post*, 21 February 1815
9. ibid.
10. *Freeman's Journal*, 20 February 1815
11. ibid.
12. ibid.
13. *The Examiner*, 1 October 1815
14. ibid., 20 April 1817
15. *Saunders' News-Letter*, 1 March 1815
16. *Freeman's Journal*, 24 April 1815
17. ibid., 2 March 1815
18. *Saunders' News-Letter*, 2 March 1815
19. *Dublin Evening Post*, 2 March 1815
20. *Freeman's Journal*, 27 March 1815
21. ibid.
22. ibid., 29 March 1815
23. *Dublin Evening Post*, 11 May 1815
24. ibid., 23 May 1815
25. ibid., 11 May 1815
26. *Freeman's Journal*, 31 January 1815
27. *Saunders' News-Letter*, 20 February 1815
28. *Freeman's Journal*, 3 May 1815
29. ibid., 19 August 1815
30. ibid., 24 August 1815
31. ibid., 8 November 1815
32. *The Morning Chronicle*, 18 August 1815
33. *Freeman's Journal*, 3 November 1815
34. *The Morning Chronicle*, 16 November 1815
35. *Saunders' News-Letter*, 13 November 1815
36. ibid., 14 November 1815
37. ibid., 8 December 1815
38. *Freeman's Journal*, 12 December 1815
39. Baron-Wilson, Mrs C., *Our Actresses*, London 1844, 2 vols, I, p. 107
40. *Freeman's Journal*, 13 December 1815
41. Kelly, p. 209
42. *Freeman's Journal*, 8 January 1816
43. ibid., 9 January 1816
44. Boaden, James, *Memoirs of the Life of John Philip Kemble*, London 1825, 2 vols, II, p. 126
45. *Dublin Evening Post*, 9 January 1816
46. *Freeman's Journal*, 9 January 1816
47. ibid., 7 February 1816
48. ibid., 6 April 1816
49. ibid., 7 February 1816
50. ibid., 19 March 1816
51. ibid., 3 April 1816
52. ibid., 9 April 1816
53. *Dublin Evening Post*, 13 January 1816
54. *Freeman's Journal*, 18 May 1816
55. *The Times*, 16 April 1806
56. *The Monthly Mirror*, June 1810
57. *The Harmonicon*, June 1831
58. *The Morning Post*, 28 March 1806
59. *The Examiner*, 17 June 1810
60. *The Harmonicon*, June 1831
61. *Freeman's Journal*, 15 September 1817

62. *Dublin Examiner*, December 1816, p. 143
63. *Freeman's Journal*, 31 January 1815
64. ibid., 28 May 1816
65. ibid., 27 May 1816
66. ibid., 16 September 1816
67. ibid., 29 May 1817
68. ibid., 10 June 1817
69. *Dublin Chronicle*, 21 June 1816
70. Baptie, David, *A Handbook of Musical Biography*, R/Boethius, Kilkenny 1986, and Baptismal Register
71. *The Universal British Directory of Trade and Commerce*, London 1790 and 1791, p. 298
72. *The Morning Chronicle*, 4 October 1813
73. ibid.
74. Baron-Wilson, I, p. 277
75. ibid., p. 278
76. *The Morning Chronicle*, 24 September 1813
77. ibid., 4 October 1813
78. Baron-Wilson, I, p. 286
79. Cox, I, pp. 61, 62
80. Saxe Wyndham, Henry, *Annals of Covent Garden Theatre from 1732 to 1897*, London 1906, 2 vols, II, pp. 273, 274
81. *New Monthly Magazine*, October 1833, p. 190
82. *The Morning Chronicle*, 10 November 1813
83. *Dublin Evening Post*, 11 July 1816
84. *Freeman's Journal*, 13 August 1816
85. ibid., 15 August 1816
86. *Dublin Evening Post*, 20 August 1816
87. ibid.
88. ibid., 22 August 1816
89. ibid., 15 August 1816
90. ibid., 20 August 1816
91. Parke, II, p. 111
92. *Dublin Chronicle*, 26 August 1816
93. *Dublin Evening Post*, 31 August 1816
94. ibid.
95. *Freeman's Journal*, 7 September 1816
96. *The Era Almanack*, London 1888, p. 67
97. *Dublin Evening Post*, 1 February 1816
98. *Freeman's Journal*, 16 March 1816

Chapter 9 An Irish Lark and that extraordinary Child, Master Balfe. 1817- 1818

1. *Freeman's Journal*, 24 December 1804
2. Gilbert, I, pp. 354, 355
3. *Life of Theobald Wolfe Tone*, Ed by his Son, 2 vols, Washington 1826, I, p. 197
4. Gilbert, II, p. 245
5. ibid.
6. *Dublin Evening Post*, 6 May 1817
7. ibid., 8 May 1817
8. Gilbert, II, p. 245
9. *Freeman's Journal*, 19 December 1816
10. ibid., 23 December 1816
11. *Dublin Examiner*, December 1816, pp. 143, 144
12. *Dublin Evening Post*, 11 January 1817
13. ibid., 15 May 1817
14. *Ramsey's Waterford Chronicle*, 26 July 1817
15. *The Examiner*, 19 October 1817
16. *The British Stage*, March 1817, p. 56
17. *Dublin Evening Post*, 23 January 1817
18. ibid., 16 January 1817
19. *Saunders' News-Letter*, 30 January 1817
20. *The Monthly Mirror*, October 1810, p. 313
21. ibid.
22. *The British Stage*, July 1817, p. 160
23. *Dublin Evening Post*, 31 July 1817
24. *The Examiner*, 2 July 1809
25. ibid., 12 July 1818
26. ibid., 7 July 1816
27. ibid.
28. ibid., 28 July 1816
29. *The Morning Chronicle*, 21 November 1816
30. *The Quarterly Musical Magazine*, VIII, 1826, p. 141
31. *Freeman's Journal*, 15 December 1817
32. ibid.
33. *The Examiner*, 25 May 1817
34. *Saunders' News-Letter*, 9 January 1818
35. *Freeman's Journal*, 9 January 1818
36. ibid., 12 January 1818
37. ibid., 8 January 1818
38. ibid., 13 January, 1818
39. ibid., 9 January 1818
40. *The Monthly Mirror*, January 1811, p. 5
41. Baron-Wilson, II, p. 224
42. ibid., p. 225
43. *The Examiner*, 4 July 1819
44. ibid., 5 November, 1815
45. ibid., 1 September 1816
46. *Freeman's Journal*, 20 January 1818

47. ibid., 2 February 1818
48. ibid., 23 May 1818
49. *Dublin Evening Post*, 26 May 1818
50. ibid.
51. ibid.
52. *Freeman's Journal*, 25 May 1818
53. *Dublin Evening Post*, 26 May 1818
54. ibid.
55. *Saunders' News-Letter*, 26 May 1818
56. *Dublin Evening Post*, 26 May 1818
57. *Wexford Herald*, 1 February 1810
58. ibid., 1 December 1814
59. ibid., 13 May 1816
60. ibid., 4 December 1815
61. ibid., 5 December 1814, 21 December 1815, 13 May 1816, 16 June 1817
62. ibid., 15 November 1813
63. *Wexford Conservative*, 10 July 1833
64. *Freeman's Journal*, 27 May 1817
65. *Saunders' News-Letter*, 18 June 1817
66. ibid., 25 June 1817
67. *Freeman's Journal*, 22 January 1818
68. ibid., 13 January 1818
69. ibid., 2 February 1818
70. ibid., 5 March 1818
71. *Dublin Evening Post*, 24 March 1818
72. ibid., 31 March 1818
73. *Freeman's Journal*, 17 April 1818
74. ibid., 13 May 1818
75. ibid.
76. *Dublin Evening Post*, 26 May 1818
77. *Freeman's Journal*, 5 June 1818
78. Vocal score at British Library (H.123a)
79. *Dublin Evening Post*, 11 June 1818
80. *Freeman's Journal*, 9 June 1818
81. ibid., 11 June 1818
82. Dowden, Wilfred S., (Ed.), *The Letters of Thomas Moore*, Oxford 1964, 2 vols, I, p. 232
83. *Dublin Evening Post*, 8 January, 1818
84. ibid., 19 March 1818
85. ibid., 12 February 1818
86. ibid., 3 March 1818
87. ibid., 14 March 1818
88. *Freeman's Journal*, 27 July 1818
89. *Saunders' News-Letter*, 17 August 1818
90. *Freeman's Journal*, 17 August 1818
91. *Saunders' News-Letter*, 17 August 1818
92. *Freeman's Journal*, 8 September 1818
93. ibid.
94. *Dublin Evening Post*, 8 September 1818
95. *Freeman's Journal*, 8 September 1818
96. *Saunders' News-Letter*, 8 September 1818
97. ibid., 14 September 1818
98. *The Private Theatre of Kilkenny*, pp. 96, 97, 103-106
99. *Saunders' News-Letter*, 22 October 1818
100. *Freeman's Journal*, 22 October 1818
101. ibid., 12 November 1818
102. *Dublin Weekly Register*, 14 November 1818
103. *Freeman's Journal*, 14 November 1818
104. ibid.
105. *Dublin Evening Post*, 17 November 1818
106. ibid.
107. *Dublin Weekly Register*, 21 November 1818
108. *Saunders' News-Letter*, 13 November 1818
109. *Dublin Weekly Register*, 12 December 1818
110. *Freeman's Journal*, 8 December 1818
111. *Saunders' News-Letter*, 2 December 1818
112. *Freeman's Journal*, 8 December 1818
113. ibid.
114. *Saunders' News-Letter*, 5 December 1818
115. *Freeman's Journal*, 8 December 1818

Chapter 10 The Original Don Giovanni 1819-1820

1. *Dublin Evening Post*, 8 December 1818
2. *Freeman's Journal*, 4 December 1818
3. *Dublin Evening Post*, 8 December 1818
4. *Saunders' News-Letter*, 10 December 1818
5. *Dublin Evening Post*, 10 December 1818
6. ibid., 15 December 1818
7. ibid., 17 December 1818
8. *Freeman's Journal*, 23 January 1819
9. ibid., 20 January 1819
10. ibid., 8 February 1819
11. ibid.
12. ibid., 26 January 1819
13. *Saunders' News-Letter*, 13 April 1818
14. *Freeman's Journal*, 13 April 1818
15. *The British Stage*, IV, December 1819
16. *Freeman's Journal*, 4 March 1819
17. *Saunders' News-Letter*, 5 March 1819
18. ibid., 8 March 1819

19. ibid., 17 November 1814
20. ibid., 20 March 1819
21. ibid., 22 March 1819
22. *Freeman's Journal*, 15 April 1819
23. ibid.
24. ibid., 16 April 1819
25. ibid., 17 April 1819
26. ibid.
27. ibid., 19 April 1819
28. ibid., 20 April 1819
29. ibid., 21 April 1819
30. ibid.,
31. ibid., 22 April 1819
32. ibid.
33. ibid.
34. ibid., 23 April 1819
35. *The British Stage*, IV, January 1820
36. *Saunders' News-Letter*, 27 September 1819; *Freeman's Journal*, 29 September 1819
37. *Freeman's Journal*, 23 September 1819
38. *Saunders' News-Letter*, 25 September 1819
39. *Dublin Evening Post*, 23 September 1819
40. Pereíra Peíxoto, p. 159
41. Carlo Gatti, II, p. 26
42. *Allgemeine Musikalische Zeitung*, 25 December 1816
43. William C. Smith, p. 143
44. Ebers, John, *Seven Years of the King's Theatre*, London 1828, p. 392
45. *The Harmonicon*, June 1830
46. *The Examiner*, 20 April 1817
47. *The Anti-Gallican Monitor*, 27 April 1817
48. *The Morning Chronicle*, 3 February 1817
49. *The London Magazine*, June 1822
50. ibid., February 1822
51. Ebers, p. 392
52. *The Times*, 9 March 1825
53. *The Examiner*, 23 March 1817
54. *The British Stage*, V, April 1821; *The London Magazine*, April 1821
55. *The Harmonicon*, April 1827
56. *Allgemeine Musikalische Zeitung*, 21 December 1814
57. Fitzsimons, Edward, *Letters from France and the Netherlands in the Summers of 1820 and 1821*, Dublin 1821, pp. 204, 205
58. *The Harmonicon*, February 1833
59. *The Times*, 1 March 1819
60. *The British Stage*, III, April 1819
61. [*Gold's*] *London Magazine*, February 1820
62. Ebers, p. 394
63. *The Morning Chronicle*, 19 January 1818
64. *Allgemeine Musikalische Zeitung*, 8 January 1823
65. *The Examiner*, 25 January 1818
66. *The London Magazine*, February 1826
67. *Allgemeine Musikalische Zeitung*, 16 July 1823
68. ibid., 13 May 1835
69. *New Monthly Magazine*, March 1818
70. *The London Magazine*, August 1820
71. *The Morning Chronicle*, 25 July 1820
72. ibid., 26 March 1813
73. Ebers, p. 394
74. *New Monthly Magazine*, November 1816
75. *The Public Ledger*, 13 January 1817
76. Martorelli, Giulio Cesare, (comp.), *Indice o sia Catalogo dei teatrali Spettacoli musicali italiani di tutta l'Europa*, Rome 1822, p. 79
77. ibid.
78. *The Harmonicon*, April 1823
79. Stendhal, *Life of Rossini*, (translated Richard N. Coe), London 1956, p. 417
80. *Allgemeine Musikalische Zeitung*, 12 January 1825
81. *The Harmonicon*, November 1826
82. ibid., March 1827
83. *Allgemeine Musikalische Zeitung*, 5 September 1832
84. *The London Magazine*, November 1820
85. *Allgemeine Musikalische Zeitung*, May 1836
86. *Saunders' News-letter*, 28 September 1819
87. ibid., 29 September 1819
88. *Freeman's Journal*, 28 September 1819
89. *Allgemeine Musikalische Zeitung*, 13 March 1839
90. *The Harmonicon*, June 1826
91. Levey, R. M. and O'Rorke, J., *Annals of the Theatre Royal, Dublin*, Dublin 1880, pp. 73, 74
92. Ebers, pp. 56, 57
93. Personal communications
94. *Saunders' News-letter*, 30 September 1819
95. ibid.
96. *Freeman's Journal*, 4 October 1819
97. *Saunders' News-Letter*, 4 October 1819

98. ibid., 6 October 1819
99. ibid., 7, 8 October 1819
100. *Freeman's Journal*, 9 October 1819
101. *Saunders' News-Letter*, 11 October 1819
102. ibid., 13 October 1819
103. ibid., 15 November 1819
104. *The Dublin Evening Post*, 6 February 1817
105. ibid., 26 November 1818
106. ibid., 1 December 1818
107. Gilbert, II, p. 244
108. ibid.
109. ibid., p. 245
110. *Saunders' News-Letter*, 27 November 1818
111. Gilbert, II, p. 248
112. ibid., pp. 249, 250
113. *Saunders' News-Letter*, 14 March 1820
114. *Dublin University Magazine*, June 1852, p. 679
115. *The Globe*, 30 March 1820
116. ibid., 31 March 1820
117. *Saunders' News-Letter*, 4 May 1820
118. ibid., 8 May 1820
119. *The British Stage*, V, November 1821
120. Mss book of plays and their casts 1817-1820 at The Royal Irish Academy, Dublin
121. *Dublin University Magazine*, June 1852, p. 679
122. Gilbert, II, pp. 254, 255
123. ibid., pp. 252, 253
124. *Saunders' News-letter*, 11 September 1821
125. Gilbert, II, p. 254
126. ibid. (reported), p. 239
127. *Dublin Evening Post*, 8 November 1834

Appendix C
1. *Freeman's Journal*, 8 August 1814
2. *Gemme d'Antichità*, London 1864, No 68
3. *Saunders' News-Letter*, 11 August 1814
4. *Freeman's Journal*, 9 August 1814
5. ibid., 15 August 1814
6. *Saunders' News-Letter*, 10 August 1814
7. ibid., 10 [11] August 1814
8. ibid.
9. *Freeman's Journal*, 13 August 1814, [*Saunders' News-Letter*, 12 August 1814]
10. *Freeman's Journal*, 15 August 1814
11. ibid., 6 September 1814
12. ibid., 8 September 1814
13. *Saunders' News-Letter*, 10 September 1814
14. *Freeman's Journal*, 10 September 1814
15. ibid., 12 September 1814
16. ibid.
17. ibid.
18. ibid., 13 September 1814
19. ibid., 16 September 1814

Bibliography

MANUSCRIPT
Royal Irish Academy, Dublin. MS book of plays and their casts: 1817-1820

MICROFILM
Minute Book of the Incorporated Musical Fund Society, The National Library, Dublin

NEWSPAPERS AND PERIODICALS

Dublin
The Correspondent
The Cyclopaedian Magazine, and Dublin Monthly Register
The Dublin Chronicle
The Dublin Evening Post
The Dublin Examiner; or, Monthly Journal of Science, Literature and Art
The Dublin Journal
The Dublin University Magazine
The Dublin Weekly Register
The Freeman's Journal
The Hibernia Magazine and Monthly Panorama
The Hibernian Journal; or, Chronicle of Liberty
The Irish Dramatic Censor
The Monthly Panorama
Saunders' News-Letter
Walker's Hibernian Magazine; or, Compendium of Entertaining Knowledge

London
The Alfred
L'Ambigu
The Anti-Gallican Monitor
Le Beau Monde; or, Literary and Fashionable Magazine
La Belle Assemblée; or, Bell's Court and Fashionable Magazine
The British Press; or, Morning Literary Advertiser
The British Stage and Literary Cabinet
The Cabinet; or, Monthly report of polite literature
The Courier
The Court Journal
The Daily Advertiser, Oracle, and True Briton
The Dramatic Censor
The Era Almanack: dramatic and musical
The European Magazine and London Review
The Examiner
The Gentleman's Magazine
The Globe
Gold's London Magazine and Theatrical Inquisitor
The Harmonicon
Johnson's Sunday Monitor, and British Gazette
Knight's Quarterly Magazine
The London Magazine
The Monthly Mirror
Morning Advertiser
The Morning Chronicle
Morning Herald
The Morning Post
The Musical World
The National Register
The New Monthly Magazine and Universal Register
The Pilot
The Public Ledger
The Quarterly Musical Magazine and Review
The St James's Chronicle
The Statesman
The Sun
The Times
The Universal Magazine

Miscellaneous
Allgemeine Musikalische Zeitung (Leipzig)
Berlinische Musikalische Zeitung
Caledonian Mercury (Edinburgh)
The Edinburgh Advertiser
Gazette nationale: ou le Moniteur universel (Paris)
The General Advertiser; or, Limerick Gazette
Journal de Paris
Journal des Luxus und der Moden (Weimar)
The Musical Quarterly (New York)
La Quotidienne (Paris)
Ramsey's Waterford Chronicle
Revue et Gazette Musicale de Paris
Rivista Musicale Italiana (Turin)
The Wexford Conservative
The Wexford Herald
The Wexford Independent

BOOKS AND PRINTED ARTICLES

Adolphus, John, *Memoirs of John Bannister, Comedian*, 2 v, London 1839

Baptie, David, *A Handbook of Musical Biography*, 1st ed London 1883; rep, Kilkenny 1986

Baron-Wilson, Mrs C. [Margaret]. *Our Actresses; or, Glances at Stage Favourites, past and present*, 2 v, London 1844

Barrett, William Alexander, *Balfe: his Life and Work*, London 1882

Barrington, Sir Jonah, *Personal Sketches of his own Times*, 3 v, London 1827-32

[Berry, Mary] *Extracts of the Journals and Correspondence of Miss Berry from the year 1783 to 1852*, Lady Theresa Lewis (Ed.), 3 v, London 1865

BIBLIOGRAPHY

Blunden, Edmund Charles, *Leigh Hunt's "Examiner" Examined*, London 1928

Boaden, James, *Memoirs of the Life of John Philip Kemble, Esq.*, 2 v, London 1825

Bottenheim, S. A. M., *De Opera in Nederland*, Amsterdam 1946

Burney, Charles, *A General History of Music*, 4 v, London 1776-89

The Cabinet of Irish Literature, C. A. Read and K. T. Hinkson (Eds), 4 v, London 1902

Calendar of Ancient Records of Dublin, Dublin 1916

Carmena y Millán, Luis, *Crónica de la Ópera Italiana en Madrid*, Madrid 1878

Carr, Bruce, 'Theatre Music: 1800-1834', in *The Athlone History of Music in Britain*, vol V: *The Romantic Age*, Ed. Nicholas Temperley, London 1981

Carse, Adam, *The Orchestra from Beethoven to Berlioz*, Cambridge 1948

[Charlotte Augusta, Princess] *Letters of the Princess Charlotte 1811-1817*, A. Aspinall (Ed.), London 1949

Choron, A. E. and F. Fayolle, *Dictionnaire historique des Musiciens*, 2 v., Paris 1810

Clark, William Smith, *The Irish Stage in the County Towns, 1720 to 1800*, Oxford 1965

Clayton, Ellen Creathorne, *Queens of Song*, 2 v, London 1863

C[okayne], G. E. *The Complete Peerage of England, Scotland, Ireland*, 12 v, The Hon. Vicary Gibbs, a.o. (Ed.), London 1910-59

Corder, Frederick, 'The Works of Sir Henry Bishop', *The Musical Quarterly*, IV, New York 1918

Cox, John Edmund, [Rev] *Musical Recollections of the last Half Century*, 2 v, London 1872

[Croker, John Wilson], *Familiar Epistles to Frederick J—s, Esq., on the present state of the Irish Stage*, 5th ed., Dublin 1806

Curiel, Carlo Leone, *Il Teatro S. Pietro di Trieste, 1690-1801*, Milan 1937

Derwent, Lord [G. H. Johnstone], *Rossini and some forgotten Nightingales*, London 1934

Detcheverry, Arnaud, *Histoire des Théâtres de Bordeaux*, Bordeaux 1860

Dictionary of National Biography, Sir L. Stephen and Sir S. Lee (Eds.), 63 v, London 1885-1900; 22 v, London 1973

Ebers, John, *Seven Years of the King's Theatre*, London 1828

Edgcumbe, Richard, 2nd Earl of Mount Edgcumbe, *Musical Reminiscences of an old Amateur*, 4th ed. London 1834

Edwards, Henry Sutherland, *The Lyrical Drama*, 2 v, London 1881

Enciclopedia dello Spettacolo, 9 v, Rome 1954-62

Escudier, Marie and Léon, *Vie et Aventures des Cantatrices Célèbres*, Paris 1856

[Farington, Joseph], *The Farington Diary*, James Greig (Ed.), 8 v, London 1922-28

Fenner, Theodore, *Leigh Hunt and Opera Criticism; the "Examiner" Years 1808-21*, Kansas 1972

Ferrari, Giacomo Gotifredo, *Aneddoti piacevoli e interessanti occorsi nella vita di Giacomo Gotifredo Ferrari da Roveredo*, 2 v, London 1830

Fétis, François Joseph, *Biographie universelle des Musiciens*, 8 v, 2nd ed. Paris 1860-65; *Supplement*, A. Pougin (Ed.), 2 v, Paris 1878-80

Fitzsimons, Edward, *Letters from France and the Netherlands in the summers of 1820 and 1821*, Dublin 1821

Fonseca Benevides, Francisco da, *O Real Theatro de S. Carlos de Lisboa*, Lisbon 1883

Gardiner, William, *Sights in Italy, with some account of the present state of music and the sister arts in that country*, London 1847

Gatti, Carlo, *Il Teatro alla Scala*, 2 v, Milan 1964

[Genest, John], *Some Account of the English Stage from the Restoration in 1660 to 1830*, 10 v, Bath 1832

Giazotto, Remo, *Giovan Battista Viotti*, Milan 1956

Gilbert, Sir John T., *A History of the City of Dublin*, 3 v, Dublin 1861

Gordon, Pryse Lockhart, *Personal Memoirs; or, Reminiscences of Men and Manners at home and abroad*, 2 v, London 1830

Grove's Dictionary of Music and Musicians, eds. 1-5, London 1878-1961

The New Grove Dictionary of Music and Musicians, Stanley Sadie (Ed.), 20 v, London 1980

Gutierrez, Beniamino, *Il Teatro Carcano (1803-1914)*, Milan 1914

Hazlitt, William, *A View of the English Stage*, ed. W. Spencer Jackson, London 1906

Herbert, J. D., [Dowling], *Irish Varieties for the last fifty years...* London 1836

Heywood, Joseph, 'Recollections of the Life of Joseph Heywood, and some of his Thoughts about Music' in *The Cornhill Magazine*, London December 1865

Highfill, P. H. (jnr), K. A. Burnim and E. A. Langhans, *A Biographical Dictionary of Actors, Actresses, Musicians, Dancers, Managers and Other Stage Personnel in London 1660-1800*, Illinois 1973–

Leigh Hunt's Dramatic Criticism 1808-1831, L. H. and C. W. Houtchens (Eds), New York 1949

[Jerningham, (Hon, Frances) Lady], *The Jerningham Letters (1780-1843)*, 2 v, ed. Egerton Castle, London 1896

[Jordan, Dorothy], *Mrs Jordan and her Family, being the unpublished correspondence of Mrs Jordan and the Duke of Clarence, later William IV*, A. Aspinall (Ed.), London 1951

BIBLIOGRAPHY

[Kelly, Michael], *Reminiscences*, Roger Fiske (Ed.), London 1975

Kenney, Charles Lamb, *A Memoir of Michael William Balfe*, London 1875

King, Alec Hyatt, *Musical Pursuits. Selected Essays*, London 1987

Kurze Lebensbeschreibung der Madame Angelica Catalani aus ganz authentischen Quellen gezogen, Berlin 1816

Lamb, Charles, 'Imperfect Sympathies', in *Elia* (first series), London 1823

Lassels, Richard, *An Italian Voyage; or, A compleat Journey through Italy*, 2 pts., 2nd ed. 1698

Levey R. M. and J. O'Rorke, *Annals of the Theatre Royal, Dublin, from its opening in 1821 to its destruction by fire, February 1880*, Dublin 1880

Loewenberg, Alfred, *Annals of Opera 1597-1940*, 3rd ed., rev. H. Rosenthal, London 1978

The London Stage 1660-1800, 11 v, Illinois 1960-68

[Mackintosh, Matthew], *Stage Reminiscences ... by an old stager*, Glasgow 1866

Manferrari, Umberto (Ed.), *Dizionario Universale delle Opere Melodrammatiche*, 3 v, Florence 1954-55

Martorelli, G. C. (Ed.), *Indice o sia Catalogo dei teatrali Spettacoli musicali italiani di tutta l'Europa, incominciando dalla Quaresima 1821*, Rome 1822

[Moore, Thomas], *The Letters of Thomas Moore*, Wilfred S. Dowden (Ed.), 2 v, Oxford 1964; *Memoirs, Journal and Correspondence of Thomas Moore*, Ed. The Rt Hon. Lrd John Russell, M.P., 8 v, London 1853-56

[Moscheles, Ignaz], *Aus Moscheles' Leben*, Charlotte Moscheles (Ed.), 2 v, Leipzig 1872-3

Nicoll, Allardyce, *A History of English Drama: 1660-1900*, 6 v, 4th ed., Cambridge 1952-59

Northcott, Richard, *The Life of Sir Henry R. Bishop*, London 1920

Oxberry, Catherine E. (Ed.), *Oxberry's Dramatic Biography and Histrionic Anecdotes*, 7 v, London and Bristol 1825-27

Parke, William Thomas, *Musical Memoirs*, 2 v, London 1830

Payer von Thurn, Rudolf (Ed.), *Joseph II. als Theaterdirektor. Ungedruckte Briefe und Aktenstücke aus den Kinderjahren des Burgtheaters*, Vienna 1920

Pereíra Peíxoto d'Almeida Carvalhaes, Manoel, *Marcos Portugal - na sua musica dramatica*, Lisbon 1910

The Private Theatre of Kilkenny, Kilkenny 1825

Radiciotti, Giuseppe, *Teatro, Musica e Musicisti in Sinigaglia*, Milan 1893

Rinaldi, Mario, *Due Secoli di Musica al Teatro Argentina*, Florence 1978

Robson, William, *The Old Play-Goer*, London 1846

Rogge, Hendrik Cornelis, 'De Opera te Amsterdam', in *Oud-Holland*, Amsterdam 1887

Ruders, C. I., *Portugisisk Resa*, 3 pt, Stockholm 1805-89

Sainsbury, John S., *A Dictionary of Musicians from the Earliest Times*, 2 v, 1st ed. London 1824, rep. New York 1971

Sartori, Claudio, *Catalogo unico dei libretti italiani a stampa fino all'anno 1800*, 19 v, Milan 1973-85

Smith, William Charles, *The Italian Opera and Contemporary Ballet in London: 1789-1820*, London 1955

Sorge, Giuseppe, *I Teatri di Palermo nei secoli XVI, XVII and XVIII*, Palermo 1926

Soubies, Albert, *Le Théâtre-Italien de 1801 à 1913*, Paris 1913

Speyer, Edward, *Wilhelm Speyer der Liederkomponist 1790-1878*, Munich 1925

Stendhal (Marie Henri Beyle), *Life of Rossini*, trans. Richard N. Coe, London 1956

Stieger, Franz (Ed.), *Opernlexikon*, 11 v, Tutzing 1975-83

Stockwell, La Tourette, *Dublin Theatres and Theatre Customs, 1637-1820*, Tennessee 1938

Strickland, Walter George, *A Dictionary of Irish Artists*, 2 v., Dublin/London 1913

O'Sullivan, Michael John, *A Fasciculus of Lyric Verses*, Cork 1846

Thurner, Auguste, *Les Reines du Chant*, Paris 1883

[Tone, Theobald Wolfe]. *Life of Theobald Wolfe Tone*, Ed. by his son, William Theobald Wolfe Tone, 2 v, Washington D.C. 1826

[Trench, R. C.], *The Remains of the late Mrs Richard Trench*, edited by her son, the Dean of Westminster, London 1862

The Universal British Directory of Trade and Commerce, London 1790-91

Warburton, J., Rev. J. Whitelaw and Rev. R. Walsh, *History of the City of Dublin*, 2 v, London 1818

Watson, Ernest B., *Sheridan to Robertson*, Cambridge, Mass. 1926

W. A., (Ed.), *I Giudizj dell'Europa intorno alla Signora Catalani*, 2nd ed, Milan 1816

White, Eric Walter, *The Rise of English Opera*, London 1951; A Register of *First Performances of English Operas and Semi-Operas*, London 1983

Wiel, Taddeo, *I Teatri Musicali Veneziani del Settecento*, Venice 1897

Willis's Rooms (London), *Collection of Programmes of six concerts* [1810] (with MS notes by Sir G. T. Smart), British Library C.61.g.18

Wyndham, Henry Saxe, *The Annals of Covent Garden Theatre from 1732 to 1897*, 2 v, London 1906

Index

Abbey Street theatre, 232
Abbott, Mr, 229
Abercorn, Marquess of, 34n
Act of Parliament (Ireland), 225
Actresses, contemporary opinion of, 11
Adam, Adolphe, 138
Adami, Anna, *see* Corri, Mrs Haydn
Adams, The Misses, 24, 71, 125
Addison, Mrs John, née Elizabeth Willems, *see also* Nunn, 233-4
Adolfo e Chiara (Pucitta), 137
Adolphe et Clara; ou, Les Deux Prisonniers (Marsollier), 28
Adopted Child, The (Attwood), 16, 233
Agnese (Paer), 214
'Ah! perfido' (Beethoven), 167
Ailiffe, Mr, 246
Alcina (Handel), 33
Alday, Paul, 33
Alessandria, Italy, 217
Alessandro nell' Indie (D. Corri), 219
Ali d'Amore, L' (Rauzzini), 42
Alley, Mr, 66n
Alphonso (?), 27
Amanti consolati, Gli (Sarti), 80
Ambigu-Comique, Théâtre de l', 28
Ambrogetti, Giuseppe, 211-2, 214-5, 218, 220, *222*, 224, *227*, 248
America, United States of, 180
Amilie, (Rooke), 190
Aminta (?Tenducci), 5
Amor contrastato, L' (Paisiello), 108
Amours de Bastien et Bastienne, Les, 4
Amsterdam, 68, 106, 111-2, 137, 140
'An Cailín', 33
Ancona, 37
Andreozzi, Gaetano, 128, 165, 244
Anfossi, Pasquale, 37, 42
Angelica, 117, 119
Angellini, Signor, 71
Angiolini, Pietro, 71
Anzico and Coanza (Stevenson), 206, 248
Apothecaries Hall (Dublin), 230
Apparizione di San Michele Archangelo nel Monte Gargano, L' 105
Argyle Rooms (London), 112
Argyle Street Fashionable Institute, 68
Arias, interpolation of, 59, 97, 114, 117, 119, 121
Armide (Gluck), 215
Arne, Michael, 7, 158-9
    Thomas Augustine, 7, 158, 171

Arnold, Samuel, 23, 43, 96-7, 184, 243
    Samuel James, 125, 149
Artaserse (Cimarosa/Portugal), 139
Artaxerxes (T. A. Arne), 120, 155, *157*, 170, 195
Arte contro l'Arte, L', see *Furbo contro il Furbo, Il*
Astley's Theatre (Dublin) 86n
Athenaeum Club (London), 26
Atkinson, Joseph, 18, 233
Atter, Mr, 66n
Attwood, Master, 208
    Mr, 20
    Thomas, 16, 19, 23, 102, 233, 239, 246
Aubry, Mr, 66n
Audiences, comments on, 85
Augustus Frederick, Prince, 139
Aungier Street Theatre (Dublin), 4, 6
Austria, 200
'Austrian Trumpet, The', 93
Autié, Léonard, 42
Ayres, James, 5
Aylett, Miss, 183, 206
Ayrton, William, 116

Bacchanali di Roma, I (Nicolini), 38
Baily, Mr, 13
Balance of payments (Ireland 1811), 103
Balassi, Francesco, 111, 113, 117, 240-1
Balfe, Michael William, 3, 189-193, 229
    William, 189, 191
Ballet, 69, 71, 119, 125, 156, 205
Bannister, John, 100, 234
Banti, Brigida, 17
Barber of Seville, The (Rossini/Shield), 101
Barbier, Jane, 7
Barbiere di Siviglia, Il (Paisiello), 42
    (Rossini), 155, 212n, 218, 223
Barnes, Thomas, 42
Barnett, George, 3, 183, 185
Barons de Felsheim, Les, 149
Barret, Mr, 66n, 134n
Barrett, William Alexander, 189, 190
Barrington, Sir Jonah, 6, 49
Barry, Spranger, 229
Barrymore, William, 150, 242
Barthélemon, François Hippolyte, 139, 219
Bartleman, James, 138
Bartley, Mr 237
Bartoli, Mr, 17
Barton, James, 66n, 134n, 190-1, 198-9, 229, 246
    R., 134n
Baseggio, Signor, 72

278

INDEX

Bassi, Luigi, 211
Bate Dudley, Rev. Sir Henry, 6
Bath (Somerset), 89, 169, 180
Baudouin d'Aubigny, Jean Marie Théodore, 162
Baumgarten, Carl Friedrich, 19
Bayly, Thomas Haynes, 190n, 193
'Bay of Biscay, The', 83
Beaumont Hospital (Dublin), 129
'Beauty in tears', 165
Becher, William Wrixon, MP, 132-3
Bedford, 6th Duke of, 34
    Duchess of, 34, 35n, 57
Bedford, Master, 243
    Mr, 244
Bedouins, The; or, The Arabs of the Desert (Stevenson), 20, 234
Beethoven, Ludwig van, 58n, 167, 224
Beggar's Opera, The (Gay), 7 , 22, 25, 95, 146, 167, 170, 173, 180-1, 200n, 218
Begrez, Pierre Ignace, 116, 211, 215, 220, 224, 248
Belfast, 47, 62, 105, 143, 195
Bellamy, Thomas Ludford, 18, 233
Bellchambers, Mrs, 162, 244-5
Bellew, Sir Edward, Bt, 142
Bellini, Vincenzo, 217
Benelli, Monsieur, 40n
Beresford, John Claudius, 148
    Sir John Poo, 176
Berlin (Germany), 69, 140
Bernardin de Saint-Pierre, Jacques Henri, 21
Bertinotti-Radicati, Teresa, 105-7, 113-4, *115*, 117-20, 123, 125, 170, 240-1
Betty, William Henry West ('Master Betty'), 77
Bianchi, Angiola, 125, 138, 250
    Francesco, 16, 53, 138
    J. M. C., 16
Bianchi-Lacy, Jane, 134, 138, 250-2
Billington, Elizabeth, 118, 154, 167, 170, 172
Birch, Samuel, 16, 233
Bird, Mr, 134n
Birmingham (England), 47
Bishop, Sir Henry Rowley, 3, 84, 129-30, 150, 154, 162, 163, 165, 173, 179, 182, 223, 238, 242-7, 251
Blaise et Babet; ou, La Fête au Village (ballet), 71
Blanchard, Master, 233
    Mrs, 233
Blewitt, Jonathan, 131-2, 148, 150, 154, 155-6, 159, 166, 242-4
Boaden, James, 41
Boadicea (Pucitta), 218
Boieldieu, François Adrien, 154, 243
Bologna (Italy), 86, 105n, 107
Bond, Mr, 66n, 134n

Bondocani, Il; or The Caliph Robber (Attwood and Moorehead), 23
Boodle's Club (London), 28
Border Feuds; or, The Lady of Buccleuch (Stevenson), 240
Botturini (or Butturini), Mattia, 241
Bouilly, Jean Nicolas, 28
Bourke, D, 66n
    Joseph, The Very Rev., 132
Bowden, A. W., 199
    Mr, 66n, 134n
Braham, John, as actor, 92-3
    as composer, 31, 32, 96, 97, 149, 236-7, 239, 242
    as teacher, 216
    at Covent Garden, 89, 91
    at Drury Lane, 89, 91
    at King's Theatre, 89, 91, 215
    Charles Lamb on, 89, 91
    early life of, 89
    his fees, 91
    his vocal technique, 91
    his voice assessed, 91-2, 93, 99, 178-9
    in concerts, 99, 169
    in Dublin, 88-94, 96-7, 99-100, 178-9, 239, 246
    in oratorio, 88, 99
    on the Continent, 89
Brandon, James, 98
Brett, Frances R., 19n
Bristol, 169
Broad, Mr, 134n
    Mrs, 209
Broadhurst, Mr, 160
Brooke, Henry, 21
    Honor, 20
Brother and Sister (Bishop and Reeve), 173, 245
Buck, Adam, 39
'Buffo' singing, styles compared, 137-8,
Buona Figliuola, La (Piccinni), 42, 121
Buonaiuti, Serafino, 78, 237
Burgess, Mr (clarinet player), 66n
    Mr (horn player), 134n
    Mr (singer), 244-7
    Mrs, 242, 244-6
Burgtheater (Vienna), 6, 241, 248
Burney, Charles, 42
Burning of Moscow, The (Code/Stevenson), *145*
Burns, Robert, 186
Burton, Mr, 234
Butturini, Mattia, *see* Botturini, M.
Byrne, Edward, sen., 175
    Edward, jun., 175, 210
    Mary, 7, 175-6, 177, 178-9, 201, *202*, 206, 208-11, 246, 248
    Mr, 233-5

Byron, 6th Lord, 131

Cabinet,The, (Reeve, Moorehead, Davy and Braham), 31, 32, 96, 97, 146, 173, 236
Cahir (Caher), Lord, 131
Caigniez, Louis Charles, 162
Calcraft, John William, 230
Callan, Mr, 19n, 233
Callcott, John Wall, 252
Calypso and Telemachus (Galliard), 7
Calzolaio, Il; o La Bacchetta portentosa (Portugal), 141, 254
Cambridge, Duke of, 68, 112
Camden, Earl, 11
Campbell, Counsellor, 149
Campion, Miss, (later Mrs F. E. Jones), 11
Capel Street Theatre, 4, 6, 93, 95, 156
Capricciosa pentita, La (Fioravanti), 135
Caravan, The; or, The Driver and his Dog (Reeve), 156, 243
Caravita, Giuseppe, 237
Carey, Henry, 5
Carlton House (London), 110
Carmichael, Thomas, 6
Carroll, Mr, 235-6
Cartagenova, Giovanni Orazio, 109n
Carter, Charles Thomas, 34, 252
Carty, Mr, 201
Casanova, Giovanni Giacomo, 5
Cassidy, Miss, 154
Castil-Blaze, François Henri Joseph, 49
Castle of Andalusia, The (pasticcio: Arnold), 96, 128, 146, *177*, 181
Castlereagh, Lord, 49
Castle Spectre, The (Lewis/Kelly), 27
Catalani, Adelina, 38n
    Agostino, 37, 38
Catalani, Angelica, 39, 44, 50, 65, 67, 147, 219
    and Alberto Cavos, 38
    "God Save the King", 54, 62, 80, 142, 252, 254, 256
    and "Rule Britannia", 251, 253-4
    and Miss Hughes, 138, 154, 171
    and Frederick Jones, 41, 51, 55
    and her husband, 40, 41, 49, 51
    and Michael Kelly, 49, 53, 55, 75-6
    and Napoleon, 40
    and Catherine Stephens, 172
    as an actress, 76-77
    as teacher, 216-7
    "created" by Mme Tussaud (Dublin), 57
    début in Venice, 38
    early life, 37-40
    fashions inspired by, 57, 72
    gifts bestowed on, 40
    her attraction on the wane, 54
    her charity performances, 62, 141-2, 255
    her marriage, 40
    her children, 40
    her excessive embroidery of arias, 59
    her fees, 40, 41, 51, 55
    her "graces of the shawl", 79
    her indispositions, 48, 56, 75, 143, 256
    her vocal range, 45
        L. Spohr's opinion of, 45
    her vocal technique, 45-7
    her voice assessed, 45-7, 143-4
        in decline, 143-4
    in Belfast, 62
    in Birmingham, 47
    in Bordeaux, 40
    in Cork, 141
    in Dublin, 41-3, 47-9, 51-3, 55, 56-65, 72-80, 123, 134, 140-4, 237-8, 250-6
        at Crow Street Theatre, 56-9, 61-2, 64, 71-80, 141-4, 254-5
        at the Rotunda, 51, 53, 140
        at St Werburgh's Church, 134, 140
        her concert appearances in, 51-4, 250-6
        her social life in, 49, 62, 80, 140
    in Edinburgh, 62
    in Florence, 38
    in Hamburg, 143
    in Handel's music, 47, 140, 250, 255-6
    in Leipzig, 143
    in Limerick, 141
    in Lisbon, 38, 40, 41
    in London, 40, 103, 111, 143
    in Madrid, 40
    in Milan, 38, 45
    in Munich, 137
    in Paris, 40, 137
    in Rome, 38, 46
    in Trieste, 38
    in travesti rôles, 61
    Patti (Adelina), compared to, 46
    racehorse named after, 57
    ticket prices raised for, 40, 73
        reduced for, 51, 62
Catalani, Benizio, 38n
    Guglielmo, 66n, 68
    Maria Antonia, 38n
Catholic Emancipation, 129, 152, 175
Catholic University of Ireland, 230
Catley, Ann, 7
Cauvini, Angiolina, 111, 113, 116-7, 120, 240-1
    Carlo, 111, 113, 116-7, 119-20, 240-1
Cavalli, Pietro Francesco, 219
Cavan Militia, 189

INDEX

Cavos, Alberto, 38,
Chalmers, William, 21
Charity performances, 128, 155, 173, 198-9, 224
Charlemont, Earl of, 192
    Countess of, 205
Charlotte, Princess, 110, 112
    Queen, 130
Che Originali!, see *Fanatico per la Musica, Il*
Cheese, Miss, 142, 169, 256
Cherry, Andrew, 19n, 83-84, 95, 238
'Cherry ripe', 181
Cherubini, Louis Charles Zénobie Salvador Marie, 100, 102, 138, 148, 239, 244
Chiappari, Giuseppe, 118
Chiodi, Giuseppe, 134, 137, 141-3, 250-6
Ciabattino, Il, ossia il Diavolo a quattro, see *Calzolaio, Il*
Cimarosa, Domenico, 38, 52n, 57-8, 68, 81, 105, 110, 120, 137-8, 139, 149, 167, 218, 252-3
Clairon, Mlle, 76
Clanricarde, Lord, 5
Clanwilliam, Lord, 5
Claremont, William, 146
Clarence, Duke of, later: William IV, 48
Clarendon Street Chapel (Dublin), 173
Clayton, Ellen Creathorne, 54n
Clemenza di Tito, La (Mozart), 76, 91, 167, 253
Cleopatra (Nasolini), see *Morte di Cleopatra, La*
Clifford, Henry, 98
Clifton, John Charles, 166, 245
Clitennestra (Zingarelli), 38
Cloken, Mr, 66n
Cobb, James, 21, 24, 28, 234-5
Cobham, Thomas, 247-8
Code, Henry Brereton, 102, 145, 200
Coffey, Charles, 141n
Cogan, Philip, 5, 58, 64, 149
Colin and Flora; or, Rural Sports (dance), 24
Collini, Signora, 116
Collins, William, 229
Colman, George, the Younger, 28, 97, 100, 102, 183, 235, 239-40
Comédie-Italienne (Paris), 21
Comerford, John, 50
Concerts, 17, 51-4, 99, 169, 250-6
Connor, Charles, 240, 242-5
    Mrs, 244
Continent, The, 6, 9, 17, 143, 217
Conway, William Augustus, 240, 245
Cooke, B., 66n, 134n
    Bartlett, 23
    E., 66n
    Grattan, 80
    R., 66n

Cooke, Thomas Simpson,
    and Catholic party tunes, 36
    as composer, 26, 28, 29, 31, 34, 96, 100, 149, 150, 235-6, 242-3, 246
    as father, 80
    as instrumentalist, 26, 56
    as leader (conductor), 23, 52, 54, 57, 64-6, 99, 126
    as musical director, 56, 131
    as singer, 54, 56, 126-7, 242-4
    Mrs T. S., 29, 33, 34, 52-3, 54, 57, 61, 80, 126, 128, 132, 236, 238-40, 242-4; *see also* Howells, Fanny. W., 242
Cooper, Martin, 4
Copenhagen (Denmark), 69
Coppola, Giuseppe, 105
Corbett, Mr, 100
Corelli, Arcangelo, 26
Cork (Ireland), 12, 63, 103, 105n, 141-2, 197-8, 209, 257-60
Cornelys, Miss, 19n
Corri, Angelina, 211, 217
    Domenico, 83, 96, 112, 216, 219, 238, 242
    [-Paltoni] Frances (Fanny), 216-7, 219-20, 223, 248
    Haydn, 112, 119-20, 219, 221
    Mrs Haydn, 219
    Natale, 216
    Rosalie, 211, 216-20, 223, 248
    Sophia *see* Mme Dussek
Corry, James, 192
Corsair, The; or, The Pirate's Isle (Blewitt), 131, 242
Così fan tutte (Mozart), 33, 107, 110-2, 114, 116, 120, 124, 215, 220, 223, 241, 249-50
Cottin, Mme, 84
Countess of Salisbury, The (Hartson), 82n
Covent Garden Theatre, 7, 28, 31, 36, 47, 63, 86, 89, 95, 98, 99, 102-3, 125, 130, 144, 148, 154-5, 170, 172-3, 182, 194, 196, 201, 218, 228, 233-40, 242-3, 245-7
Cowper, Lord, 41
Cox, Rev. John Edmund, 92, 170
Coyle, James, 13
Coyne, Mr, 27, 233-4
Cramer, Miss, 71
Crampton, John, 83-4, 103, 152, 226, 228
Cranfield, Miss, 68
Creation, The (Haydn), 252, 255
Crescentini, Girolamo, 37, 140, 251
Creswell, Mrs, 233-4
Crispin, rival de son Maître (Le Sage), 78
Crivelli, Gaetano, 109n, 116
Croker, John Wilson, MP, 24, 26-7, 163
Crotch, William, 45
Crouch, Anna Maria, 6
Crow, William, 13n

Crow Street Theatre,
  and R. Daly, 9, 11, 102
  opened by F. E. Jones, 14
  and bond holders, 75, 225-6, 229, 232
  and Catholic Party tunes, 35-6, 129, 210
  and George III, 88
  and marshal law, 24
  and morals, 225
  artists' salaries at, 27, 61, 153
  auditorium of, 12-13, 14, 174, 203
  ballet at, 69, 71-2, 79, 119, 148, 154
  box office receipts, 13, 17, 61, 73, 80, 123, 156, 174, 179, 203
  capacity of, 13
  censorship at, 11
  charity performances at, 128, 155, 191
  chorus: inadequacy of, 64, 117, 201
  command performances at, 61, 87, 121, 150, 173
  "complimentary" tickets for, 12, 58, 79, 121, 211, 229
  costumes and scenery at, 77, 83, 165-6, 174, 201
  disturbances at, 31-2, 35, 61, 71, 95, 131, 149-50, 152, 181, 187-8
  exterior described, 13, 132, 134
  M. Fitzgerald as trustee, 162, 194
  free admission to, 75, 121, 130, 229
  "gold tickets", 176
  Italian opera seasons at, 4, 64-80, 103, 105-20, 125, 211, 220
  lighting at (oil lamps), 20, 77, 174
    (gas), 207
  managed by: T. L. Bellamy, 18
    J. Crampton, 83, 153
    W. Farren, 162, 174, 194, 196
    four trustees, 152
    J. G. Holman, 23
    A. Rock, 153
  management criticised, 152-3, 195
    defended, 162
  opera concerts at, 56-62, 141, 250, 252-4
  orchestra at, 20, 23, 33, 66, 134n., 148, 153, 159, 201, 219, 221
  patents for, 12, 224-6, 228
  patrons' attire, suggestions for, 16, 85
    carriages, instructions to, 14
  "peace celebrations" at, 132-3
  peace officers in, 181
  pickpockets in, 29, 32
  police in, 32, 194, 204, 209
  prompters at, 185, 187
  public taste at, 3, 85, 100, 158, 178-9, 192
  refreshments served at, 16, 131
  refurbishing of, 12-13, 14, 23, 85-6, 147, 161-2, 174, 203-5
  rules and regulations at, 14, 16, 148, 201, 257-60
  seat prices at, 58, 62, 72-3, 141, 211
  servants, restrictions for, 14
  shares, issued on, 58
  soldiers in, 194
  stage management at, 75, 185
  stage, size of, 12
  standard of productions at, 64, 77, 83-4, 118
  structural alterations to, 85
  Wexford Rebellion and, 16
  women at, 32, 93, 130, 148, 153, 204, 211
  working conditions at, 123
  "its end is nigh", 224
  F. E. Jones ousted as patentee, 225-6, 228
    his total expenditure on, 230
  H. Harris succeeds, 228
  principal creditor objects, 228
  finally closed as Theatre Royal, 229-30
  falls into neglect, 230
  1798 season, 14
  1799 season, 17
  1800 season, 20
  1801 season, 21
  1802 season, 23
  1803 season, 23-4
  1804 season, 24-5, 27-8
  1805 season, 28
  1806 season, 29, 31-2
  1807 season, 34-6, 47-62
  1808 season, 63-83
  1809 season, 86-99
  1810 season, 99-100
  1811 season, 4, 100-2, 105-23
  1812 season, 125
  1813 season, 125-9
  1814 season, 129-148, 250-6
  1815 season, 153-163
  1816 season, 165-175
  1817 season, 178-181
  1818 season, 182-200
  1819 season, 4, 201-224
  1820 season, 229
Crozier, Mr, 66n
Cubitt, Mr, 66n, 134n
Cunningham, Mr, 189
Curfew, introduction of, 16, 24
Curioni, Alberico, 116
Curran, John Philpot, 49, 192
Currencies, Irish v. English, 51n
Curtis, Miss, 229
Cymon (M. Arne), 158
  (Bishop), 158n
  (Stevenson), 158, 244

INDEX

Dalayrac, Nicolas Marie, 28
Dalton, Edward Tuite, 83, 103, 124, 152, 226, 228
Daly, Denis Bowes, MP, 152
   Richard, 9, 11-12, 102, 129, 229, 230
   Mr, 134n
Dama Soldato, La (Orlandi), 135
Da Ponte, Lorenzo, 96, 241, 248
D'Arcy, Mr, 186, 191
Dauberval, Jean, 119
David, Félicien César, 192n
Davidson, Miss, 27, 234
Davis, Mr, 66n
   Mr, 234
   Mrs, 236, 239
Davy, John, 31, 83, 96, 97, 185-6, 236, 247
Dawson, Mrs, 234
   Timothy, 4n
   William, 4
Day, Francis, 156
Dean, Winton, 7
'Dear little shamrock, The', 83, 95
De Lihu, Annette, 205
   Victorine, 205
Delvecchio's Music Shop, 148
Demofoonte (pasticcio), 42
Denmark Street Chapel (Dublin), 141-2, 255
Dennett, The Misses, 125, 126, 132, 154, *164*
Descombes, Jean Charles François Maurice, 35
Deux Journées, Les (Cherubini), 100
Devil to pay, The (Coffey), 141n
Deville, Giuseppe, 211, 215-6, 220, 248
   Paolo, 63
Devil's Bridge, The (Braham/Horn/Corri) 125, 181, 242
Devonshire House (London) 49
Dibdin, Charles, 93, 173
   Charles Isaac Mungo, 245
   Thomas John, 19, 20, 23, 28n, 31, 36, 97, 102, 233-4, 236-7, 239
Dickons, Maria, 95, *101*, 103, 172, 240
Didactic aims, 85, 225
Didone (Paisiello), 75, 77, 237, 249
Dilettante, Il (pasticcio), 139
Dimond, William, 31, 173, 236, 240, 245
Dinorah (Meyerbeer), see *Pardon de Ploërmel, Le*
Dizi, François Joseph, 119-120
Dog of Montargis, The, see *Forest of Bondy, The*
'Dog Riots' (Dublin), 149-50
Donaghadee, 47
Donegall, Marchioness, 105
Don Giovanni (Mozart), 116, 154, 167, 182, 211-2, 214, 218, 220-4, 246, 248-9
Donizetti, Domenico Gaetano, 84, 154
Donne cambiate, Le, see *Calzolaio, Il*

Donzelli, Domenico, 92, 109n
Dorset, Duchess of, 129, 140, 150, 167
Downes, Mr, 234
Doyle, Mr, 75
Drogheda (Ireland), 132
Drury Lane Theatre, 24, 36, 56, 82-3, 86, 89, 91, 93, 95, 99, 100, 156, 165, 176, 179, 181, 183, 217, 233-6, 238-9, 241-6
Dublin Castle, 13
   map of, *15*
   Musical Fund, 17
Due Nozze ed un sol Marito (P. C. Guglielmi), 109, 113, 119-20, 141, 143-4, 240, 249, 254
Duenna, The (Linley), 96, 144, 170, 172-3
Duff, John R, 84, 238
   Mrs Mary Ann, 239
Duignan, Mr, 123
Dunn, Mr, 234
Dussek, Mme, 114
Dutton, Thomas, 21, 43
Dyke, Mary Ann, 238
   Miss, 239
   Mr, 156, 165
   Mrs, 234, 239

Ebers, John, 116, 221
Edgeworth, Maria, 97, 206
Edinburgh (Scotland), 83, 105, 117, 216, 219, 229
Edwards, Henry Sutherland, 88
Edwin, Elizabeth Rebecca, 163, 235, 244
   John, sen., 163
   John, jun., 163
Edwin and Angelina (Stevenson and Clifton), 166, 245
Egbert and Ethelinda; or, The Draw Bridge (Blewitt), 166, 244
Elena e Malvina (Schira), 38n
Elgin Marbles, 26
Elisabeth; ou, les Exilés de Sibérie (Mme Cottin), 84-5
Elisabetta, Regina d'Inghilterra (Rossini), 215
Elisir d'amore, L' (Donizetti), 111
Elliston, Robert William, 31, 236
Emmet rising (1803), 11, 24
Emmet, Robert, 24
English Fleet in 1342, The (Braham), 32, 36, 97, 237
English Opera House, 183; see also Lyceum Theatre
Equestrian displays, 160, *231*
Ereby, Miss, 165
Escapes, The; or, The Water Carrier (Cherubini/Attwood), 100, 239
Essex, 5th Earl of, 173-4
Événemens imprévus, Les (Grétry), 102
Exile, The; or, The Deserts of Siberia (Mazzinghi and Bishop), 84-5, 238

Fagan, Patrick, 7
Fallon, John, 66n
False Alarms; or, My Cousin (King and Braham), 97, 98-9, 239
Familiar Epistles...on the present state of the Irish Stage (J. W. Croker), 24, 26-7, 163
Fanatico per la Musica, Il (Mayr), 56, 57, 59, 61, 72-3, 77, 80, *108*, 114, 119-20,121, 142, 144, 167, 223-4, 237, 240, 248-9, 252, 255
Farinelli, Giuseppe, 120
Farington, Joseph, 6
Farrel, Mr, 13
Farren, Percy, 162n, 239-40, 243, 245
 William, 155, 162, 166, 172, 174, 179, 187-8, 194, 196, 239-47
Fashions of the day, 16, 56-7, 72, 85, 112-3
Faulkner's Hotel (Dublin), 212
Favart, Madame, 4
Favières, Edmond Guillaume François de, 21
Fawcett, Captain, 225
 John, 102, 246
Federici, Francesco, 105, 118, 241, 249
 Vincenzo, 118
Fees and salaries, 27, 40, 43, 68, 71, 91, 110, 123, 197, 212, 215, 216, 218
Femme à deux maris, La (Guilbert de Pixérécourt), 28
Fenice, Teatro La, *see* Venice
Feramors (Rubinstein), 192n
Ferlendis, Alessandro, 135, 136, 139, 143, 148, 253
 Camilla, 134-5, *136*, 137, 141-3, 148, 253-6
 Giuseppe, 135-7
Ferrara (Italy), 105, 107, 118
Ferrari, Giacomo Gotifredo, 42, 58n, 59n, 78-9, 237
Fétis Concert, Paris, 219
Feudal Times (Kelly), 61
Fidelio (Beethoven), 28
Field, John, 135n
Fielding, Henry, 87
Figaro (burletta), (Mozart), 169
Fille mal gardée, La (ballet), 119
Fingall, Earl, 142
Fioravanti, Valentino, 73, 78, 135, 237, 249, 251-2
First Attempt, The; or, The Whim of the Moment (T. Cooke a.o.), 34, 96, 236
Fishamble Street Music Hall (Dublin), 9, 11
 Theatre, 14n, 28, 121, 144, 146, 156, 160
Fiske, Roger, 165n
Fitzgerald, Maurice, MP, 194, 228
Fitzsimons, Edward, 206, 216, 248
Five Lovers, The (Cooke), 29, 235
Flauto Magico, Il, see *Zauberflöte, Die*
Florence (Italy), 69, 89, 107, 113
Fontainbleau (Shield a.o.), 180
Foote, Mr, 31

Foppa, Giuseppe, 141n
Ford, Miss, 229, 247
Forest of Bondy, The; or, The Dog of Montargis (Blewitt and Cooke), 149-50, 152, 242
Forty Thieves, The (Kelly and Cooke), 61, 100, 239
Franz II,(I) Emperor of Austria, 200
Frascatana, La (Paisiello), 5, 79-80, 238, 249
Frederick the Great; or, The Heart of a Soldier (Cooke), 149, 243
Freemasons (Ireland) 155n, 191
Freemasons' Hall (London), 167
Freischütz, Der (Weber), 181
Frizzi, Benedetto, 46
Fry, Mr, 246
Fullam, Michael, 131, 155, 235-40, 243-8
Fulton, Mrs, 242
Furbo contro il Furbo, Il (Fioravanti), 78-9, 112, 237, 249

Gaelic Athletic Association, 9
Gain, Mr, 66n
Gainford family, The, 189
Gale, Mr, 134n
Galindo, P., 233-4
 Catherine, 235
Galliard, Johann Ernst, 7
Galuppi, Baldassare, 53
Garat, Dominique Pierre Jean, 215
Garbett, Mr, 134, 250, 252
Garcia, Manuel, 111, 116
Garrick, David, 100n, 158, 188, 244
'Garryowen', 35
Gas lighting, 20
Gaudry, Richard, 233-4
Gavan, Michael, 62
Gaven, Mr, 244
Gay Deceivers, The (Kelly), 102, 240
Gayton, Miss, 79n
Gazza Ladra, La (Rossini), 162, 215
Geeson, W., 218
 Mrs, *see* Corri, Rosalie
General Asylum (Dublin), 173
Generali, Pietro, 250
Genoa (Italy), 69, 89, 118
George III, 88, 225
George IV (as Prince Regent), 20, 110, 112, 134
 (as King), 232
Gerbini, Luigia, 134, 139-40, 142, 148, 149, 250, 252, 256
 Signor, 149
Germany, National Anthem of, 200
Gherardini, Giovanni Michele Secondario Crisippo, 163
Giardini, Felice, 97

INDEX 285

Gilardoni, Domenico, 84
Giordani, Nicolina, 4
    Tommaso, 4, 17, 23, 34
Giornovichi, Giovanni Maria, 23
Gipsy Prince, The (Kelly), see *Gypsy*
Giroux, "the three Misses", 238
    Mr, 238
Giulietta e Romeo (Zingarelli), 110
Giulio Sabino (Sarti), 58n
Glasgow (Scotland), 105, 179, 183
Glenelg, Lord, *see* Grant, Charles
Glover, Mr, 66n, 134n
    Mr, jun., 134n
Gluck, Christoph Willibald Ritter von, 19, 215
Goldsmid, Abraham, 89
Goldsmith, Oliver, 166
Gomery, Robert, 239
Goodwin, Mr, 234
Gordon, 4th Duke of, 35n
    Pryse Lockart, 106-7
'Gott erhalte Franz den Kaiser' (Haydn), 200
Goulding, Phipps, D'Almaine & Co., 21
'Graces of the Shawl', 79
Grant, Charles, 226, 230
    Master, 236
    Miss, 236
    Mr, 236
Grassini, Giuseppina Maria Camilla, 28n
'Gratias agimus tibi' (P. A. Guglielmi), 250-1, 256
Grattan, Henry, 49, 80
Gray, Mr, 134n
Green, Mrs, 206
Gregory, George, 152
    William, 226
Grétry, André Ernest Modeste, 102
Greville, Colonel Harry F., 68
Grey, Miss, 246
    Mr, 66n
Griglietti, Elizabeth Augusta, 148, 166-9, 173, 176, 179, 245-6
Griselda (Paer), 117, 120, 253
Guerini, Vincenzo, 33
Guglielmi, Pietro Alessandro, 58, 250
    Pietro Carlo, 109, 113, 223, 240, 249, 254
Gustavus Vasa (Kelly), 103, 240
Gutierrez, Beniamino, 118
Guy Mannering (Bishop), 207, 246
Gypsy Prince, The (Kelly), 53, 61, 120, 124

Hales (D'Hèle), Thomas, 102
Halliday, Joseph, 189
Hamburg (Germany), 89
Hamerton, J., 233
Hamilton, W. H., 155, 243

Hamlet (Shakespeare), 132
Hammersley, Miss, 182, 246-8
    Mr, 201
Handel, George Frederick, 16, 19, 99, 138, 168, 250
Hanover Square concerts, 162, 167
Hardwicke, 3rd Earl of, 20, 24, 36, 129
    Countess of, 21, 24, 36
Hargrave, Mr, 235
Harrington, Earl of, 49
    Countess of, 72
Harris, Captain George, 228
    Henry, 228-9, 232
    Thomas, 98
Harrowby, Lady, 49
Hartson, Hall, 82
Haunted Tower, The (Storace), 3, 96, 99, 167, 175-6, 195, 201-2, 229
Haydn, Franz Joseph, 21, 99, 120, 149, 250-2, 255
Haymarket Theatre (London), 23, 31, 86, 93, 144, 183, 218
Hayward & Co., Thomas, 167n
Hazlitt, William, 21n, 29, 133, 155, 164, 180, 182, 185, 243n
Henderson, John, 165
Henley, Mr, 66n
Henn, Mr, 75
Henry, Lady E., 49
Hermit, The (Goldsmith), 166
Heureuse Erreur, L' (Patrat), 173
Hewett, Mr, 245
Heywood, Joseph, 88
Hill, James, 238, 241
His Majesty's Theatre (London), 235-6, 240
Hitchcock, Sarah, 236-7
Hodson, George Alexander, 165, 179, 186, 195-7, 199, 208, 244-7
    Mrs, 196, 244
Hogan, Ita Margaret, 7
Hogarth, William, 170
Holcroft, Thomas, 28, 102, 239
    Mrs Thomas, 28
Holland, Mr, 97
Holman, Joseph George, 23
    Mr, jun., 82n
Holyhead (Wales), 48, 62, 173
Hook, James, 35, 89, 236
    Theodore Edward, 34, 236
Holy Cross College (Ireland), 9
Horn, Charles Edward, 124n, 125, 167, 180-3, 185-8, 192, 195, 197-9, 201, 241-2, 246-8
    Karl Friedrich, 180
Horncastle, Mr, 146
Hortense, Queen of Holland, 206
Howard, Samuel, 19

Howells, Fanny, 27, 235 *see also* Mrs T. S. Cooke
Huddart, Thomas, 238
Hughes, Miss (later Mrs Gattie), 134, 138, 154-6, *157*, 159, 243-4, 250-3, 256
   Mr, 234
Hungarian Cottage, The; or, The Brazen Bust (Bishop), 130
Hunt, James Henry Leigh, 21n, 24, 43, 47, 92, 95, 124, 167-8, 176, 180, 217
Hunter of the Alps, The (Kelly), 31, 236
Huntley, Mr, 242-4

Idomeneo (Mozart), 114
Ifigenia in Aulide (Mosca), 38
Incledon, Charles Benjamin, 21, *22*, *25*, 28n, 31-2, 36, 103, 144, 146, 160, 172, 186, 234, 236-7, 240
Incorporated Musical Fund Society (Dublin), The, 128
Indian Settlers, The (ballet), 205
Ingleby, Mr, 156
Inkle and Yarico (Arnold), 32, 97
Innkeeper's Daughter, The (Cooke and Barton), 185, 246
Invisible Girl, The (J. Hook), 34, 236
Irish balance of payments (1811), 103
Irish clothes promoted, 61
Irwin, Eyles, 20-1, 234
Isaacs, John, 243n
Isabella (Garrick), 100n
Isouard, Nicolò, 155-6
Israel in Egypt (Handel), 88
Italiana in Algeri, L' (Rossini), 205
Ivrea (Italy), 212n

Jackson, John, 171
Jack the Giant-Queller (Brooke), 21
Jamaica, 89
Jean de Paris (Boieldieu), 154; see also *John of Paris*
Jenkins, Florence Foster, 186
Jenkinson, William, 134n
Jocasto, Mrs, 234
John of Paris (Boieldieu, Bishop, Blewitt and Pucitta), 154-6, 159, 243
Johnson, Mr, 155, 182-3, 233-6, 238-40, 242-8
   Mrs, 236
   Samuel, 188
Johnston, Henry Erskine, 24n, 86, 88, 123, 125
   Miss, 243-4
   Mr, 233, 244
   Mrs, 237
Johnstone, John Henry, 24, 235
   Miss, 245-7
Jolie, Dr, 149
Jones, Charles Horatio, 232

Jones, Frederick Edwards ('Buck'), 9
   opens private theatre, 9, 11
   petitions to open public theatre, 11
   Crow Street Theatre property assigned to, 12
   Government subvention received by, 12
   patent obtained by, 11
   petitions to Parliament, 17, 225
   and bond ticket holders, 225-6, 229, 232
   and T. L. Bellamy, 18
   and John Braham, 91
   and Mary Byrne, 208-11
   and Angelica Catalani, 41, 51, 55
   and J. W. Croker, 24, 26-7
   and Richard Daly, 11-12, 102-3
   and Charles Incledon, 146
   and Michael Kelly, 105
   and lease of Cork & Limerick theatres, 12, 103
   and Mozart "adapted", 118-20
   and Royal Hibernian Theatre, 123
   and his shareholders, 58, 103
   and theatre disturbances, 149-50, 152, 194
   as actor, 9
   as manager of Drury Lane Theatre, 51, 82
   death of his mother, 78
   defends his actions, 161
   his expenditure on Crow Street Theatre, 13, 17, 230
   his generosity to artists, 32
      to charities, 99, 199
   his home (The Red House: "Clonliffe"), *10*, 194, 232
   his management criticised, 152-3, 160, 195, 225-6
   his patent questioned, 225
      relinquished, 194, 197
   imprisoned for debt, 228n.
   in dispute with artists or stage personnel, 123, 125, 198-9, 208-11
   in litigation, 75, 100, 194
   in negotiation with W. Farren, 162, 194
   refurbishes Crow Street Theatre, 85, 147, 203, 225
   resumes management, 83, 85, 153
   seeks renewal of patent, 225-6
   sells Crow Street Theatre share, 83
   transfers interest in Crow Street Theatre, 83, 228
   withdraws from management, 51, 82-3, 152, 162
   seeks compensation, 230, 232
   death of, 232
Jones, Frederick, jun., 196n, 197, 232n
Jones, I., 244
   Nicholas, 236, 238-42, 244-5
   Richard, 235-6
   Richard Talbot, 232
Jordan, Dorothy, 48-9, 62

INDEX 287

Joseph II, Emperor, 6
Josephine, Empress, 89
Jovial Crew, The (Bates), 185

Kais (Braham and Reeve), 96
Kärntnerthor Theater (Vienna), 215
Kavanagh, Mr, 134n
Kean, Edmund, 174
'Keep those tears for me' (Moore), 96
Kelly, Frances (Fanny) Maria, 179, 183, *184*, 185, 195, 246-7
  Joseph, 6
  Luke, 183n
  Lydia Eliza, 179-80, 183-4, 197, 246-8
  Mark, 179
  Mary, 183n
Kelly, Michael William (1762-1826), 6, 42, 69, 82, 104, 179, 183-4
  and Catalani, 48, 49, 52, 53, *55*, 57, 61, 64, 75-6, 78, 80
  and Tom Cooke, 56
  as composer, 20, 28, 43, 79, 100, 102, 103, 120
  as critic, 32n.
  as pianist, 41, 43
  as singer, 41, 43, 53, 54, 61, 64, 75, 78, 80, 120, 165, 237-8
  as stage manager, 41, 52, 105, 111-3, 118, 120, 123
  his "Reminiscences" corrected, 63, 112, 121, 165
  in backstage discord, 61
  "laid up with gout", 57, 75
  leaves Dublin, 62
  Mr (singer), 234
  Mr ('cellist), 66n
  Mr (architect), 86
  Mr (stage machinist), 156
  Mrs, 234
Kemble, Charles, 130, 182, 214, 244
  John Philip, 89, 91, 92, 98, 165
Kemp, Mr, 201
Kenney, Charles Lamb, 189, 190
  James, 28, 97, 167, 235, 239
  James, senior, 28
Kent, Mr, 239
Kilkenny Private Theatre, 124, 126n, 132, 152, 197
Killeen, Viscount, 142
Kilwarden, Lord, 11
King, Sir Charles Simeon, 4
  Alec Hyatt, 167n
  Matthew Peter, 28, 97, 125, 167, 235, 239, 241
  Mr, 75
  Mr (singer), 234
  Mrs G., 234
  William, Archbishop of Dublin, 4-5
King Lear (Shakespeare), 3

King's Theatre (London), 5, 31n, 40, 42-3, 51, 64, 68-9, 71, 73, 75, 76-9, 86, 88-9, 91, 95, 103, 105, 107, 109, 111-4, 116-7, 121, 135, 137-40, 144, 148, 155, 167-8, 175, 183, 205, 211-2, 215-8, 223, 238
Knight, Thomas, 20, 233
Kotzebue, August Friedrich Ferdinand von, 20
Kreutzer, Rodolphe, 21, 244
Krifft, William B. de, 6

Lablache, Luigi, 111
La Costa, Mme, 71
Lacy, John, 134, 138, 245, 252
  Michael Rophino, 54
La Feuillade, Mme, 128
La Finilarde, Mme, 128
Lalla Rookh; or, The Minstrel of Cashmere (Horn), 192, 247
Lamb, Charles, 88, 185
Lampe, John Frederick, 5
Lancashire Chorus Singers, The, 140, 201
Lanza, Gesualdo, 169-70, 172, 218
La Rosa, Rosina, 128
La Touche, Mr, 229
La Touche's Bank, 11
Lavenu, Lewis, 219
Lay of the Last Minstrel, The (Scott), 102
Lazenby, George, 168
  Mrs, 168, 186, 246-7
Le Clercq, Mr, 148, 154
Lee, Mr, 235, 238-9
Lees, Mr, 155n
Le Fanu, Alicia, 49
Lefroy, Mr Sergeant, 229
Legislation, 26 Geo. III c.57 (1786), 225
Leinster, Duke of, 134, 155, 191
  Duchess of, 205
Leipzig (Germany) 143
Lennox, Charles, *see* Richmond, Duke of
Leoni, Michael, 4, 89
Lesage, Alain René, 78
'Let fame sound the trumpet' (Shield), 32
Levasseur, Nicholas Prosper, 219
Lewis, Matthew Gregory ('Monk'), 125, 185, 241, 247
  Mr, 239
Libertine, The (Mozart/Bishop), 182, 246
Libretti disparaged, 92
Limerick (Ireland), 12, 63, 83, 103, 141-2, 195, 197, 257-60
Lincoln's Inn Fields Theatre (London), 5
Lindley, Robert, 135, 139, 251-3
Lindsay, Mr, 27, 235
Linley, Thomas, 96, 127
Lionel and Clarissa (C. Dibdin), 166
Lisbon (Portugal), 40, 71, 107, 135

Liverati, Giovanni, 175
Liverpool, 63, 105, 149, 175
Livigni, Filippo, 238
Livorno (Italy), 41, 105
Lock and Key (Shield), 144, *177*
Locke, Miss, 238
Loder, John David, 134, 139, 253
Lodi (Italy), 218
Lodoiska (Cherubini) 163, 165
    (Kreutzer), 163, 165
    (Storace), 165, 166
    (pasticcio), 244
    La (Mayr), 38
Logier, Johann Bernard, 149
Lord, Mr, 165
Lord Lieutenants (Ireland), 9, 11, 20, 24, 34-6, 61, 78, 96-7, 121, 128-9, 130, 134, 140, 150, 167, 173, 194, 198, 200n, 204-5, 224, 232
Lord of the Manor (Bishop a. o.), 167
Louis Bonaparte, 106, 111
Love in a Blaze (Stevenson), 18, 21, 233
Love in a Village (pasticcio), 146, 162, 170, 172, 176, 181, 195, 206
Love Laughs at Locksmiths (Kelly), 28, 235
'Lover's Mistake, The' (Bayly/Balfe), *193*
Lupino, Georgina, 71
Lyceum Theatre (London), 124, 126, 149, 167-8, 179, 180, 241-3, 247
Lying Valet, The (Garrick), 16
Lyons, Miss, 241
Lyric Novelist, The, 95

M.P.; or, The Blue Stocking (Moore), 124-5, 167, 181, 241
McCauley, Miss, 236
McCulloch, Miss, 239
    Mrs, 236, 238-9, 243-4
    Mrs Charles, 243-4
McDonnell, Mrs, 246
McGouran, Rev., 142
McKenna, Mr, 86
McKeon, Mr, 242
McNally, Mr, 119
    Leonard, 5, 179, 246
MacNamara, Mrs, 235
Madden Mr, 234
Madrid (Spain), 139
Maestro di Cappella, Il (Cimarosa), 68, 81
Magic Flute, The, see *Zauberflöte, Die*
Magicien sans Magie, Le (Isouard), 155
Magician without Magic, The (Blewitt), 156, 243
Mahmoud (Storace), 89
Mahon, Gilbert, 19n
    Mrs Gilbert, 19n, 34
    John, 33, 134n, 149

Mahon, Robert, 4
    William, 33
Maid and the Magpie, The (Bishop), 3, 162, 163, *184*, 185, 244
'Maid of Lodi, The', 119
Maid of the Mill, The (Arnold a.o.) 173
Maison à vendre, La (Dalayrac), 28
Manchester, 105, 170
Manchester Chorus Singers, 140n
Manners, Lady Jane, 205
Manni, Niccola, 111, 113, 117, 240-1
Mansell, Mr, 235
Mansfield, Sir James, 98
Mansion House Committee (Dublin), 198
Mara, Gertrud Elizabeth, 58
Marchesi, Francesco, 41
    Luigi, 38
Mardyn, Mrs, 243
Maria I, Queen of Portugal, 40
Maria Luisa, Queen of Spain, 40
Maria Theresa, Empress of Austria, 200n
Marinari, Gaetano, 86, 97, 130, 132, 156, 165, 182, 201, 203, 206
Marsollier des Vivetières, Benoît Joseph, 28
Martial law (Dublin 1803), 24
Martinelli, Peter, 21
Martin y Soler, Vicente, 53, 97
Mary's Abbey (Dublin), 4
Mason, John Monck, 14
    Thomas Monck, Captain, RN, 14
    William Monck, 4n, 14n
    Mrs, 84, 238
Matrimonio Segreto, Il (Cimarosa), 58, 137-9, 167
Matrimony (King), 28, 235
Maurice, Charles, *see* Descombes, J.C.F.M.
Maximilian Joseph, Archduke of Austria, 199-200
Mayhew, Mr, 229
Mayr, Johann Simon, 4, 38, 52, 56-9, 73, 119, 138, 167, 223, 237, 248-50
Mayseder, Joseph, 190
Mazzinghi, Joseph, 20, 21, 24, 28, 84, 233-5, 238
Meadows, T. P., 190
Meath Charitable Loan Society, 99
Meath, Countess of, 49
    Earl of, 49, 152
Meglier, Mr, 66n
Méhul, Étienne Nicolas, 28
Mellish, Mr, 66n
Melodrame, 3, 130, 162
Mengozzi, Bernardo, 53
Merchant of Venice, The (Shakespeare), 16, 230
Merlin, Comtesse, 111
Messiah (Handel), 18, 34, 128, 134, 140, 250, 255-6
Metastasio, Pietro, 237

INDEX

Metheringham, Mr, 66n, 134n
Metherington, Mr, 134n
Mexican Festival, The (ballet), 79
Meyer, Mr, 201
Meyerbeer, Giacomo, 149
Miarteni, Maria, 63, 68-9, 72, 81, 237-8
    Nicolò, 63, 68, 72, 78, *81*, 112, 237-8
Michael, Grand Duke of Russia, 200n
Midas (pasticcio), 146-7
Milan (Italy), 38, 64, 69, 89, 107, 118, 128, 212, 217
Miller and his Men, The (Bishop), 130, 150, 242
Minstrel's Summer Ramble, A, 186, 187, 189
'Mirth and Minstrelsy...', 146
Mitridate (Sarti), 54, 59
Mock Doctor, The (Fielding), *87*
Molière, 87
Molinara, La (Paisiello), 58n, *108*, 178n
'Molly Astore', 33
Monck Mason, J. *see* Mason, J. M.
Montgomery, H., 208-9, 211, 245-6
Moore, Miss E., 242
    Mr, 236
    Thomas, 17, 24, 53, 96, 124-5, 131, 181, 192, 251
Moorehead, John, 23, 28, 31, 96, 236
Moran, Mr, 165
Morandi, Pietro, 37, 38
Morelli, Giovanni, 41-2, 52, 53, 57-8, 61-2, 68
Morgan, Lady, *see* Owenson, Sydney
    Sir Thomas Charles, 34n
Mori, A, 218, 220
    Maria, 218
    Marietta, 218
    Nicolas, 135, 139, 211-2, 219-21, 223-4
    Rosina, 219
    Signora, 211, 223-4, 248
Morigi, Andrea, 42
Morrison's Great Rooms (Dublin), 192
    Hotel (Dublin), 200n
    Tavern (Dublin), 229
Morte di Cleopatra, La (Nasolini), 52, 56, 59, 61
Morte di Mitridate, La (Nasolini), 53
    (Portugal), 40, 79-80, 141, 238, 249, 254
Morte di Semiramide, La (Portugal), 38, 41, 52, 56, 57, 59, 61, 67, 70, 72, 74, 76-7, 141, 144, 237, 249, 253-4
Mortellari, Michele, 64, 72
    Michele C., 64, 66n
Morton, Thomas, 179, 245
Mosca, Giuseppe, 38
Moscheles, Ignaz, 66
Mount Edgcumbe, Richard, 42, 47, 69, 92, 109, 111, 136, 139
Mount Melleray Abbey (Ireland), 223
Mount of Olives, The (Beethoven), 224
Mountain, Rosoman, 28n, 93, 95-6, 97, 166

Mountain, John, 95
Mountjoy, Lord, 9
Mouth of the Nile, The (Attwood), 19, 233
Mozart, Wolfgang Amadeus (1756-91), 4, 33, 53, 79, 120, 149, 251
    Clemenza di Tito, La, 76, 91, 253
    Così fan tutte, 33, 107, 110-2, 114, 116, 120, 124, 215, 220, 223, 241, 249-50
    Don Giovanni, 116, 154, 167, 182, 211-2, 214, 218, 220-4, 246, 248-9
    Idomeneo, 114
    Nozze di Figaro, Le, 43, 103, 111-2, 116-7, 205, 211-2, 214-6, 220, 223-4, 248-9
    Zauberflöte, Die, 59n., 112, 117, 120, 167, 183, 218, 223, 250, 252-3
Mulligan, Mr, 66n, 134n
Munden, Joseph Shepherd, 24, 233, 235
Munich (Germany), 105
Musicamania, La, see *Fanatico per la Musica, Il*
My Spouse and I (Whitaker), 166, 245

Naldi, Antonio, 109
    Carolina, 110, 119
    Giuseppe, 105, 107, 109-10, 113-4, 116-20, *122*, 143, 181, 218, *222*, 240-1
    Signora, 110, 119-20
Naples (Italy), 64, 69, 78n, 105, 108, 118, 219, 237
Napoleon I, 40, 49, 89, 134
Napoleon III, 206
Nasolini, Sebastiano, 52, 53, 56, 65, 149
Naufrage, Le (Lafont), 18
Naval Pillar, The (Moorehead), 28n
Nelson, Mr, 134n
Nelson monument (Dublin), 33
Netley Abbey; or, The Sailor in his Glory (Shield a. o.), 3, 19, 131
Neville, Mr, 242
New Concert Rooms, (Dublin), 33-4
New English Opera, 185, 243n
Newenham, Mr, 233
New Rooms, Hanover Square (London), 138
Newry (Ireland), 83
Nichols, Mr, 242
Nicolini, Giuseppe, 38, 119
    né Nicola Grimaldi, 4-5, 7
Nobility's Concerts (London), 138
Nora, Mme, 71, 79
Norma (Bellini), 217
Norman, John, 234, 236
Norton, Miss S., 242, 245
No Song, No Supper (Storace), 200n
Novello, Vincent, 66
Nozze di Figaro, Le (Mozart), 43, 103, 111-2, 116-7, 205, 211-2, 214-6, 220, 223-4, 248-9

Nunez, Giovanni Battista, 46
Nunn, Mrs, 29, 93, 95-6, 97, 235-7, 239
Nurmahal; oder, Das Rosenfest von Caschmir (Spontini), 192n

O'Callaghan, Mr, 181, 183, 186, 246-7
Occasional Oratorio (Handel), 16
Of Age tomorrow (Kelly/Paisiello), 20, 234
Oh! This Love!; or, The Masqueraders (King), 167
O'Keeffe, John, 124
Olympic Pavilion (Dublin), 160
O'Moran, Mme, 128
One o'Clock; or, The Knight and the Wood Daemon (Kelly and King), 125, 241
O'Neill, Eliza, 132, *133*, 241-2
  John, 132
Onorati, Bernardino, 37
Opera Agreement (1792), 167n
Opera dress fashions, 56-7, 72, 112-3
  glasses, 57
  production, 64, 77, 83-4, 152-3, 156, 159, 174
Opera House Theatre (Dublin), 28n
Opie, John, 98
  Mrs Amelia, 214
Oratorios, concerts of, 17-18, 128, 134, 140, 167n, 250-2, 255
Orazi e i Curiazi, Gli (Cimarosa), 38, 52n, 105, 253
Orchestral conductors and leaders, 65-6
Ord, Mr, 229
O'Reilly, Myles, 194, 226
Orfeo ed Euridice (Gluck), 19
Orlandi, Ferdinando, 135
O'Rourke, Mr, 186
  William Michael, 190, 191, 201
Ossory, The Very Reverend Dean of, 132
O'Sullivan, Michael John, 131-2, 183, 192, 242, 247
Otello (Rossini), 216
Othello (Shakespeare), 232n
Otto Mesi in due ore, ossia Gli Esiliati in Siberia (Donizetti), 84
Otway, Thomas, 132
Oulton, Walley Chamberlain, 23
Out-Post, The (Stevenson), 206, 247
'Overture à la chasse' (Cooke), 31
Owenson, Robert, 34, 236
  Sydney, 34, 49, 96, 192, 236

Padlock, The, (C. Dibdin), 176
Paer, Ferdinand, 59n, 69, 117, 120, 137, 214, 253
Paganini, Nicolò, 58n
Paisiello, Giovanni, 4, 19, 20, 41, 42, 57-8, 61, 75, 79, 108, 119, 149, 178, 223, 237-8, 249
Palermo (Italy), 118, 241
Paltoni, Giuseppe, 217

Panormo, Percival, 208-9, 224
Pantheon, The (London), 42, 66, 86, 111, 148, 169
Pardon de Ploërmel, Le (Meyerbeer), 149
Paris, 21, 38n, 40, 40n, 42, 69, 89, 110, 111, 116, 119, 135, 137, 139, 140, 143, 147, 165, 183, 212, 215, 216, 219, 218-9
Parke, William Thomas, 19, 79, 83, 109
Parleur éternel, Le (Maurice), 35
'Partant pour la Syrie' (Queen of Holland/Drouet), 205
Passions, The (Collins), 229
Pasta, Giuditta, 110
'Pastoral opera' genre, 3
Pastorella Nobile, La (P. A. Guglielmi), 58
Patents, 11-12, 194, 197, 224-6, 232
Patrat, Joseph, 173
Patrician Society, The (Dublin), 191
'Patrick's Day' 35-6, *see also* 'St Patrick's Day'
Patriot, The; or, Hermit of Saxellen (Stevenson), 102, 251
Paul and Virginia (Mazzinghi and Reeve), 21, 234
Pavia (Italy), 128, 218
Payne, Mr, 97, 239
Peace celebrations (Dublin), 132-3
Pearce, William, 19, 131
Peel, Sir Robert, 225
Peeping Tom of Coventry (O'Keeffe), 124
Pelham, Thomas, 11
Pellegrini, Felice, 147
Pendennis (Thackeray), 132n
Penelope (Cimarosa), 218
Peninsula War, The, 190n
Percy's Private Theatre (Dublin), 28n
Pergolesi, Giovanni Battista, 251
Peterson, Mr, 201
Phedra (Radicati), 119
Phelps, Mr, 114
Philadare, Mme, 128
Philharmonic Society Concert (London), 110, 205, 211
Philipps, Thomas, 26-7, 29, *30*, 31-3, 36, 52-3, 57, 61, 162, 165, 169, 172-3, 174, 178, 183, 234-8, 244-5
Piccinni, Nicolò, 64, 121
Pickpockets, 29
Pie voleuse, La; ou, La Servante de Palaiseau (Caigniez and Baudoin d'Aubigny), 162
Pigault-Lebrun, Charles Antoine Guillaume, 149
Pigott, Mr, 224
Pindar, Peter, *see* Wolcot, Dr John
Piovano, Francesco, 42
Pitman, Mr, 234, 236
Pixérécourt, René Charles Guilbert de, 28, 150
Pizarro (Sheridan), 97
Plagiarism, Leigh Hunt on, 43

INDEX 291

Pleyel, Ignace Joseph, 23
Plunkett, Luke, 186
Pobje, Mr, 100n
Pocock, Isaac, 130, 154, 162, 182, 185, 242-4, 246-7
Popular Tales (Edgeworth), 97, 206
Porpora, Nicole Antonio, 219
Portpatrick, 47
Portugal, Marcos Antonio da Fonseca, 4, 38n, 40-1, 52, 56-7, 59, 61, 65, 67, 70, 72, 74, 76-7, 79-80, 141, 144, 237-8, 249, 253-4
Postillon de Lonjumeau, Le (Adam), 138
Potts, James, 83
Power family, The, (Kilkenny), 132
Power, William, 125
   Mr, 246-7
Prague, 69, 211, 220, 248
Presle, Mlle, 69, 71
Prince Regent, *see* George IV
Principe di Taranto, Il (Paer), 253
Pro-Cathedral (Dublin), 220
Pucitta, Vincenzo, 128, 137, 149, 154, 175, 218, 223, 243, 250-2
Pugnani, Giulio Gaetano Gerolamo, 107, 139
Puppet theatre, 4
Purcell, Henry, 158
Putnam, Mr, 236-7
Puzzi, Giovanni, 205-6

Queen of Carthage, The, and the Prince of Troy (Shield/Hodson), 179, 246
Queen of Spain; or, Farinelli in Madrid (Lampe), 5

Radcliffe, Mr 235
Radicati, Felice Alessandro, 105, 107, 111, 115, 119-20
Rae, Mr, 84, 238
Rainville, Herr, 89
Raising the Wind (Kenney), 28
Ramah Droog; or, Wine does Wonders (Mazzinghi/Reeve), 24, 235
Ramsgate (Kent), 170
Ratchford, Mr, 234
Rauzzini, Venanzio, 42, 89, 95, 180
Rawling, Mr, 233
Reck, Padge, 6
Reeve, William, 20, 21, 24, 31, 96, 97, 156, 173, 233-6, 239, 243, 245
Reggio Emilio (Italy), 118
Reid, Mr, 244, 248
Re Teodoro in Venezia, Il (Paisiello), 119
Review, The (Arnold), 197
Reynolds, Frederick, 84, 156, 238, 243
   Sir Joshua, 76
Rich and Poor (Horn), 185, 247
Richard the Third (Shakespeare), 229-30

Richmond, Duchess of, 35, 61, 73, 86, 96-7
   4th Duke of, 29, 35, 61, 73, 86, 97, 121, 128-9
Righi, Signor, 48
Rimini (Italy), 69
Rinaldo (Handel), 7
Ritorno di Serse, Il (Portugal), 52, 65, 253
Rizzio, David, 146
Robertson, Henry, 106
Robinson, Mr, 66n, 134n
Rob Roy Macgregor; or, Auld Lang Syne (Davy), 185-6, 188, 247
Robson, William, 95
Rock, Anthony, 125, 148, 159, 160
   Mary, 125, 201, 241-2, 244-5, 247
Rode, Jacques Pierre Joseph, 124, 191
Roles, surprising casting of, 116-7
Roman Catholic Committee, 142, 175
Romberg, Bernhard Heinrich, 253
Rome (Italy), 38
Romeo and Juliet (Shakespeare), 132
Romero, Emanuele, 216n
   Signor, 211, 215-6, 220, 224, 248
Ronconi, Giorgio, 109n
Rooke, William, 190
Rose, Signora, 128
Rosenberg, Franz Xavier Wolf, Count, 6
Rosina (Shield), 3, 173, 180, 229
Rossi, Gaetano, 237
   Signor, 71
Rossini, Gioacchino Antonio, 34, 54, 101, 147, 154, 162, 163, 180, 205, 215-7, 223
Rossmore, Lady, 205
Rotunda (Dublin), 33-4, 51, 53, 56, 60, 128, 134, 140, 148, 168, 180, 190-1, 198-9, 205, 224, 228, 251, 253
Rousseau, Jean Jacques, 71
Rovedino, Carlo, 41, 42-3, 52-3, 57, 61, 62, 64, 72, 125, 237-8
   Tommaso, 217
Rowswell, Mr, 244, 246
Royal Academy of Music (London), 139
Royal Amphitheatre (Dublin), 28n, 86
Royal College of Surgeons (Ireland), 5-6
Royal Exchange (Dublin), 190
Royal Hibernian Theatre (Dublin), 86, 102, 123, 125
Royal Oak, The (Kelly), 120
Royalty Theatre (London), 89
Rubini, Giovanni Battista, 109n
Rubinstein, Anton, 192n
'Rule Britannia', 251, 253-4
Ruling Passion, The (Cogan), 5
Russia, 112
Ryder, Thomas, 12

S, W. jun., 247

Sacchini, Antonio Maria Gasparo Gioacchino, 21, 53, 57-8, 149, 252
St Albin, Mr, 183, 206
St Andrew's Church (Dublin) 131
St James's Theatre (London), 138
St Laurence's [Hospital] (Dublin), 129
St Mary's Lane Chapel (Dublin), 256
St Michael's and St John's Chapel, 224
St Michan's Parish Chapel (Dublin), 141, 142, 255
St Patrick's Day, 145
'St Patrick's Day', 129, 210
St Patrick's Hall (Dublin), 13
St Petersburg, 69, 71, 135
St Pierre, M, 69, 71-2, 79, 119, 154, 239
    Master, 165, 206
    Miss, 165, 206
    Mme, 71
St Werburgh's Church (Dublin), 134, 140, 163n, 168, 251
Salle Cléry (Paris), 40n
Salmon, Eliza, 33, 180
Samson (Handel), 33, 256
Samuellini, Mr, 201
Sarti, Giuseppe, 53-4, 57-9, 80
Saunders, Mr, 134n
Savage, Mr, 156, 165
Savoi, Gasparo, 5
Schiavi per amore, Gli (Paisiello), 41
Schira, Francesco, 38n
Schmidt, Johann Georg, 20
Scott, Sir Walter, 102, 154, 186, 246
Seagrave, Mr, 165
Seapoint House, Blackrock, 160
Semiramide, see *Morte di Semiramide, La*
Seymour, Mr, (scenic artist), 13, 21, 85-6
    Mr, 234
Shadwell, Thomas, 182
Shakespeare, William, 3, 132, 180, 230
Shakespeare Gallery (Dublin), 155
Shakespeare Tavern (Dublin), 14
Sharpe, Miss, 240
Shaw, Mr, 162
Sheridan, Richard Brinsley, 43, 82, 97, 112, 127
    Thomas, 32, 82
Shield, William, 19, 32, 95, 101, 103n, 131, 144, 173, 177, 179, 246
Short, T., 243
    Mr, 155n, 243-4
Siboni, Giuseppe Vincenzo Antonio, 63, 69, 72-3, 75-6, 78-80, 237-8
    Signora, 73
Siddons, Mrs (Sarah), 133, 214
Siege of Belgrade, The (Storace), 93, 96, 99, 126, 166, 180, 181

Simon, Mlle, 183, 223
    M., 183, 206
    M. jun., 206
Simpson, Mr, 238
Sinclair, John, 144, 146-7, *151*
Sipper, Mr, 66n
Sixty-third Letter, The (Arnold), 23
Slave, The (Bishop), 179, 245
Sloman, Mr, 239-40
Smart, Sir George Thomas, 180, 217
Smith, John Christopher, 5
    Miss, 238, 240
    Mr, 234
    Mr (double bass player), 66n
    Mr (organist), 134n
Smock Alley Theatre (Dublin), 4, 6, 11, 23, 157, 224
Smythson, Mr, 245-7
    Mrs, 245-6
Soane, George, 246
Sografi, Antonio Simeone, 238
Soult, Nicholas Jean de Dieu, Marshal, 189n
Spagnoletti, Paolo, 64, 66, 72, 79
Spanish Dollars, The; or, The Priest of the Parish (Davy), 83
Sparks, Mrs, 234
Sparre, Count de, 110
    Countess de, 111, *see also* Naldi, Caroline
Spermaceti oil, 20
Speyer, Wilhelm, 66
Spohr, Louis, 45, 66
Spontini, Gaspare Luigi Pacifico, 192n, 215, 218
Spray, John, 33-4, 114, 134, 250-2
Stanford, Sir Charles Villiers, 192n
Stanley, John, 5
Stansbury, George Frederick, 134, 138, 250, 253
State lottery, 130
Steibelt, Daniel, 149
Stephens, Catherine, 169, *171*, 172, 175, 176, 178, 195-200, 245
    Edward, 169
    Edward jun., 198
    Mr, 29
Stevenson, Sir John Andrew, 18, 20, 21, 62, 102, 125, 158-9, 206, *213*, 233-4, 240, 244, 247, 251
    Olivia, 125
Stewart, Mr, 233-5
    Mrs, 27, 235-6, 238-9, 242
Stockholm (Sweden), 140
Storace, Nancy, 89, 91
    Stephen, 21, 89, 93, 96, 99, 165, 202, 244
Strasbourg, 220
Stretch's Puppet Theatre, 4
Strickland, Walter George, 100n
Strinasacchi, Teresa, 212n

INDEX

Sullivan, M. J., *see* O'Sullivan, M. J.
Sussex, Duke of, 68
Sutcliffe, Miss, 246
Swift, Jonathan, 29
   Theophilus, 29, 235

Talbot, Earl, 198, 200n, 205, 225-6, 232
   Countess, 198, 204
   Richard Wogan, 152
   Montague, 34, 195, 235, 242
Tamburini, Antonio, 109n
Tamerlano (Handel), 7
Tancredi (Rossini), 110, 180, 215, 217
Tasca, Luigi, 42
Taylor, James, 152, 194, 208
   William, 51
Teatro alla Scala (Milan), 38, 39, 212, 217
   Argentina (Rome), 38
   Carcana (Milan), 118
   del Cocomero (Florence), 41, 240
   della Pergola (Florence), 38, 42
   de los Caños del Peral (Madrid), 40, 139
   La Fenice (Venice), 38, 212n
   San Benedetto (Venice), 237
   San Carlino (Naples), 105
   San Carlo (Naples), 237
   San Moisè (Venice), 141n
   San Pietro (Trieste), 38, 42
   San Samuele (Venice), 37, 41, 78n, 237-8
   Santa Cecilia (Palermo), 118, 241
   São Carlos (Lisbon), 38, 40-1, 109n, 125, 135, 138-9, 218, 237-8
   Zagnoni (Bologna), 86
Temple, Earl, 13
Tenducci, Giusto Ferdinando, 5
   Dorothea, 5
Teraminta (Smith), 5
   (Stanley), 5
Terry, Daniel, 246
Tett, Mr, 134, 250
Thackeray, William Makepeace, 132n
Theatre Royal, Hawkins Street (Dublin), 4n, 138, 147, 219, 228, 230, 232n
Theodora (Handel), 251
Thirty Thousand; or, Who's the richest?, 97-8, 239
Thompson, Mr, 242-4, 247-8
   Mr (viola player), 66n
'The Thorn' (Shield), 32
Three Weeks after Marriage (Murphy), 229
Tinney, Mr, 134-5, 250
Tipperary County, 5
Tomkinson, Mr, 134n
Tottola, Andrea Leone, 78, 237
Touring, exigencies of, 48

Townsend, John, 229
Tramezzani, Diomiro, 111, 116-7, 125, 138
Tramore (Co. Waterford), 57
Travellers at Spa, The (musical monologue), 95
Travellers, The; or, Music's Fascination (D. Corri), 83, 96, 238
Travers, John, 252
Trento, Vittorio, 66
Trieste (Italy), 38, 42
Trimlestown, Lord, 142
Trinity College Dublin, 9, 11, 24
Trionfo dell' amor fraterno, Il (Winter), 58, 96
Tritto, Giacomo Domenico Mario Antonio Pasquale Giuseppe, 42
Turin (Italy), 78, 105
Turnpike Gate, The (Mazzinghi and Reeve), 20, 233
Tussaud, Madame Anne Marie, 57
Tutore accorto, Il (?), 142, 255
Tyrconnell, Lord, 112
Tyrer, Miss, 114

Una cosa rara (Martin y Soler), 97
Une Folie (Méhul), 28
Union, Act of, 57, 58, 96, 129
Urbani, Pietro, 32-4, 114, 148, 149, 168, 224
   Signora, 33, 169

Valabrègue, Paul, 40, 41, 49, 51, 55, 140
Veiled Prophet of Khorassan, The (Stanford), 192n
Venice (Italy), 89, 139
Venice Preserved (Otway), 132
Vera Costanza, La (Anfossi), 37
Vercelli (Italy), 212n
Vergine del Sole, La (Andreozzi), 128
Verona (Italy), 212
Vestale, La (Pucitta), 128, 252
   (Rauzzini), 103n
   (Spontini), 215, 218
Vestris, Armand, 107
   Mme, 218
Viaggiatori felici, I (Anfossi), 42
Vicende amorose, Le (Tritto), 42
Vicenza (Italy), 216
Vienna (Austria), 69, 89, 105, 215, 241, 248
Viganoni, Giuseppe, 64
Villanella rapita, La (F. Bianchi), 138
Viotti, Giovanni Battista, 42, 139, 140, 219
Virgin of the Sun, The, see Vergine del Sole, La
Virgina (F. Federici), 105-6, 118
Voice ranges, nomenclature, 109n, 143
   and casting of roles, 116-7

Waldré, Vincent, 13
Walker, Mr, 66n, 134n

Wallace, Mr, 196, 247
Walsh, Dr Robert, 176
Walstein, Miss, 238, 240, 242
Ward, Charles, 100, 239
   Mr, 247
Warren, Mr, 235, 247
Warsaw (Poland), 69
Waterford (Ireland), 7, 176
   Bishop of, 176
   Marquis of, 176
Waterloo, victory at, 155
Waterman, The (C. Dibdin), 3
Waters, Edmund, 51, 68
Weather, inclemency of, 18, 130, 149
Webb, Miss, 233-4
   Mr, 160
Webbe, Samuel, the Elder, 250
Weber, Carl Maria von, 181
Weichsel, Charles, 64
   Mrs Frederica, 7
Weidner, Johann Carl, 134
Wellington, Duke of, 134, 190n
Welsh, Thomas, 144, 170, 180
West Indian, The (Cumberland), 209
Westmeath, Earl of, 9
Westmorland, Earl of, 9
Weston, Mr, 235-6
Wexford, 6, 189, 191
   Assembly Rooms, 189
   Rebellion, 16
Weyman, David, 33, 114
Whitaker, John, 166, 245-6
   Miss, 248
White, Robert, 198
Whitworth, Viscount, 129, 130, 134, 140, 142, 150, 167, 194, 205, 225
Wife of Two Husbands, The (Mazzinghi and Cooke), 27-8, 235
Wildfang, Der (Kotzebue), 20
Wild Irish Girl, The (S. Owenson), 34
Wilkinson, Tate, 93
Williams, Master, 146, 160
   Mr, 155, 233-9, 241-8

Williams, Mrs, 233
   Mrs I, 240-1
Willis, Isaac, 191
   Mrs, 191, 208
Willis's Rooms (London), 112
Willman, Henry, 32, 134, 191, 252, 255-6
   Thomas Lindsay, 134-256
Wilson, Rev. Thomas, 11
   Mr, 66n
Wilton, Mr, 234-5
Winter, Peter, 57-8, 96
Wivell, Abraham, 104, 227
Wolcot, Dr John, 58
Wolfe, Arthur, *see* Kilwarden, Lord
Wood, Mr, 209
Wood Demon, The (Kelly), 120
Woodman, The (Shield), 103n
Worcester (England), 197
Wordsworth, William, 186
Worsdale, James, 5
Wranitsky, Anton, 251
Wright, Mr, 170

Xerse (Cavalli), 219

Yaniewicz, Felix, 16, 48, 63, 191
Yates, Mr, 246
   Mrs, 246-7
York, Duke of, 112, 228
Young, Miss, *see* Stephens, Catherine
   Miss, 156
   Mr, 214
Younger, Mr, 238-40, 242, 244-5, 247
Youth, Love and Folly (Kelly), 61

Zafforini, Filippo, 13, 21, 23, 76, 86, 123, 125
Zaira (F. Federici), 105-6, 118, 120, 241, 249
   (Portugal), 38
Zamboni, Luigi, 107
   Signor, 38
Zauberflöte, Die (Mozart), 59n, 112, 117, 120, 167, 183, 219, 223, 250, 252-3
Zingarelli, Niccolò Antonio, 38, 65, 110